Poverty and Social Exclusion

Poverty and inequality remain at the top of the global economic agenda, and the methodology of measuring poverty continues to be a key area of research. This new book, from a leading international group of scholars, offers an up to date and innovative survey of new methods for estimating poverty at the local level, as well as the most recent multidimensional methods of the dynamics of poverty.

It is argued here that measures of poverty and inequality are most useful to policy-makers and researchers when they are finely disaggregated into small geographic units. *Poverty and Social Exclusion: New Methods of Analysis* is the first attempt to compile the most recent research results on local estimates of multidimensional deprivation. The methods offered here take both traditional and multidimensional approaches, with a focus on using the methodology for the construction of time-related measures of deprivation at the individual and aggregated levels. In analysis of persistence over time, the book also explores whether the level of deprivation is defined in terms of relative inequality in society, or in relation to some supposedly absolute standard.

This book is of particular importance as the continuing international economic and financial crisis has led to the impoverishment of segments of population as a result of unemployment, bankruptcy and difficulties in obtaining credit. The volume will therefore be of interest to all those working on economic, econometric and statistical methods and empirical analyses in the areas of poverty, social exclusion and income inequality.

Gianni Betti is Associate Professor in Statistics and Economics and member of the Research Centre for Income Distribution at the University of Siena, Italy. He has worked for several projects for the World Bank and European Commission, and has been closely involved with the development of the EU Statistics on Income and Living Conditions.

Achille Lemmi is Full Professor in Statistics and Economics and Director of the Research Centre for Income Distribution at the University of Siena, Italy. He is Associate Editor of *Statistics in Transition* and member of the Editorial Board of *Journal of Economic Inequality*. His life-long research interest has been the study of income distribution, inequality and poverty.

Routledge advances in social economics
Edited by John B. Davis
Marquette University

This series presents new advances and developments in social economics thinking on a variety of subjects that concern the link between social values and economics. Need, justice and equity, gender, cooperation, work poverty, the environment, class, institutions, public policy and methodology are some of the most important themes. Among the orientations of the authors are social economist, institutionalist, humanist, solidarist, cooperatist, radical and Marxist, feminist, post-Keynesian, behaviouralist and environmentalist. The series offers new contributions from today's most foremost thinkers on the social character of the economy.

Publishes in conjunction with the Association of Social Economics.

Previous books published in the series include:

1 **Social Economics**
Premises, findings and policies
Edited by Edward J. O'Boyle

2 **The Environmental Consequences of Growth**
Steady-state economics as an alternative to ecological decline
Douglas Booth

3 **The Human Firm**
A socio-economic analysis of its behaviour and potential in a new economic age
John Tomer

4 **Economics for the Common Good**
Two centuries of economic thought in the humanist tradition
Mark A. Lutz

5 **Working Time**
International trends, theory and policy perspectives
Edited by Lonnie Golden and Deborah M. Figart

6 **The Social Economics of Health Care**
John Davis

7 **Reclaiming Evolution**
A Marxist institutionalist dialogue on social change
William M. Dugger and Howard J. Sherman

8 **The Theory of the Individual in Economics**
Identity and value
John Davis

9 **Boundaries of Clan and Color**
Transnational comparisons of inter-group disparity
Edited by William Darity Jnr. and Ashwini Deshpande

10 **Living Wage Movements**
Global perspectives
Edited by Deborah M. Figart

11 **Ethics and the Market**
Insights from social economics
Edited by Betsy Jane Clary, Wilfred Dolfsma and Deborah M. Figart

12 **Political Economy of Consumer Behaviour**
Contesting consumption
Bruce Pietrykowski

13 **Socio-Economic Mobility and Low-Status Minorities**
Slow roads to progress
Jacob Meerman

14 **Global Social Economy**
Development, work and policy
Edited by John B. Davis

15 **The Economics of Social Responsibility**
The world of social enterprises
Edited by Carlo Borzaga and Leonardo Becchetti

16 **Elements of an Evolutionary Theory of Welfare**
Assessing welfare when preferences change
Martin Binder

17 **Poverty in the Inner City**
Overcoming financial exclusion
Paul Mosley and Pamela Lenton

18 The Capability Approach
Development practice and public policy in the Asia-Pacific region
Edited by Francesca Panzironi and Katharine Gelber

19 Poverty and Social Exclusion
New methods of analysis
Edited by Gianni Betti and Achille Lemmi

Poverty and Social Exclusion

New methods of analysis

Edited by Gianni Betti and Achille Lemmi

First published 2014
by Routledge
2 Park Square, Milton Park, Abingdon, Oxon OX14 4RN

Simultaneously published in the USA and Canada
by Routledge
711 Third Avenue, New York, NY 10017

Routledge is an imprint of the Taylor & Francis Group, an informa business

British Library Cataloguing in Publication Data
A catalogue record for this book is available from the British Library

Library of Congress Cataloging in Publication Data
Poverty and social exclusion: new methods of analysis/edited by
Gianni Betti and Achille Lemmi.
pages cm
Includes bibliographical references and index.
1. Poverty–Measurement. 2. Poverty–Social aspects. 3. Marginality, Social. I. Betti, Gianni. II. Lemmi, Achille.
HC79.P6P6829 2013
305.5′69–dc23

2013001757

ISBN: 978-0-415-63634-6 (hbk)
ISBN: 978-0-203-08517-2 (ebk)

Typeset in Times New Roman
by Wearset Ltd, Boldon, Tyne and Wear

Printed and bound in the United States of America by Publishers Graphics, LLC on sustainably sourced paper.

Contents

Illustrations

Figures

Tables

Contributors

Gianni Betti is Associate Professor in Statistics and Economics and member of the 'Centro Interdipartimentale di Ricerca sulla Distribuzione del Reddito' (C.R.I.DI.RE. – 'C. Dagum'), University of Siena, Italy. He has worked for several projects for the World Bank and European Commission, and has been closely involved with the development of the EU Statistics on Income and Living Conditions (EU-SILC).

Walter Bossert is Professor of Economics and CIREQ Research Fellow at the University of Montreal. He has published articles on social choice theory, welfare economics, public economics, decision theory, game theory, theories of rational choice, economic index numbers and issues involving ethics and economics. Professor Bossert is a co-author of *Population Issues in Social Choice Theory, Welfare Economics, and Ethics*, and of *Consistency, Choice, and Rationality*. He serves as a member of the editorial boards of *Economics and Philosophy*, *Mathematical Social Sciences* and *Social Choice and Welfare*.

German Caruso is a PhD candidate at the University of Illinois at Urbana Champaign and a visiting professor at Universidad de San Andrés. He specializes in poverty and development economics.

Lidia Ceriani is lecturer at the Department of Policy Analysis and Public Management, Bocconi University. Her research interests are in the fields of Income Distribution, Poverty Measurement and Welfare Effects of Privatization Policies.

Satya R. Chakravarty is a Professor of Economics at Indian Statistical Institute, Kolkata, India. He has articles published in *Econometrica*, *Journal of Economic Theory, Games and Economic Behavior*, *International Economic Review*, *Economic Theory*, *Social Choice and Welfare*, *Journal of Development Economics*, *Canadian Journal of Economics*, *Mathematical Social Sciences*, *Economics Letters*, *Theory and Decision*, *Review of Income and Wealth*, *Journal of Economic Inequality*; and books published by Springer-Verlag, Cambridge University Press, Anthem Press and Avebury. He is a member of the Editorial Board of *Journal of Economic Inequality*, a co-editor of *Economics E-Journal* and also a member of Advisory Board for the Book

Series 'Economic Studies in Inequality, Social Exclusion and Well-Being', Springer-Verlag. He is a fellow of the Human Development and Capability Association, an external adviser of the World Bank, and has been awarded the Mahalanobis Memorial Prize by the Indian Econometric Society. He has also been a grantee of the Ford Foundation, the German Research Foundation and the French Ministry of Higher Education. His major research interests are: measurement of inequality, well-being, poverty, polarization, etc., and voting games.

Ray Chambers is Professor of Statistical Methodology and a Professorial Fellow in the National Institute for Applied Statistics Research Australia, located at the University of Wollongong, Australia. He is a Fellow of the American Statistical Association and an elected member of the International Statistical Institute. He is the International Representative on the Board of the American Statistical Association 2011–2013, and President of the International Association of Survey Statisticians 2011–2013. His research interests include analysis and design for sample surveys, robust methods in statistical modelling, small area inference and statistical analysis using combined data sources.

Conchita D'Ambrosio is Associate Professor of Economics at Università di Milano-Bicocca. Her research focuses on income and wealth distributions, deprivation, polarization and social exclusion. She has been member of the editorial board of the *Review of Income and Wealth* since 2001 and managing editor of the same journal since 2007. She is also a member of Advisory Board for the Book Series 'Economic Studies in Inequality, Social Exclusion and Well-Being', Springer-Verlag. She has articles published in *Economica, International Economic Review, Social Choice and Welfare, Economics Letters, Review of Income and Wealth, Social Indicators Research, World Development.*

Steve Donbavand is a postgraduate researcher at the University of Southampton where he is currently working on the application of small area models within Poverty Mapping and is particularly interested in the performance of different methodologies under various population assumptions and the development of accurate MSE estimation techniques for outlier robust semi-parametric methods. More generally, his research interests centre around improving the accuracy of the measurement of social phenomena and as such has previously carried out research on behalf of a wide range of organizations, including the Department for Environment, Food and Rural Affairs (Defra, UK), the University of Sunderland and various NGOs.

James E. Foster is Professor of Economics and International Affairs at The Elliott School of International Affairs at The George Washington University, and the Director of the Institute for International Economic Policy. He is also a research associate to the Oxford Poverty and Human Development Initiative (OPHI), Queen Elizabeth House (QEH), Department of International Development at the University of Oxford. His areas of expertise include development economics, inequality and poverty, economic theory and policy.

Francesca Gagliardi is research fellow at the Department of Economics and Statistics, University of Siena. By the end of the PhD in Applied Statistics, she has been working on poverty measures from various points: cross-sectional and longitudinal, multidimensional, absolute or relative, national level or local and small area level. She has also been working on methods for variance estimation.

Caterina Giusti is Researcher in Statistics in the Department of Economics and Management of the University of Pisa. Her research interests include multi-level models, imputation for nonresponses in sample surveys, small area estimation models.

Naama Haron is a social economist who recently received her PhD from Bar-Ilan University, Israel. Chapter 4 of the book is based on part of her dissertation.

Stephen Haslett is Professor of Statistics in the Institute of Fundamental Sciences, Massey University, New Zealand. He has published widely in both theoretical and applied statistics journals, and been involved in small area estimation of poverty and malnutrition projects in Bangladesh, Bhutan, Cambodia, Nepal, Philippines and Timor-Leste.

Stephan Klasen is Professor of Economics at the University of Göttingen in Germany, where he also coordinates the Courant Research Center 'Poverty, equity, and growth in developing and transition countries'. He holds a PhD from Harvard University and has since held positions at the World Bank, King's College (Cambridge, UK) and the University of Munich. His research focuses on analysing poverty and inequality issues in developing countries.

Achille Lemmi is Full Professor in Statistics and Economics and Director of C.R.I.DI.RE. – 'C. Dagum' at University of Siena, Italy. He is Associate Editor of *Statistics in Transition* and member of the Editorial Board of *Journal of Economic Inequality*. His life-long research interest has been the study of income distribution, inequality and poverty.

Stefano Marchetti is Research Fellow in Statistics at the Department of Economics and Management, University of Pisa. He holds a PhD in Applied Statistics from the Department of Statistics, University of Florence (2009). During 2008 he spent a visiting period of three months at the Cathie Marsh Centre for Census and Survey Research, University of Manchester, UK. He teaches Statistics in graduate and post-graduate courses of the University of Pisa. He has research interests in survey sampling methodology, resampling methods, small area estimation, with special focus on M-quantile models and on poverty mapping. He has participated in research supported by the Ministry of the Education, University and Research and the European Union's Seventh Framework Programme.

Isabel Molina is Associate Professor in the Department of Statistics at Universidad Carlos III de Madrid. She is the author of many papers in small area estimation and in particular several papers in poverty estimation. She received the 'Best Paper Award 2010' of the Statistical Society of Canada for the joint paper with J.N.K. Rao on small area estimation of poverty indicators published in the *Canadian Journal of Statistics*, 2010.

Catherine Porter is a British Academy Postdoctoral Research Fellow in the Department of Economics and Corpus Christi College, University of Oxford. She is an applied micro-economist working on poverty and wellbeing issues in developing countries. Particular areas of interest are chronic poverty measurement, risk and shocks, health, children, social protection and aid effectiveness, and she has worked or published in several developing countries including Ethiopia, The Gambia, Peru, Uganda and Vietnam.

Monica Pratesi is Full Professor of Statistics in the Department of Economics and Management of the University of Pisa. She is member of the Council of the International Association of Survey Statisticians. She was scientific coordinator in many Italian and European research projects. Her research interests include nonresponses in telephone and web surveys, small area estimation models, inference for elusive populations.

Natalie Naïri Quinn is a Domus Fellow and Tutor in Economics at Lady Margaret Hall, University of Oxford. She works on poverty and welfare evaluation in developing countries and has a particular interest in the ethical and mathematical foundations of methods of evaluation. She is also interested in issues of public service delivery in developing countries. She has worked in Sierra Leone, Tanzania and Namibia.

Jon N.K. Rao is Professor Emeritus and Distinguished Research Professor in the School of Mathematics and Statistics at Carleton University. He has published numerous papers in small area estimation including poverty estimation. He is the author of the book *Small Area Estimation* (Wiley 2003). He received the Gold Medal of the Statistical Society of Canada in 1993 and Honorary Doctorate in Mathematics from the University of Waterloo in 2008.

Nicola Salvati is Researcher in Statistics in the Department of Economics and Management of the University of Pisa. His research interests include survey sampling, model-assisted and design-based inference, robust regression, quantile and M-quantile regression, multilevel models, geographically weighted regression, spatial statistics, applications of small area models in poverty mapping, new technologies in survey methodology (computer assisted telephone surveys, electronic data interchange, Internet surveys).

Vincenzo Salvucci is resident adviser for the University of Copenhagen in Maputo, Mozambique. Starting from his PhD thesis at the University of Siena, his research focuses on different aspects of the Mozambican economy mainly related to poverty analysis.

Maria Emma Santos is Assistant Professor at Departamento de Economía, Universidad Nacional del Sur (UNS) and Research Fellow at the Instituto de Investigaciones Económicas y Sociales del Sur (IIES) – UNS and Consejo Nacional de Investigaciones Científicas y Técnicas (CONICET), Bahía Blanca, Argentina. She is also Research Associate at the Oxford Poverty and Human Development Initiative (OPHI), Department of International Development, University of Oxford. Her main research interests are the measurement, determinants and analysis of multidimensional and chronic poverty, income inequality and education.

Jacques Silber is Professor Emeritus of Economics at Bar-Ilan University, Israel. He holds a PhD (1975) in Economics from the University of Chicago. He was the editor of a *Handbook on Income Inequality Measurement* (Kluwer) and of several other books, and the author with Y. Flückiger of *The Measurement of Segregation and Discrimination in the Labor Force* (Physica-Verlag). He was the founding editor and first Editor-in-Chief of the *Journal of Economic Inequality* (Springer). He has published more than 100 articles in international academic journals.

Walter Sosa-Escudero is Associate Professor at the Department of Economics of Universidad de San Andres, Argentina. He specializes in econometric theory, and applications to welfare analysis. His work has been published in the leading journals of his fields. He is the president of the Argentine Economic Association.

Marcela Svarc is Assistant Professor at the Department of Mathematics of Universidad de San Andres. She holds a PhD in Mathematics from Universidad de Buenos Aires. Her work on cluster methods has been published in leading journals like the *Journal of the American Statistical Association*.

Nikos Tzavidis is Associate Professor at the University of Southampton and has previously held posts as Assistant Professor and Associate Professor at the Institute of Education, University of London and at the University of Manchester. He is specializing in small area estimation, quantile regression, robust estimation for random effects models and in applications of multilevel models. He has published research papers in the *Journal of the Royal Statistical Society Series B*, *Biometrika*, *Survey Methodology* and *Computational Statistics and Data Analysis* and his research is supported by the British Academy, the Economic and Social Research Council in the UK and the European Union's Seventh Framework Programme.

Vijay Verma is Visiting Professor at University of Siena. He has worked for development of official statistics worldwide, and has written on sample survey methodology and income and poverty analysis.

Gaston Yalonetzky is Lecturer in Economics at the University of Leeds, UK and Research Associate at the Oxford Poverty and Human Development Initiative. He holds a DPhil (2008) in Economics from the University of Oxford. He is an Associate Editor of Oxford Development Studies and has served as Economics PhD Director at Leeds University Business School. His research interests span distributional analysis and wellbeing measurement. Some of his work has been published in special journal issues on these topics.

1 Introduction

Gianni Betti and Achille Lemmi

Over the last five years the University of Siena's Research Centre on Income Distribution "C. Dagum" and its research group have been awarded several research grants funded by international bodies such as the European Union, the World Bank and Eurostat. In particular, the SAMPLE project, funded by the EU under the seventh FP, is the main source of theory and results for this book (www.sample-project.eu/). The reason that prompted us to put together this collective volume is that measures of poverty and inequality are most useful to policymakers and researchers when they are finely disaggregated, i.e. when they are estimated for small geographic units, such as cities, municipalities, districts or other "local" administrative partitions of a country. This book is the first attempt to bring together the most recent research results on local estimates of multidimensional poverty (deprivation).

Moreover, one of the book's significant innovations lies in its analysis of poverty dynamics using both traditional and multidimensional approaches. The focus will be on using this methodology for the construction of *time-related measures of deprivation at individual and aggregated levels*. Regarding the analysis of persistence over time, a related issue dealt with concerns whether the level of deprivation is defined purely in terms of the relative disparity in society, or in relation to some supposedly absolute standard. The book explores the implications of this choice on the analysis of the persistence of deprivation in society.

These research themes constitute a consistent and relevant part of the recent economic, econometric and statistical methods and empirical analyses in the international scientific literature. They also play a seminal role in policymaking, given the importance of inequality and poverty reduction, and of improving living conditions within a framework of sustainable economic development.

Moreover, the present dramatic international economic and financial crisis emphasizes the role of social equity in policymaking, in order to effectively combat the increasing impoverishment of relevant segments of population, due to unemployment, bankruptcy and difficulties related to bank credit.

The book we propose contains a selection of the most original research from the abovementioned projects, complemented by chapters proposed by outstanding researchers in the field of poverty. The chapters have been chosen taking into account three criteria: (i) scientific quality; (ii) capacity to represent the key issues in current scientific research; (iii) existence of a sort of scientific "fil rouge" among the contributions, with the aim of proposing a homogeneous, useful and up-to-date book.

In particular, the main themes are related to: (i) innovation in the theory and methods regarding poverty and social exclusion; (ii) poverty mapping and its policy implications; (iii) worldwide empirical applications in applied economic analysis.

The book boasts a good balance between theoretical/methodological and empirical chapters. The group of authors constitutes a mixture of very renowned researchers in the fields of economic and statistical approaches to poverty measurement, and a group of young and very promising scholars. The case studies are always applications of the economic theories proposed and appeal to an international audience; they cover the Americas, Asia, Africa and Europe.

The book is made up of 16 chapters. After this Introduction, the others are divided into three Parts, respectively devoted to: Part I, Multidimensional poverty; Part II, Longitudinal and chronic poverty; Part III, Small area estimation methods. Each Part begins with an Overview chapter written by leading researchers on that topic.

Chapter 2 by Jacques Silber and Gaston Yalonetzky on "Measuring multidimensional deprivation with dichotomized and ordinal variables" is more than an overview of Part I on Multidimensional poverty and deprivation. The purpose of this chapter is to provide an introduction to the measurement of multidimensional poverty when dealing with counting and ordinal variables. The authors first present a general framework for measuring multidimensional poverty with ordinal variables. A distinction is made between individual and social poverty functions. At the individual level they stress the difference between the identification function and the "breadth of poverty" function and review the properties of such an individual poverty function, combining "identification" and "breadth of poverty" elements. When aggregating these individual deprivation functions they then make a distinction between the case in which the social poverty function is an average of the individual poverty functions and that of alternative aggregation methods. This chapter also discusses the issue of inequality in the distribution of deprivation among the poor, looks at alternative identification functions and discusses the issues of weights, robustness and partial ordering. Finally, a simple empirical illustration of the different approaches is presented.

Chapter 3 by Walter Sosa-Escudero, German Caruso and Marcela Svarc on "Poverty and the dimensionality of welfare" reviews recent methods for quantifying the dimensionality of welfare and its relationship with deprivation. The authors discuss two alternative strategies based on factor analytic methods and on variable selection after cluster analysis. Unlike latent variable methods, variable selection strategies are immediate to interpret and resample, since they choose variables originally in the data set. The advantages and disadvantages of both strategies are discussed, as well as some recent empirical applications of these methods. The methods are shown to be capable of summarizing an initially large list of variables into a few new variables (as in factor analytic methods) or a subset of the original ones (as in feature selection/cluster methods), which can serve the purpose of characterizing the poor. These methods can contribute to the conceptual search for relevant dimensions of welfare, or provide confirmatory analysis of alternative, probably multidisciplinary, studies aimed at isolating relevant factors for poverty analysis.

Chapter 4 by Naama Haron on "Income, material deprivation and social exclusion in Israel" aims at illustrating the state of economic poverty, material deprivation and social exclusion in Israel. The author examines the correlation between these phenomena and attempts to identify and characterize the afflicted population. The research presented herein is based on Israeli societal survey data of 2007. The Central Bureau of Statistics has carried out this survey since 2002, permitting the development of an index uniquely adapted to these needs. In order to maintain clarity and create as clear a distinction as possible between economic poverty, material deprivation and social exclusion, the definition of the paradigm of social quality is followed (Berman and Phillips, 2000). This paradigm serves as a superstructure for understanding and organizing the various levels of an individual's life in the society in which he/she resides, including the absence of economic security and social exclusion.

Chapter 5 by Gianni Betti, Francesca Gagliardi and Vincenzo Salvucci deals with "Multidimensional and fuzzy measures of poverty and inequality at national and regional level in Mozambique". This chapter provides a step-by-step account of how fuzzy measures of non-monetary deprivation and also monetary poverty can be calculated in Mozambique. For non-monetary deprivation, meaning dimensions or groupings of initial items of deprivation are identified using explanatory and confirmatory factor analyses, and a weighting system is applied for the aggregation of individual items into the dimension they represent. An application is conducted on Mozambique using 2008–2009 data: estimates are provided at national level and also disaggregated at provincial level. Standard errors are provided using a recent methodology based on Jack-knife Repeated Replication (Verma and Betti, 2011).

Chapter 6 by Vijay Verma and Francesca Gagliardi "On assessing the time-dimension of poverty" constitutes an overview of Part II on Longitudinal and chronic poverty. In the chapter, various issues are addressed concerning the time-dimension of poverty, in particular measures of poverty trend over time at the aggregate level and measures of persistence or otherwise of poverty at the micro level. The main measurement problem in the assessment of poverty trends is the definition of the poverty threshold and its consistency over time. The main measurement problem in the assessment of persistence of poverty concerns the effect of random measurement errors on the consistency of the individuals' observed poverty situation at different times.

Chapter 7 by Walter Bossert, Lidia Ceriani, Satya R. Chakravarty and Conchita D'Ambrosio looks at "Intertemporal material deprivation". The purpose of this chapter is to add intertemporal considerations to the analysis of material deprivation. The authors employ the EU-SILC panel data set, which includes information on different aspects of wellbeing over time. EU countries are compared based on measures that take this additional intertemporal information into consideration. Following the path of material deprivation experienced by each individual over time, a picture that differs from the annual results is obtained. Since the measurement of material deprivation is used by the EU member states and the European Commission to monitor national and EU progress in the fight against poverty and social exclusion, the results suggest that time cannot be neglected. Countries should not only be compared based on their year by year results, but additional information needs to be gained by following individuals over time and producing an aggregate measure once time is taken into account.

Chapter 8 by James E. Foster and Maria Emma Santos on "Measuring chronic poverty" proposes a new class of chronic poverty measures, which does not require resources in different periods to be perfect substitutes when identifying the chronically poor. The authors use a general mean to combine the resources of a person into a permanent income standard, which is then compared to a poverty line to determine when a person is chronically poor. The parameter of the general mean allows for varying degrees of substitutability over time, from perfect substitutes to perfect complements. The decomposable CHU poverty measure with the same parameter is applied to the distribution of permanent income standards to measure overall chronic poverty. Each measure has a convenient expression in terms of a censored matrix and satisfies a host of properties, including decomposability. This chapter provides an interesting empirical application of the new measures using panel data from urban areas in Argentina.

Chapter 9 by Catherine Porter and Natalie Naïri Quinn on "Measuring intertemporal poverty: policy options for the poverty analyst" presents an analytical review of the recent technical literature on intertemporal poverty measurement (also known as chronic poverty measurement or lifetime poverty measurement). Individual measures of wellbeing (or the lack of) are aggregated both over time and across people, to compare wellbeing over more than one time period. The chapter has a practical emphasis. The main aims are to make the intertemporal poverty measures which have been introduced in the technical literature accessible to applied practitioners conducting poverty analysis and to offer advice on which may be appropriate in alternative circumstances. Different measures have different properties which reflect alternative normative principles or judgements. First, the authors present intuitive motivations for and explanations of these properties; then they relate these properties to practical considerations, such as measurement error in data, as well as normative principles. The chapter then gives a comprehensive review of the properties satisfied by several of the recently suggested measures, in order to understand when they might be appropriate.

Chapter 10 by Stephan Klasen on "measuring levels and trends in absolute poverty in the world: open questions and possible alternatives" critically analyses changes in the Global Poverty numbers generated by the World Bank in 2008. While they have little impact on observed poverty trends and while there are good reasons to believe that the previous numbers were based on weak foundations, the new numbers on levels of poverty in the developing world create new uncertainties and questions. In particular, there are conceptual and empirical issues involved in using one (updated) ICP round to update all poverty numbers, as well as questions regarding the adjustment of the international poverty line. This chapter reviews these issues and finds that we cannot be very certain about levels of absolute poverty in the world and that the current method for generating absolute poverty numbers is problematic and should possibly be abandoned. At the same time, poverty trends are much less affected by these methodological issues. This chapter also discusses potential alternatives to the current methods and highlights their strengths and weaknesses. Unfortunately, there is no readily available alternative to the current method, although such an alternative could be developed, with some difficulty.

Chapter 11 by Ray Chambers and Monica Pratesi on "Small area methodology in poverty mapping: an introductory overview" proposes the use of small area estimation methods for estimating poverty at local level; these methods constitute a set of advanced statistical inference techniques that can be used for the measurement of poverty and living conditions by survey practitioners, researchers in private and public organizations, official statistical agencies and local governmental agencies. In particular, the estimates produced using small area estimation methods are well suited to mapping geographical variations in these conditions. The aim of the chapter – and the entire Part III of the book – is to provide the reader with an overview of small area estimation methods that focus on poverty measurement.

Chapter 12 by Stephen Haslett concerns "Small area estimation of poverty using the ELL/PovMap method, and its alternatives". Small area estimation of poverty-related variables, such as poverty incidence, gap and severity is an important analytical tool used to target the delivery of food and other aid in developing countries at a finer level. Although there is a wide variety of small area estimation methods, the principal method used for poverty estimation (and hence poverty mapping) at small area level is that of Elbers, Lanjouw and Lanjouw (2003), commonly referred to as ELL. The ELL method differs from many other small area methods in that it uses both survey and census data at household level, and a model based on the survey data to make household level predictions for all census observations, which are then aggregated. When the model is correct, using the ELL method can give considerable improvements in terms of accuracy over small area estimation methods based on surveys alone. However, when it is incorrect the estimated standard errors can be severe underestimations. This chapter explores the ELL method and its underlying assumptions, considers model diagnostics and the utility of the World Bank's PovMap software that implements ELL, and outlines both variations of and alternatives to ELL.

Chapter 13 by Isabel Molina and Jon N.K. Rao on the "Estimation of poverty measures in small areas" describes methods for the estimation of poverty indicators at local level. The methodology described here is generally applicable to practically any poverty indicator, but for the sake of illustration they focus on the class called FGT poverty measures. This class includes *poverty incidence*, or the proportion of individuals living in poverty in a small area, *poverty gap*, which measures the area mean of the relative distance from the poverty line of each individual, and *poverty severity*, which measures the area mean of the squared relative distance: large values of this measure point to areas with severe levels of poverty. A comparison of the different estimation procedures through model-based and design-based simulation studies is provided.

Chapter 14 by Nicola Salvati, Caterina Giusti and Monica Pratesi on "The use of spatial information for the estimation of poverty indicators at the small area level" considers a model-based perspective and presents different models that use geographical information. The authors apply these spatial models to data from the EU-SILC 2008 survey to estimate various indicators of poverty and living conditions for each Local Labour System in three Italian regions: Lombardy in the North, Tuscany in Central Italy and Campania in Southern Italy. The choice of these three regions, out of the 20 existing in Italy, is motivated by the geographical differences that characterize the country. In particular, since each of the three regions can be considered as representative of the corresponding geographical area of Italy (Northern, Central and Southern/Insular Italy), this allows us to investigate the so-called "north–south" divide.

Chapter 15 by Nikos Tzavidis, Stefano Marchetti and Steve Donbavand on "Outlier robust semi-parametric small area methods for poverty estimation" reviews small area estimation methodologies for poverty indicators, with specific emphasis on outlier robust estimation. The World Bank and Empirical Best Prediction approaches are presented alongside an outlier robust, semi-parametric approach based on the M-quantile model. The World Bank and M-quantile approaches are empirically compared using Monte-Carlo simulation. Using data from the EU survey of income and living conditions (EU-SILC) and Census micro-data from Italy the authors apply the M-quantile approach to derive head count ratio and poverty gap estimates for provinces.

Finally, Chapter 16 by Gianni Betti and Achille Lemmi on "poverty and social exclusion in 3D: multidimensional, longitudinal and small area estimation" is not merely a conclusion of the book. It seeks to connect the three Parts of the book, or the three dimensions of poverty and social exclusion: Multidimensionality, Longitudinal poverty and Small area estimation. As such it represents the "fil rouge" of the whole book, and examines and proposes methods that combine these three dimensions.

References

Berman Y., Phillips D. (2000), Indicators of social quality and social exclusion at national and community level. *Social Indicators Research*, 50(3), pp. 329–350.

Elbers C., Lanjouw J.O., Lanjouw P. (2003), Micro-level estimation of poverty and inequality. *Econometrica*, 71(1), pp. 355–364.

Verma V., Betti G. (2011), Taylor linearization sampling errors and design effects for poverty measures and other complex statistics, *Journal of Applied Statistics*, 38(8), pp. 1549–1576.

Part I

Multidimensional poverty

2 Measuring multidimensional deprivation with dichotomized and ordinal variables

Jacques Silber[1] *and Gaston Yalonetzky*

2.1 Introduction

The multidimensional nature of poverty enjoys broad consensus. However there is an ongoing debate regarding the suitability of bringing together the multiple indicators of deprivations into one composite index. Those against lament the potential loss of information and prefer a dashboard approach (see, for example, Ravallion, 2011). Those in favour stress that a composite index may be unavoidable when the purpose is to quantify the incidence of multiple deprivations within the same individuals (Yalonetzky, 2012).

Even among those in favour of composite indices of poverty there is a lively debate as to the best way to combine the information from different variables. For instance, Ravallion (2011) advocates the traditional monetary approach whereby the different variables (e.g. consumption items) are summed using market or shadow prices as weight, and then the aggregate (household) consumption metric is compared against a monetary poverty line, both for the purpose of identifying the (monetary) poor and of measuring the severity of their poverty experience. But as stressed by Thorbecke (2007) the drawback of such an income based approach is that some (non-monetary) attributes cannot be purchased because markets do not exist or because they operate very imperfectly. Relying on income as the sole indicator of well-being is hence problematic since such an approach cannot take into account basic dimensions of poverty such as life expectancy, literacy, the provision of public goods, the extent of freedom.

In this chapter we emphasize the multidimensionality of poverty but our focus is on the challenges and limitations posed by measuring multidimensional poverty when only ordinal variables are available. The study of multidimensional deprivations emphasizing the idea of counting was in fact introduced by Townsend (1979). Atkinson (2003) then compared an analysis of multidimensional deprivation based on the concept of social welfare with one derived from a counting approach. The question is: what can we actually "say", "measure" when we have only ordinal or dichotomized indicators? The answer is that we can identify who is poor (in a multidimensional sense) and we can say something about the "breadth" of poverty as in number of deprivations. We can also say something about the inequality in deprivation distribution, i.e. whether deprivations are concentrated among the same people or not. As we will see, there are different ways of taking these elements into account.

By contrast, what is it about poverty that we cannot say when we have ordinal variables? We cannot say much about the "depth" of poverty because any scaling of an ordinal variable is bound to be arbitrary, so the actual distance between the value of the variable and its respective deprivation line does not really make much sense. However, we can still do some dominance analysis (partial orderings), whereby we derive conditions whose fulfilment guarantees the robustness of pairwise ordinal comparisons (i.e. ordinal as in: "A>B" but not knowing by how much) for a broad range of possible scalings (i.e. possible individual poverty functions).

Although there is by now a vast theoretical and empirical literature on the measurement of multidimensional poverty in the case of continuous variables, there is only a limited amount of work focusing on counting or on ordinal variables, despite the fact that the dissemination of dichotomized and qualitative variables in the data sets of the main statistical agencies largely justifies emphasizing such a topic.[2] In fact the purpose of this chapter is to introduce the reader to the measurement of multidimensional poverty when dealing with counting and ordinal variables. We do not provide a comprehensive survey of the literature on this topic but instead focus our attention on selected topics, comparing on one hand individual and aggregate measures of deprivation, and, on the other hand, approaches based on the expected utility theory and on the dual approach.

More precisely we first present (Section 2.2) a general framework allowing us to measure multidimensional poverty with ordinal variables. We make a distinction between individual and social poverty functions. At the individual level we also stress the difference between the identification function and the "breadth of poverty" function and review the properties of such an individual poverty function combining "identification" and "breadth of poverty" elements. When aggregating these individual deprivation functions we then make a distinction between the case where the social poverty function is an average of the individual poverty functions and that of alternative aggregation methods. We also discuss the issue of inequality in the distribution of deprivations among the poor. In Section 2.3 we next look at alternative identification functions while in Section 2.4 we discuss the issues of weights, robustness and partial ordering. Section 2.5 presents a simple empirical illustration of the approaches described in previous sections while concluding comments are given in Section 2.6.

2.2 Multidimensional poverty measurement with ordinal variables: a general framework[3]

2.2.1 Notation

Consider the matrix of attainments X. Its N rows have information on the attainment of an equal number of individuals over D variables that relate to ordinal indicators of well-being. A typical element of X is x_{nd} ($\in \mathbf{N}_{++}$). In the matrix, X_n is the row of attainments in each variable for individual n, and X_d is the column showing the distribution of variable d over the population. The variable-specific deprivation lines are in the vector $Z=(z_1, z_2, \ldots, z_D)$.

It is also useful to define a vector of variable-specific weights: $W=(w_1, w_2, \ldots, w_D)$, such that $w_d>0 \, \forall \, d \in [1, D]$ and

$$\sum_{j=1}^{D} w_j = 1 .$$

The positive, real-valued scalar cut-off that participates in the identification of the multidimensionally poor is k, such that $0 \leq k \leq 1$.

2.2.2 *The individual poverty function*

Sen (1979) explained that social poverty measurement is done in two steps. The first step defines the individual (or household) poverty function, which involves the identification of the poor; then the second step generates the social poverty function by aggregating the individual poverty functions.

When applying the counting approach, it is best to follow a three-step procedure. The first step identifies individual deprivations in each variable. The second step identifies the multidimensionally poor and produces the individual multidimensional poverty function. Finally, the third step generates the social poverty function through aggregation as mentioned above.

In this section we describe how the first two steps work in the measurement of multidimensional poverty with ordinal variables and using the counting approach. In the first step each attainment variable, x_{nd}, is compared against its respective deprivation line z_d. If $x_{nd}<z_d$ then person n is said to be deprived in variable d.[4]

In the second stage we formulate an individual poverty function that identifies the multidimensional poor and is sensitive to the number of deprivations among the poor.

The identification function

In the counting approach the identification of the poor rests on an individual deprivation score that we call the (real-valued) counting function and define as:

$$c_n : \mathbf{N}_{++}^{D} \times \mathbf{N}_{++}^{D} \times [0,1]^{D} \rightarrow [0,1] .$$

More specifically, we define the following counting function:

$$c_n \equiv \sum_{d=1}^{D} \mathbb{I}(x_{nd} < z_d) w_d \tag{2.1}$$

According to (2.1), c_n is the weighted sum of deprivations suffered by person n. In the counting approach this deprivation score is compared against the multidimensional cut-off k. If $c_n \geq k$ then person n is considered multidimensionally poor, or simply poor. Hence the poverty identification function of the counting approach is the following:

$$\psi(X_n; Z, W, k) = \mathbb{I}(c_n \geq k) \tag{2.2}$$

If $k \leq \min\{w_1, w_2, \ldots, w_D\}$, then $\exists d{:}x_{nd} > z_d \rightarrow \psi = 1$; i.e. the presence of one deprivation suffices to consider n as poor. This is the *union approach* to poverty identification in a multivariate context. Even though this approach yields the highest proportion of poor people, it is used by most of the theoretical literature on multidimensional poverty measurement, albeit often implicitly, as shown below.

On the other extreme, if $k=1$, then $\forall d{:}x_{nd} > z_d \leftrightarrow \psi = 1$; i.e. n is considered poor if and only if the person is deprived in each and every well-being variable. This is the *intersection approach* to poverty identification. It yields the lowest proportion of poor people and is expected to detect the most indigent people.

The general formula in (2.2) was proposed by Alkire and Foster (2011). It also includes *intermediate* identification approaches, between union and intersection, when $\min\{w_1, w_2, \ldots, w_k\} < k < 1$.

A simple graphical representation of the various ways of classifying different identification functions is given in Figure 2.1. Basically either they are discrete (e.g. Alkire–Foster) or they are "fuzzy". If the latter are stretched towards the corners of the unitary square (in the positive quadrant) then we get the union and intersection approaches in the limit.

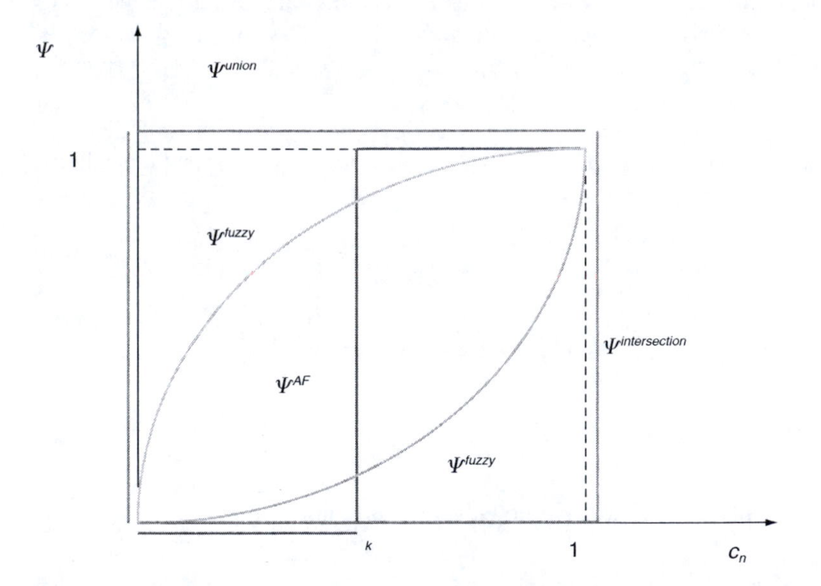

Figure 2.1 Some possible shapes of the identification function ψ.

Such a fuzzy function was suggested by Rippin (2010). For instance, she uses $\psi^{Rippin} = [c_n]^{\gamma}$, so that the identification function can be concave or convex depending on whether γ is between 0 and 1 or whether it is higher than 1. Now this identification is fuzzy because unless $c_n = 0$ or $c_n = 1$, everybody can be "somewhat poor", e.g. $\psi = 0.8$ would mean that someone is "80 per cent poor".

How does one choose between a concave and a convex function? That evidently depends on whether it is assumed that the dimensions are complements or substitutes. If they are substitutes, then deprivation in one dimension may be overcome by having no deprivation in another dimension so that as long as an individual is not deprived in all dimensions his overall deprivation score will be equal to zero in the case of perfect substitution (the "intersection" case) or smaller than one when dimensions are imperfect substitutes (the case of a convex identification function). If on the contrary the deprivation dimensions are complements, as soon as an individual is deprived in one dimension, he must suffer from some overall deprivation. If it is assumed that the deprivation dimensions are perfect complements, then one obtains the "union" case while if they are imperfect complements one gets the more general case of a concave identification function. Note that this implies, as stressed by Rippin (2010), that the Alkire–Foster (2011) approach assumes implicitly that up to some critical point the dimensions are substitutes while they are complements beyond this critical point (see Figure 2.1).

One can also transform any "fuzzy" identification function into a dichotomized identification function, e.g. to move from ψ^{Rippin} to ψ^{AF}. How? Just consider a cut-off value $m \in [0, 1]$ and identify the poor if and only if $\psi^{fuzzy} \geq m$. Otherwise, if $\psi^{fuzzy} < m$, the person is non-poor. Since ψ^{fuzzy} is a continuous function that increases monotonically with c_n, then the choice of m implicitly defines a threshold k for the individual count c_n that solves the following implicit function: $\psi^{fuzzy}(k) = m$. Therefore the new dichotomized function works exactly like ψ^{AF}.

Note that besides being scale invariant, the non-fuzzy versions of ψ fulfil a key property called poverty consistency (Lasso de la Vega, 2010). If an identification function satisfies this property and identifies a person with $c_n = m$ as poor, then it identifies as poor any person for whom $c_n > m$. Lasso de la Vega (2010) shows that ψ^{AF} is the only "non trivial dichotomized identification function" (p. 6) that satisfies poverty consistency, but her results are easily extendable to any non-fuzzy ψ.

The function defining the breadth of poverty

Both the univariate and the multivariate poverty measurement literature require the individual poverty function not only to identify the poor, but also to capture the intensity, or breadth, of the poverty experience. In the case of continuous variables, the magnitude of the poverty gap between the attainment and the poverty line is of usual interest. However such measurement is meaningless with ordinal variables. Yet with the latter it is still possible and meaningful to make the individual poverty function depend on the number of deprivations. Therefore the individual poverty function can be defined as the product of an identification function and a function g that captures the breadth of the poverty experience. More precisely we express the individual poverty function p_n as

$$p_n(X_n;Z,W,k) = \psi(X_n;Z,W,k)g(X_n;Z,W) \tag{2.3}$$

where g, which measures the breadth of poverty, depends on the number of deprivations: g may in fact be considered as a deprivation score, like c_n. For instance, in the case of the adjusted headcount ratio from the Alkire–Foster family, $g^{AF}=c_n$. In the case of the family of deprivation scores defined by Chakravarty and D'Ambrosio (2006), $g^{CD}=h(c_n)$, where $h'>0$ and $h''>0$ (see Figure 2.2). More generally, g is a real-valued function that maps into the interval [0, 1] and is characterized by not increasing when any attainment (e.g. x_{n1}) is increased; and by strictly decreasing when an increase, $\delta>0$, in a given variable ceases the respective deprivation, i.e. $x_{nd}+\delta>z_d>x_{nd}$.

Note that in Figure 2.2 we have also contemplated the possibility of a concave g function ($g^{concave}$). However concave breadth functions are never considered in the literature. The reason is that most authors require overall social poverty not to increase whenever inequality in deprivation counts among the poor decreases. As explained more clearly in Section 2.2.4, this property cannot be fulfilled with concave g functions.

Here an illustration will clarify the point. Imagine we have a person with three deprivations and a person with six deprivations and $g=\sqrt{c_n}$. Then the poorer person gives one deprivation to the less poor person, so now the poorer person is better-off with "just" five deprivations whereas the less poor person is

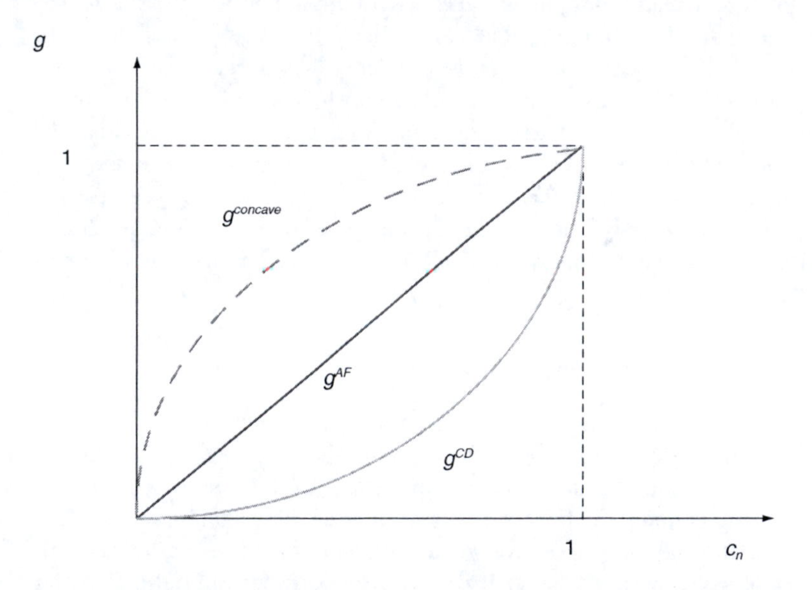

Figure 2.2 Examples of function g defining the breadth of poverty.

worse off with four deprivations. Assuming a union approach, if we compute the change in the sum of individual poverty (p_n) we get: $\Delta(p_1+p_2)=5^{0.5}+4^{0.5}-3^{0.5}-6^{0.}$ $=0.54>0$. That is overall poverty increased even though inequality among the poor was reduced. By contrast let us consider a convex function like $g=c^2$. In this case the change in poverty would be: $\Delta(p_1+p_2)=25+16-9-36=-4$. Now overall poverty decreased. That is why the literature rules out concave breadth functions.

Considering now the behaviour of ψ and g, it is easy to show that p_n satisfies the following basic properties drawn from a broader set of properties discussed by Alkire and Foster (2011):

- Normalization: $p_n=0$ if and only if person n is not poor, i.e. $\psi=0$; and $p_n=1$ if person n is deprived in every variable, i.e. $g=1$.
- Scale invariance: changes in the (arbitrary) scale of the ordinal variable do not affect p_n.
- Individual deprivation focus: if n is not deprived in d, i.e. $x_{nd}>z_d$, and n receives a transfer $\delta_d>0$, then p_n does not change.
- Individual weak monotonicity: if n receives a transfer $\delta_d>0$, then p_n does not increase.
- Individual dimensional monotonicity: if n receives a transfer $\delta_d>0$ such that $x_{nd}+\delta_d>z_d>x_{nd}$, then p_n decreases.

2.2.3 The social poverty function

The social poverty function is derived in the third stage by aggregating the individual poverty functions. There are however different ways of performing that aggregation.

The social poverty function is an average of the individual poverty functions

Here we define the social poverty function P as:

$$P(X;Z,W,k)=\frac{1}{N}\sum_{n=1}^{N}p_n(X_n;Z,W,k) \qquad (2.4)$$

Note that P inherits all the properties fulfilled by p_n and, in addition to them, it fulfils the following basic properties:

- Anonymity, also called symmetry: P does not change when the rows of X are permutated. The idea is to treat people's well-being equally.
- Principle of population, also called population replication invariance: if every person is replicated $\lambda>0$ times, then P does not change. This property enables the comparison of social poverty across populations of different size.

- Poverty focus: changes in the well-being of the non-poor that do not affect their poverty status do not change P.
- Additive decomposability: if the population is partitioned into G non-overlapping groups of people, then P can be expressed as the weighted sum of group-specific poverty indices, where the weights are the share of the population in each group, i.e.:

$$P=\sum_{j=1}^{G} s_j P_j,$$

where P_j is social poverty in group j and s_j is the proportion of the population in group j.
- Subgroup consistency: if the population is partitioned into G non-overlapping groups of people, and poverty increases/decreases in one group, but does not change in others, then overall poverty, P, increases/decreases. Subgroup consistency is implied by additive decomposability.

Table 2.1 (based on Yalonetzky, 2012) shows the family of indices of multidimensional poverty for ordinal variables available in the literature. With the exception of the family by Bossert *et al.* (2012) for $r\neq 1$, they all share the basic formulation of (2.4) and fulfil all the properties above listed. However the family of Bossert *et al.* (2012) can be expressed as $R=f(P)$, where f is continuous and $f'>0$. Hence it is easy to show that, for $r>0$, the family of Bossert *et al.* (2012) fulfils all the above properties, except for additive decomposability which can only be fulfilled when $r=1$.

Table 2.1 Explicit and implicit counting measures of multidimensional poverty for ordinal variables in the literature

Source	$\psi(X_n; Z, W, k)$	$g(X_n; Z, W)$	$P(X; Z, W, k)$
Alkire and Foster (2011)	$\mathbb{I}(c_n \geq k)$	c_n	$M^0 \equiv \frac{1}{N}\sum_{n=1}^{N}\mathbb{I}(c_n \geq k)c_n$
Bossert *et al.* (2012)	1	$[c_n]r$	$BCD \equiv \left[\frac{1}{N}\sum_{n=1}^{N}[c_n]^r\right]^{\frac{1}{r}}$
Chakravarty and D'Ambrosio (2006)	1	$h(c_n)$	$CD \equiv \frac{1}{N}\sum_{n=1}^{N}h(c_n)$
Rippin (2010)	1[1]	$[c_n]^{\gamma}c_n$	$RI \equiv \frac{1}{N}\sum_{n=1}^{N}c_n^{\gamma+1}$

Source: based on Yalonetzky (2012).

Note

1 Rippin (2010) also acknowledges the possibility of using $=\mathbb{I}(c_n \geq k)$.

Thus far none of the properties mentioned above are related to the behaviour of the indices towards the degree of inequality among the poor, which in the realm of ordinal variables, refers to the degree of association among deprivations in the population. This topic is discussed in Section 2.2.4.

Alternative aggregation methods for social poverty indices

THE AABERGE AND PELUSO (2012) APPROACH

Thus far all the social poverty indices from Table 2.1 are either means or generalized means (in the case of Bossert *et al.*, 2012) of the individual poverty functions. However there are alternative social poverty indices that can be generated with different methods of aggregation. In this section we consider one such alternative where we assume, as was done by Aaberge and Peluso (2012), that the social poverty function P is directly a function of the proportions of individuals with d deprivations, $d=1, 2,\dots, h,\dots, D$ (where D is the maximal number of deprivations)

More precisely for a number of deprivations h, let $F(h)$ be defined as $F(h)\equiv \Pr[c_n \leq h]$. Following Yaari (1987) let also Γ be a non-negative, non-decreasing, real-valued function mapping from and into the real interval [0, 1] and taking the following values $\Gamma[0]=0$ and $\Gamma[1]=1$ (for more details on Yaari's approach, see Appendix 2A).

Aaberge and Peluso (2012) propose the following social poverty function, which we normalize to lie between 0 and 1:

$$P(X;Z) = 1 - \frac{1}{D}\sum_{d=0}^{D-1}\Gamma[F(d)] \tag{2.5}$$

Note that expression (2.5) shows that if no one in the population has any deprivation,

$$\sum_{d=0}^{D-1}\Gamma[F(d)]$$

will be equal to D so that $P(X; Z)$ will be equal to 0. Conversely if everyone in the population has the maximal number D of deprivations,

$$\sum_{d=0}^{D-1}\Gamma[F(d)]$$

will be equal to 0 and therefore $P(X; Z)$ will be equal to 1.

AN EXTENSION OF THE AABERGE AND PELUSO APPROACH

For a scalar k in the real interval [0, 1] let $S(k)\equiv \Pr[c_n \geq k]$, stand for the survival function of the individual counting score c_n. Let also Γ be a non-negative, non-decreasing, real-valued function mapping from and into the real interval [0, 1], taking the values $\Gamma[0]=0$ and $\Gamma[1]=1$ and whose first and second derivatives are $\Gamma'>0$ and $\Gamma''\leq 0$.

Inspired by the work of Aaberge and Peluso (2012) we propose the following social poverty function:

$$P(X;Z)=\frac{1}{(D-m+1)}\sum_{d=m}^{D}\Gamma[S(d)] \tag{2.6}$$

where m is a positive integer in the natural interval [1, D]. For the sake of easy explanation we are omitting any uneven weights (i.e. any weight vector, W, in which $\exists i|w_i\neq 1/D$), although the family in (2.6) could be adjusted for general weighting.

The family in (2.6) takes a union approach to poverty whenever $m=1$. However by manipulating the choices of m (just like k in the case of the Alkire–Foster indices) it is possible to produce measures that identify the poor using the intersection approach, or actually any other intermediate approach. For instance, in order to measure poverty using the intersection approach we set $m=D$. For intermediate approaches we choose m such that: $1<m<D$. Interestingly, in empirical applications, when we use (2.6) with any non-union identification approach, there is no need to censor the individual deprivation count, c_n. By contrast such step is necessary, for instance, when computing the adjusted headcount ratio of the Alkire–Foster family. This operational difference is illustrated in the empirical application in Section 2.5 below.

It is easy to show that the family in (2.6) fulfils all the properties laid out in Section 2.2, with the exception of subgroup consistency. Moreover, it is also easy to show that if we impose concavity on Γ, i.e. $\Gamma''\leq 0$, then the family in (2.6) also fulfils the inequality axioms of Section 2.2.4.

2.2.4 Inequality in the distribution of deprivations among the poor.

The literature on multidimensional poverty measurement with ordinal variables has shown concern with inequality among the poor in terms of the distribution of deprivations within society. Drawing from related literatures (e.g. univariate, or focused on continuous variables) three definitions of reductions (or increases) in inequality among the poor have been proposed. In all three cases the social poverty indices are required to increase when inequality increases, or at least not to decrease (in a weak form). Hence social poverty indices are expected to penalize societies in which the burden of deprivation is shared by fewer people, thereby following an ethical tradition emphasized *inter alia* by Sen (1979) in the context of univariate continuous-variable measurement of poverty.

The first definition of change in inequality of deprivations among the poor is the rank-preserving transfer of a deprivation from the poorer to the less poor person, in which the degree of poverty is determined by the weighted number of deprivations, i.e. following a counting approach. An index that is sensitive to inequality among the poor is supposed to decrease in the presence of such a transfer, which resembles a Pigou–Dalton transfer. Rippin (2010) uses this definition and defines an axiom called “non-decreasingness under inequality increasing switch” (NDS). According to this axiom, social poverty should not decrease if a deprivation is transferred from a less poor person to a poorer person.

It is easy to show that her indices fulfil NDS if $\gamma \geq 0$. The family of Bossert *et al.* (2012) also fulfils NDS when $r \geq 1$, and the family of Chakravarty and D’Ambrosio (2006) also fulfil the same property when h is convex. The case of the Alkire and Foster (2011) family is slightly more complicated. First, such a transfer among the poor only makes sense when $k<1$. Second, if the transfer does not change the poverty status (i.e. the value of ψ) of the people involved then M^0 is not affected. That would be the case of a progressive transfer of deprivations, i.e. from the poorer to the less poor. However a regressive transfer, even among the same two people, could change the poverty status of the less poor person, depending on the relationship between k and the person’s weighted sum of deprivations. In that case, it is easy to find examples in which M^0 could *decrease* after an inequality-increasing transfer.[5] Therefore M^0 only complies with NDS when the poverty status of the people involved in the transfer does not change. Moreover, at best, M^0 remains unchanged when a progressive transfer takes place.

Chakravarty and D’Ambrosio proposed a similar axiom, called “non-decreasingness of marginal social exclusion” (NMS). According to this axiom, a deprivation increase in a poorer person should lead to higher, or at least as high, poverty than the same deprivation increase in a less poor person. In a sense, fulfilment of this property requires the individual poverty function to be quasi-convex.

Here it is worth showing that fulfilment of NDS requires the fulfilment of NMS and vice versa. Due to this axiomatic dependence, we propose:

Proposition 2.1: P satisfies NMS if and only if it satisfies NDS.
Proof.
Following Chakravarty and D’Ambrosio (2006), NMS requires that:

$$p_i(c_i + w_d) - p_i(c_i) \geq p_j(c_j + w_d) - p_j(c_j) \text{ if } c_i \geq c_j \tag{2.7}$$

where d is the variable for which there is a change in deprivation status.

Now a rank-preserving, inequality-decreasing transfer of deprivation in variable d, from i to j, yields: $c_i - w_d \geq c_j + w_d$ If p fulfils NMS we can state that:

$$p_i(c_i) - p_i(c_i - w_d) \geq p_j(c_j + w_d) - p_j(c_j) \text{ if } c_i - w_d \geq c_j + w_d \tag{2.8}$$

Because, for any $w_d > 0$, $(c_i - w_d \geq c_j + w_d) \rightarrow c_i > c_j$.

Rearranging (2.7) yields:

$$p_i(c_i)+p_j(c_j)\geq p_j(c_i+w_d)+p_j(c_j+w_d) \text{ if } c_i-w_d\geq c_j+w_d \tag{2.9}$$

Expression (2.9) is precisely the formal definition of NDS. Likewise by stating (2.9) one can work backward to demonstrate (2.7).
End of the proof.

Because of proposition 2.1, the conditions of fulfilment of NMS by the indices in Table 2.1 are the ones mentioned above for the fulfilment of NDS.

The second definition of changes in inequality among the poor is the multiplication of the column vector $C=(c_1,\ldots, c_N)'$ by a $N\times N$ bi-stochastic matrix B.[6] Both Bossert *et al.* (2012) and Alkire and Foster (2011) consider this definition. Bossert *et al.* (2012) propose a property called "S-convexity", whereby the multiplication BC' should not increase the value of P.[7] By contrast, Alkire and Foster (2011), propose a property that involves multiplying B by the achievement matrix X. They call it "weak transfers", and it is more suitable for continuous variables. Another key difference between the two properties is that Bossert *et al.* (2012) apply B to the whole population, since they take an implicit union approach to poverty identification, whereas Alkire and Foster (2011), who are more general about poverty identification, modify B so that the averaging of deprivation counts only takes place among the poor.

Just like before, it is easy to show that the indices in Table 2.1 fulfil "S-convexity" or "weak transfers" if, respectively, $\gamma\geq 0$, $r\geq 1$, h is convex. For the reasons explained above, M^0 does not fulfil "S-convexity" or "weak transfers".[8]

Finally, the third definition is proposed by Alkire and Foster (2011) and called "association decreasing rearrangement among the poor". Based on the ideas of Boland and Proschan (1988), this rearrangement is only applicable to a pair of attainment vectors X_i and X_j (corresponding to people i and j) whenever there is vector dominance between the two. Any rearrangement of attainments that breaks such vector dominance is considered valid for the definition. Based on this definition, Alkire and Foster (2011) propose the property of "weak rearrangement", whereby a poverty index should not increase after an "association decreasing rearrangement among the poor". It is easy to show that the indices in Table 2.1 fulfil this property under the same conditions as before (e.g. convexity of h). Alkire and Foster (2011) also state that M^0 fulfils "weak rearrangement". However, for the reasons explained above, it is possible to find examples for which, with certain values of k, M^0 could decrease in the event of an "association *increasing* rearrangement among the poor".

As mentioned in the previous subsection, it is also easy to show that if we impose concavity on Γ, i.e. $\Gamma'>0$ and $\Gamma''\leq 0$, then our version of the Aaberge–Peluso family, i.e. (2.6), also fulfils the inequality axioms of this section. For instance consider the fulfilment of NDS. Suppose a deprivation is transferred from a poor person with i deprivations to a poorer person with j deprivations (therefore $i<j$ and $S(i)>S(j)$). Then, if S^* stands for the survival function after the transfer, it is easy to realize that:

$$S^*(i-1)=S(i-1);$$

$$S^*(i)=S(i)-\frac{1}{N};$$

$$S^*(j)=S(j);$$

$$S^*(j+1)=S(j+1)+\frac{1}{N}.$$

Hence the change in P, ΔP, is going to be equal to: $\Delta P=\Gamma(S^*(j+1))-\Gamma(S(j+1))+\Gamma(S^*(i))-\Gamma(S(i))$. Because $\Gamma'>0$, then the right-hand side of the equation is non-negative if $\Gamma''<0$.

Now consider the fulfilment of "weak rearrangement" by our version of the Aaberge–Peluso family. If the vector of attainments of individual 1 dominates that of 2, both being poor, then it must be true that $0<c_1<c_2\leq D$. Any association-decreasing rearrangement entails a transfer of t deprivations from individual 2 to individual 1 such that: $c_1^*=c_1+t$ and $c_2^*=c_2-t$; where c^* is the deprivation score after the rearrangement, and $t<c_2-c_1$ (lest vector dominance be reinstituted but with 2 having fewer deprivations).

Then, if $c_1^*<c_2^*$, it is easy to realize that: $S^*(i)=S(i)+1/N\ \forall i\in\,]c_1, c_1^*]$; $S^*(j)=S(j)-1/N\,\forall j\in\,]c_2^*, c_2]$. Hence the change in P, ΔP, is going to be equal to: $\Delta P=\Sigma_{j\in]c_2^*,\, c_2]}[\Gamma(S^*(j))-\Gamma(S(j))]+\Sigma_{i\in]c_1,\, c_1^*]}[\Gamma(S^*(i))-\Gamma(S(i))]$. Because $\Gamma'>0$, then the right-hand side of the equation is non-positive if $\Gamma''<0$.

However, note that an association-decreasing rearrangement does not need to preserve the ranks between 1 and 2 in terms of deprivation scores. That is, it could also be the case that, after the transfer, $c_1^*>c_2^*$, which means: $c_1<c_2^*<c_1^*<c_2$ (since $t<c_2-c_1$ has to hold). In this case then the relationship between the survival probabilities after and before the transfer are the following: $S^*(i)=S(i)+1/N\,\forall i\in\,]c_1, c_2^*]$; $S^*(j)=S(j)-1/N\,\forall j\in\,]c_1^*, c_2]$. Hence the change in P, ΔP, is going to be equal to: $\Delta P=\Sigma_{j\in]c_2^*,c_2]}[\Gamma(S^*(j))-\Gamma(S(j))]+\Sigma_{i\in]c_1,c_1^*]}[\Gamma(S^*(i))-\Gamma(S(i))]$. Because $\Gamma'>0$, then the right-hand side of the equation is, again, non-positive if $\Gamma''<0$.

In summary, any transfer that breaks up vector dominance, thereby providing an association-decreasing rearrangement, cannot increase the value of our version of the Aaberge–Peluso index, independently of whether the transfer generates a reversion of ranks in terms of deprivation scores.[9]

2.3 More general identification functions and the multidimensional poverty line approach

The indices in Table 2.1 exhibit two types of relationships between the identification function, ψ, and g. Either they adopt an explicit or implicit union approach and g is a convex function of the deprivation score, c_n; or, as in the case of the Alkire–Foster family, both ψ and g depend on the same deprivation score, c_n. However these are not the only alternatives for identification, and for measurement of the intensity of multiple deprivations. Here we consider two additional alternatives.

First, we consider alternative identification functions. The counting poverty literature compares, either explicitly or implicitly, c_n against a threshold k. However, one could also define an identification approach like the following:

$$\psi(x_1, x_2) = \mathbb{I}(x_1 \leq z_1)\mathbb{I}(x_2 \leq z_2) \tag{2.10}$$

The identification function in (2.10) states that a person is deemed poor if and only if the person is deprived in variables 1 and 2, simultaneously. Note that with this approach the other variables are irrelevant for identification purposes, although they may still be considered in the measurement of the intensity of individual deprivations, i.e. they may still enter g. Likewise, with approaches like (2.10) the weighting vector is irrelevant at the identification stage.

Second, more general identification functions can bring the counting approach in closer proximity to an alternative approach to multidimensional poverty measurement based on multidimensional poverty lines. The multidimensional poverty line approach has been discussed by Duclos *et al.* (2006, 2007) in the context of partial orderings. Recently Merz and Rathjen (2012) have proposed measures of multidimensional poverty based on this approach, which are suitable for continuous variables.

The idea of this approach is to identify the poor by checking whether people's attainment bundle falls within or outside a poverty hyperspace defined by a multidimensional poverty line, which is usually a function of the individual deprivation lines. One way of doing this checking is by adopting a utility-function approach and comparing the "utility" yielded by the person's attainment vector, X_n, against a "utility" reference level stemming from evaluating the utility function using the deprivation line vector Z. This is the identification approach adopted by Merz and Rathjen (2012). Formally, if U stands for the evaluation function and fulfils some standard desirable properties, e.g. $U' > 0$, then the identification function would be:

$$\psi(x_n; Z) = \mathbb{I}[U(X_n) \leq U(Z)] \tag{2.11}$$

where $U(Z)$ is the multidimensional poverty line. Note that if, in particular,

$$U(X_n) = \sum_{d=1}^{D} w_d x_{nd},$$

and the weights are market or shadow prices, then the identification approach boils down to the traditional approach used in monetary poverty measurement. Generally, with a multidimensional poverty line approach, one could propose individual poverty indices like the following Foster–Green–Thorbecke family:

$$p_n = \mathbb{I}[U(X_n) \leq U(Z)]\left(1 - \frac{U(X_n)}{U(Z)}\right)^{\alpha}, \forall \alpha \geq 0 \quad (2.12)$$

When applying this approach to ordinal variables an important caveat to consider is that the evaluation function, U, is sensitive to the *arbitrary* scale chosen for the ordinal variables. Hence this approach is not really appropriate for ordinal variables, unless we restrict it to dichotomous variables.

2.4 Other issues in multidimensional poverty measurement with ordinal variables

2.4.1 Weights

Weight selection is one of the most important decisions in the computation of composite indices, including those measuring multidimensional poverty. The literature on weights is vast and the reader is referred to Decancq and Lugo (2013) and OECD/EU (2008) for detailed presentations. In a nutshell there are two types of weight vectors, W: exogenous weights and endogenous weights. The former include weights generated by participatory methods (i.e. a group of people chooses them), or by expert opinion (i.e. the researcher/policymaker decides). The latter include weights generated by statistical methods like principal components analysis. These weights are endogenous because they depend on X.

While the choice of exogenous weights in multidimensional poverty measurement has gained traction with the computation of the "Multidimensional Poverty Index" by the UNDP (2010), endogenous weights are also widely used both in academic and policy-oriented work. The literature discusses some of the pros and cons of resorting to the different methods of weighting. However there is not yet a proper assessment of the implications of using $W(X)$, instead of W, in terms of axiom fulfilment. It is not a priori clear that the individual and social poverty functions mentioned in this chapter fulfil the above listed axioms when weights depend on the matrix of achievements, i.e. the data.

2.4.2 Robustness and partial orderings

With several choices to make, e.g. regarding deprivation lines, weights or multidimensional cut-offs (k), the concern for the robustness of poverty comparisons based on multidimensional poverty measures is reasonable. Most of the recent literature has dealt with the robustness of ordinal poverty comparisons, i.e. whether country "A" exhibits higher poverty than "B" or whether "poverty" increased or decreased from year one to year two, regardless, in both examples, by how much. For counting measures dominance conditions have been proposed by Chakravarty and D'Ambrosio (2006), Lasso de la Vega (2010), Alkire and Foster (2011) and Aaberge and Peluso (2012). None of these approaches, though, considers full robustness of ordinal comparisons. In other words, they take for granted certain parameters, e.g. the deprivation lines, and ask whether comparisons are robust in changes to the other parameters. By contrast, Yalonetzky (2012) has recently provided full robustness conditions for ordinal comparisons for the general class of measures described by (2.4). His main result is that traditional dominance conditions are only applicable to extreme identification approaches, i.e. union and intersection.

A complementary literature looks at the sensitivity of cardinal poverty comparisons, i.e. comparisons in which the quantitative difference matters, to changes in the composite indices' parameters. One recent example of sensitivity to weighting choices is provided by Nussbaumer *et al.* (2011) who produce empirical distributions of adjusted headcount ratios (M^0) using random number generators for weights. A thorough treatment of sensitivity analysis for composite indices that is relevant for multidimensional poverty indices can be found in Saltelli *et al.* (2008).

2.5 A simple empirical illustration

Table 2.2a presents a simple empirical illustration. We assume that there are five individuals and each one has a different number of deprivations out of a total maximum of ten. Each deprivation is equally weighted. For the Alkire and Foster index we examine three possibilities for the threshold: $k=1, 5, 10$.

It is then easy to find out that the index M_0 proposed by Alkire and Foster will be equal to 0.5 when $k=1$, to 0.46 when $k=5$ and to 0.2 when $k=10$.

For the index *BCD* proposed by Bossert, Chakravarty and D'Ambrosio (see Table 2.1), assuming that $r=2$ we find that $BCD=0.615$.

For the index *CD* proposed by Chakravarty and D'Ambrosio (see Table 2.1), assuming that $h(c_n)=(c_n)^{1.2}$, it turns out that *CD* 0.4677.

Finally for the index *RI* proposed by Rippin, assuming that $\gamma=1.5$ so that $\gamma+1=2.5$, we obtain $RI=0.3414$.

If we now turn to the Aaberge and Peluso approach we find that the social deprivation index is equal to 0.7 while the "extended Aaberge and Peluso

Table 2.2a A first empirical illustration

Individual i	*Number* d *of deprivations*	*The value of* c_n *when* k = *1*	*The value of* c_n *when* k = *5*	*The value of* c_n *when* k = *10*
1	2	0.2	0	0
2	7	0.7	0.7	0
3	0	0	0	0
4	6	0.6	0.6	0
5	10	1	1	1

approach" gives a social deprivation index equal to 0.6861 when $m=1$, to 0.5872 when $m=5$ and to 0.4472 when $m=10$ (see Appendix 2B for the details of these computations).

In the second empirical illustration presented in Table 2.2b, we assume that individual 1, who had two deprivations, has now only one; while individual 2, who had seven deprivations, has now eight. In other words we made a transfer of one deprivation from a "deprived" to a "more deprived" individual. We then expect the level of social deprivation to increase.

Calculations similar to those given in the previous illustration, assuming the same values for the same functional forms for the indices *BCD*, *CD* and *RI*, give the following results.

The Alkire and Foster index M_0 is now equal to 0.5 when $k=1$, to 0.48 when $k=5$ and to 0.2 when $k=10$. When $k=1$ or 10, there was thus, as expected, no change in M_0 but M_0 increased from 0.46 to 0.48 when $k=5$.

The indices *BCD*, *CD* and *RI* are now respectively equal to 0.634, 0.474 and 0.3709 so that they all increased.

For the Aaberge and Peluso approach the social deprivation index is now equal to 0.716.

For the "extended Aaberge and Peluso approach" the index is equal to 0.6927 when $m=1$, to 0.6181 when $m=5$ and to 0.4472 when $m=10$. Note that here also, as expected, these indices are now higher than in the first empirical illustration, except in the case where m = 10 (intersection case) where no change was observed since the number of individuals with the maximum number of deprivations did not change.

In a third illustration we still assume a total number of 25 deprivations in society (as in the two previous examples) but now we take the case of extreme inequality. We assume that ten is the maximum number of deprivations for an individual. Two individuals each have ten deprivations, two others do not have any deprivation and one has five. Table 2.2c presents the basic data for this third empirical illustration.

The values of the different indices are now as follows:

For the Alkire and Foster index M_0 we find out that M_0 is equal to 0.5 when $k=1$ or 5 and to 0.4 when $k=10$.

The BCD index is now equal to 0.6708, the CD index to 0.4871 and the RI index to 0.4354.

Table 2.2b A second empirical illustration

Individual i	*Number* d *of deprivations*	*The value of* c_n *when* k = *1*	*The value of* c_n *when* k = *5*	*The value of* c_n *when* k = *10*
1	1	0.1	0	0
2	8	0.8	0.8	0
3	0	0	0	0
4	6	0.6	0.6	0
5	10	1	1	1

Table 2.2c A third empirical illustration

Individual i	*Number* d *of deprivations*	*The value of* c_n *when* k = *1*	*The value of* c_n *when* k = *5*	*The value of* c_n *when* k = *10*
1	0	0	0	0
2	10	1	1	1
3	0	0	0	0
4	5	0.5	0.5	0
5	10	1	1	1

In other words social deprivation is now even higher than in the second illustration, except, as expected, for the M_0 index when $k=1$.

For the Aaberge and Peluso approach the social deprivation index is now equal to 0.74.

For the "extended Aaberge and Peluso approach" the index is equal to 0.7036 when $m=1$, to 0.6562 when $m=5$ and to 0.6325 when $m=10$. Note that here also, as expected, these indices are now even higher than in the second empirical illustration. This is also true in the case where $m=10$ (intersection case) because now we have two individuals with the maximum number of deprivations.

2.6 Concluding remarks

This chapter sought to introduce the reader to the issues related to the measurement of multidimensional poverty when only counting or ordinal variables are available. We made a distinction not only between individual and social deprivation functions but, also, at the individual level, between the identification and the "breadth of poverty" function. We listed the properties of these individual and social deprivation functions and explained that a social deprivation function may be derived as a summation of the individual deprivation functions but we also showed that alternative aggregation methods are available. We also examined the issues of inequality in the distribution of deprivations, of weights and robustness. Finally we presented three simple empirical illustrations. This chapter includes also two appendices, one summarizing a famous paper of Yaari (1987) from which the Aaberge and Peluso (2012) approach is derived, and one giving more details on the computation of the various social deprivation functions in the three cases examined. Let us note finally that the issues analysed in this chapter are relevant whenever only counting or ordinal variables are available. The results presented could therefore be used also in studies of health and happiness since health and happiness surveys often include only counting or ordinal variables.

Appendix 2A: On Yaari's (1987) dual theory of choice under risk

In his famous article on "The Dual Theory of Choice under Risk", Yaari (1987) derived first a theorem on the expected utility theory. This theorem may be simply summarized as follows:[10]

Assume random variables defined on some given probability space with values in the unit interval and let $G_v(t) = \Pr\{v > t\}$ with $0 \leq t < 1$. Following Yaari (1987) call G the decumulative distribution function (DDF).

Yaari then defined five basic axioms from which the theorem on expected utility may be derived.

1 *Neutrality*: assume two random variables u and v with DDF's G_u and G_v. If $G_u = G_v$, then $u(v)$.
2 *Complete weak order*: the preference ordering is reflexive, transitive and connected.
3 *Continuity* (with respect to L_1-convergence).
4 *Monotonicity* (with respect to first order stochastic dominance).
5 *Independence*: if G is preferred to G' and $0 \leq \alpha < 1$, then $\alpha G + (1-\alpha)H$ is preferred to $\alpha G' + (1-\alpha)H$.

Yaari's (1987) theorem on expected utility then states that a preference relation satisfies axioms 1 to 5 if and only if there exists a continuous and non-decreasing real function ϕ, defined on the unit interval, such that for all random variables u and v, "u is preferred to v" is equivalent to $E((u) \geq E((v)$ where E refers to the expectancy.

Moreover this function φ is such that [1; φ (t)]~[t; 1] which in its simplest interpretation means that if there is a probability $((t)$ of having an income of 1 and a probability $(1-((t))$ of having an income of 0, this is equivalent to a situation where one receives with certainty an income of t.

Yaari (1987) derived then a similar theorem for what he called the "dual theory of choice". He first added the following sixth axiom.

6 *Dual independence*: this notion is equivalent to *co-monotonicity* and two variables are co-monotonic if neither of them is a hedge against the other. For Yaari (1987) this axiom can then be expressed as follows: "u is preferred to v" implies that $\alpha u + (1-\alpha)w$ is preferred to $\alpha v + (1-\alpha)w$ where $0 \leq \alpha \leq 1$.

Yaari's (1987) theorem on the dual theory of choice under risk then states that a preference relation satisfies axioms 1, 2, 3, 4 and 6 if and only if there exists a continuous and non-decreasing real function f, defined on the unit interval, such that, for all random variables u and v, u is preferred to v is equivalent to saying that

$$\int_0^1 f(G_u(t))dt \geq \int_0^1 f(G_v(t))dt. \tag{2A.1}$$

Moreover the function f is such that for all p satisfying $0 \leq p \leq 1$, $f(p)$ solves the preference equation [1; p] ([$f(p)$; 1] which means that a situation where you have a probability p of having an income of 1 and a probability $(1-p)$ of having a

0 income is equivalent to a situation where you receive with certainty an income of $f(p)$.

As stressed by Yaari, the utility U of the dual theory has two important properties.

1 U assigns to each random variable, its certainty equivalent. In other words if v (a random variable) belongs to V (the set of all random variables), then $U(v)$ is equal to that sum of money which, when received with certainty, is considered by the agent equally as good as v. Using the notations introduced previously we can therefore write that $v([U(v); 1]$.

2 U is linear in payments. In other words when the values of a random variable are subjected to some fixed positive affine transformation, the corresponding value of U undergoes the same transformation.

Yaari (1987) emphasized an important difference between the expected utility theory and his dual theory of choice under risk. Under expected utility theory, if a person has a constant absolute as well as relative risk aversion, he must be risk neutral, and as a consequence, he will always rank random variables by comparing their means. But under the dual theory of choice, there is linearity without risk neutrality being implied. In fact under the dual theory an agent would rank random variables by comparing their means if and only if the function f coincides with the identity, that is, $f(p)=p$ for $0 \leq p \leq 1$. But there is nothing which forces identity. As a consequence an agent's attitude towards wealth does not imply anything as far as his/her attitude towards risk is concerned.

Let us make a last remark by comparing the function f with the von Neumann–Morgenstern utility ϕ. In fact when we write that $[1; p]$ $([t; 1]$ we know that $f(p)$ is the value of t that solves this equivalence while $((t)$ is the value of p that solves it. Therefore $f=\varphi^{(-1)}$.

To apply Yaari's (1987) results to the topic of deprivation we have to work not with the concept of "Decumulative Distribution Functions" (DDF) as in Yaari (1987) but with the more traditional concept of "Cumulative Distribution Functions" (CDF). The intuition behind a DDF applied to the distribution of income is that if the DDF G_2 lies above the DDF G_1 for all income values (income is assumed to vary between 0 and 1), then clearly distribution 2 first order stochastically dominates distribution 1 since, whatever the income level t, there is always a higher proportion of individuals in distribution 2 having an income higher than t. If however we work with the notion of deprivation, assumed to vary also between 0 and 1, we will say that if the CDF F_2 lies above the CDF F_1, whatever the level of deprivation h, then clearly distribution 2 first order stochastically dominates distribution 1 because, whatever the level h of deprivation, there is always a higher proportion of individuals in distribution 2 with a level of deprivation smaller than h.

If we now apply to these cumulative distribution functions F a non-decreasing function $\Gamma(F)$ and assume that $\Gamma(0)=0$ and $\Gamma(1)=1$ and given that $F(0)=0$ and $F(1)=1$, we can conclude, as Yaari (1987) did in the case of risk, that [0;

$(1-F)] \sim [\Gamma\ (F);\ 1]$. In its simplest interpretation this means that if F% of the population has the maximum number of deprivations while $(1-F)$% has no deprivation at all, such a situation is equivalent, from a social welfare point of view, to one where everybody has a level of deprivation h equal to $((F)$.

Let us now reason in discrete terms and assume, as Aaberge and Peluso (2012) did, that the maximum number of deprivations is n. If we call $F(i)$ the cumulative percentage of individuals with up to i deprivations, with

$$F(i) = \sum_{h=0}^{i} q_h,$$

q_h referring to the proportion of individuals with h deprivations, Aaberge and Peluso (2012), applying Yaari's (1987) theorem on the dual theory of choice under risk and using axioms similar to those defined by Yaari (1987), concluded that stating that a cumulative distribution F_1 of deprivation counts is preferable to a cumulative distribution F_2 amounts to saying that

$$\sum_{k=0}^{n-1} (\left(\sum_{j=0}^{k} q_{1j} \right) \geq \sum_{k=0}^{n-1} (\left(\sum_{j=0}^{k} q_{2j} \right) \tag{2A.2}$$

where n is the maximum number of deprivations and q_j is the proportion of individuals with j deprivations, the subscripts 1 and 2 referring evidently to the two distributions F_1 and F_2.

Aaberge and Peluso (2012) defined then the "social deprivation score" D as

$$D = n - \sum_{h=0}^{n-1} ((F(h)) \tag{2A.3}$$

where, as assumed previously, $((0)=0$ and $((1)=1$, (being a non-decreasing function. Expression (2A.3) is easy to understand. Assume no one in the population has any deprivation. Then $F(i)=1$, $(i=1, 2, \ldots, (n-1), n$ so that

$$\sum_{h=0}^{n-1} ((F(h)) = n$$

and therefore D in (2A.3) is equal to 0. If on the contrary everyone in the population has the maximal number of deprivation, then $F(i)=0$, $(i=1, 2, \ldots, (n-1)$ and $F(n)=1$ so that

$$\sum_{h=0}^{n-1} ((F(h)) = 0$$

and hence $D=n$.

Appendix 2B: Detailed computations for the empirical illustrations of the Aaberge and Peluso (2012) approach and of the "extended Aaberge and Peluso approach"

Let us first examine the social poverty function corresponding to the Aaberge and Peluso (2012) approach. We start with the first empirical illustration and derive the social poverty function for the Aaberge and Peluso approach in Table 2B.a1.

It is then easy to see that

$$\sum_{h=0}^{10-1}\Gamma[F(h)]=3 \text{ so that } 1-\frac{1}{D}\sum_{h=1}^{10-1}\Gamma[F(h)]=1-0.3=0.7.$$

Let us now look at the second empirical illustration (Table 2B.a2).

It is then easy to see that

$$\sum_{h=0}^{10-1}\Gamma[F(h)]=2.84 \text{ so that } 1-\frac{1}{D}\sum_{h=1}^{10-1}\Gamma[F(h)]=1-0.284=0.716.$$

So here also we find that there is a greater level of social deprivation in the second than in the first empirical illustration. Let us now look at the third empirical illustration (Table 2B.a3).

It is then easy to see that

$$\sum_{h=0}^{10-1}\Gamma[F(h)]=2.6 \text{ so that } 1-\frac{1}{D}\sum_{h=1}^{10-1}\Gamma[F(h)]=1-0.26=0.74.$$

Table 2B.a.1 The Aaberge and Peluso approach in the case of the first empirical illustration

Number h *of deprivations*	*Absolute frequency of the number of deprivations*	*Relative frequency of the number of deprivations*	*Cumulative relative frequency* F(h) *of the number of deprivations*	*Value of* $((F(h))=(F(h))^2$
0	1	0.2	0.2	0.04
1	0	0	0.2	0.04
2	1	0.2	0.4	0.16
3	0	0	0.4	0.16
4	0	0	0.4	0.16
5	0	0	0.4	0.16
6	1	0.2	0.6	0.36
7	1	0.2	0.8	0.64
8	0	0	0.8	0.64
9	0	0	0.8	0.64
10	1	0.2	1	1

Table 2B.a.2 The Aaberge and Peluso approach in the case of the second empirical illustration

Number h *of deprivations*	*Absolute frequency of the number of deprivations*	*Relative frequency of the number of deprivations*	*Cumulative relative frequency* F(h) *of the number of deprivations*	*Value of* $((F(h))=(F(h))^2$
0	1	0.2	0.2	0.04
1	1	0.2	0.4	0.16
2	0	0	0.4	0.16
3	0	0	0.4	0.16
4	0	0	0.4	0.16
5	0	0	0.4	0.16
6	1	0.2	0.6	0.36
7	0	0	0.6	0.36
8	1	0.2	0.8	0.64
9	0	0	0.8	0.64
10	1	0.2	1	1

Table 2B.a.3 The Aaberge and Peluso approach in the case of the third empirical illustration

Number h *of deprivations*	*Absolute frequency of the number of deprivations*	*Relative frequency of the number of deprivations*	*Cumulative relative frequency* F(h) *of the number of deprivations*	*Value of* $((F(h))=(F(h))^2$
0	2	0.4	0.4	0.16
1	0	0	0.4	0.16
2	0	0	0.4	0.16
3	0	0	0.4	0.16
4	0	0	0.4	0.16
5	1	0.2	0.6	0.36
6	0	0	0.6	0.36
7	0	0	0.6	0.36
8	0	0	0.6	0.36
9	0	0	0.6	0.36
10	2	0.4	1	1

As expected the level of social deprivation is even higher for the third empirical illustration.

Let us now illustrate of the "extended Aaberge and Peluso approach". The maximum number D of deprivations is again assumed to be equal to ten (Table 2B.a4).

In the case where $m=1$, we derive that

$$\sum_{d=1}^{10} \Gamma[S(d)] = 6.8613.$$

Table 2B.a.4 The "extended Aaberge and Peluso approach" in the case of the first empirical illustration

Number d *of deprivations*	*Absolute frequency of the number of deprivations*	*Relative frequency of the number of deprivations*	*Value of* S(d)	*Value of* ((S(d)) *with* ((S)=√(2&S)
0	1	0.2	1	1
1	0	0	0.8	0.8944
2	1	0.2	0.8	0.8944
3	0	0	0.6	0.7746
4	0	0	0.6	0.7746
5	0	0	0.6	0.7746
6	1	0.2	0.6	0.7746
7	1	0.2	0.4	0.6325
8	0	0	0.2	0.4472
9	0	0	0.2	0.4472
10	1	0.2	0.2	0.4472

It then follows that with $m=1$ (the union approach)

$$P(X;Z)=\frac{1}{(10-1+1)}\sum_{d=1}^{10}\Gamma[S(d)]=\left(\frac{6.8613}{10}\right)=0.68613.$$

In the case where $m=10$, we find similarly that

$$\sum_{d=10}^{10}\Gamma[S(d)]=0.4472.$$

It then follows that with $m=10$ (the intersection approach)

$$P(X;Z)=\frac{1}{(10-10+1)}\sum_{d=10}^{10}\Gamma[S(d)]=\left(\frac{0.4472}{1}\right)=0.4472.$$

Finally in the case where $m=5$, we derive that

$$\sum_{d=5}^{10}\Gamma[S(d)]=3.5233\,.$$

It then follows that with $m=5$, we obtain

$$P(X;Z)=\frac{1}{(10-5+1)}\sum_{d=5}^{10}\Gamma[S(d)]=\left(\frac{1}{6}\right)3.5233=0.5872.$$

Let us now look at the second empirical illustration in the three cases examined (as far as the value of m is concerned) (Table 2B.a5).

Table 2B.a.5 The "extended Aaberge and Peluso approach" in the case of the second empirical illustration

Number h *of deprivations*	*Absolute frequency of the number of deprivations*	*Relative frequency of the number of deprivations*	*Value of* S(d)	*Value of* ((S(d)) *with* ((S)=√(2&S)
0	1	0.2	1	1
1	1	0.2	0.8	0.8944
2	0	0	0.6	0.7746
3	0	0	0.6	0.7746
4	0	0	0.6	0.7746
5	0	0	0.6	0.7746
6	1	0.2	0.6	0.7746
7	0	0	0.4	0.6325
8	1	0.2	0.4	0.6325
9	0	0	0.2	0.4472
10	1	0.2	0.2	0.4472

In the case where $m=1$, it is easy to see that

$$\sum_{d=1}^{10} \Gamma[S(d)] = 6.9268.$$

It then follows that with $m=1$ (the union approach)

$$P(X;Z) = \frac{1}{(10-1+1)} \sum_{d=1}^{10} \Gamma[S(d)] = \left(\frac{6.9268}{10}\right) = 0.6927$$

which, as expected, is a higher value than that derived from Table 2B.a.4.

In the case where $m=10$, we easily derive that

$$\sum_{d=10}^{10} \Gamma[S(d)] = 0.4472.$$

It then follows that with $m=10$ (the intersection approach)

$$P(X;Z) = \frac{1}{(10-10+1)} \sum_{d=10}^{10} \Gamma[S(d)] = \left(\frac{0.4472}{1}\right) = 0.4472.$$

Clearly in such a case the value of the social deprivation index did not change.

Finally in the intermediary case where $m=5$, we find out that

$$\sum_{d=5}^{10} \Gamma[S(d)] = 3.7086.$$

It then follows that with $m=5$, we obtain

$$P(X;Z)=\frac{1}{(10-5+1)}\sum_{d=5}^{10}\Gamma[S(d)]=\left(\frac{1}{6}\right)3.7086=0.6181$$

a value which, as expected, is higher than the one derived in this case from the data of Table 2B.a.4.

Let us finally take a look at the third empirical illustration (Table 2B.a6).

It is then easy to see that

$$\sum_{d=1}^{10}\Gamma[S(d)]=7.0355.$$

It then follows that with $m=1$ (the union approach)

$$P(X;Z)=\frac{1}{(10-1+1)}\sum_{d=1}^{10}\Gamma[S(d)]=\left(\frac{7.0355}{10}\right)=0.7036$$

which, as expected, is even higher value than the value obtained in this case in Table 2B.a.5.

Let us now examine the case where $m=10$. We observe that

$$\sum_{d=10}^{10}\Gamma[S(d)]=0.6325.$$

It then follows that with $m=10$ (the intersection approach)

$$P(X;Z)=\frac{1}{(10-10+1)}\sum_{d=10}^{10}\Gamma[S(d)]=\left(\frac{0.6325}{1}\right)=0.6325,$$

Table 2B.a.6 "The extended Aaberge and Peluso approach" in the case of the third empirical illustration

Number h *of deprivations*	*Absolute frequency of the number of deprivations*	*Relative frequency of the number of deprivations*	*Value of* S(d)	*Value of* ((S(d)) *with* ((S)=√(2&S)
0	2	0.4	1	1
1	0	0	0.6	0.7746
2	0	0	0.6	0.7746
3	0	0	0.6	0.7746
4	0	0	0.6	0.7746
5	1	0.2	0.6	0.7746
6	0	0	0.4	0.6325
7	0	0	0.4	0.6325
8	0	0	0.4	0.6325
9	0	0	0.4	0.6325
10	2	0.4	0.4	0.6325

a value of the social deprivation index which is, as expected, higher than in the two first illustrations, since we have now two individuals with the maximum number of deprivations.

Finally let us take a look at the intermediary case where $m=5$. We then see that

$$\sum_{d=5}^{10} \Gamma[S(d)] = 3.9371.$$

It then follows that with $m=5$, we have

$$P(X;Z) = \frac{1}{(10-5+1)} \sum_{d=5}^{10} \Gamma[S(d)] = \left(\frac{1}{6}\right) 3.9371 = 0.6562,$$

a value which, as expected, is even higher than the one derived in this case from the data of Table 2B.a.5.

Notes

1 Jacques Silber started working on this chapter while visiting OPHI (Oxford Poverty and Human Development Initiative) which he thanks for its very warm hospitality. He also gratefully acknowledges the financial support of the Adar Foundation of the Department of Economics of Bar-Ilan University. Both authors thank an anonymous referee for his/her very useful comments and suggestions.

2 We thank an anonymous referee for drawing our attention to this point.

3 Part of the material in this section is based on Yalonetzky (2012).

4 We do not discuss how deprivation lines are set. For a discussion see Alkire and Santos (2010).

5 For instance with equal weights, if $k=2$ and the weighted sums of deprivations for the less poor and the poorer person are, respectively, two and three. A regressive transfer of one deprivation would change the counts to one and four. Therefore the total number of deprivations between the two poor people would decrease by one, since now the remaining deprivations of the less poor person are censored out of the social count due to the person's "cessation" of poverty status.

6 A bi-stochastic matrix is a square matrix whose entries are real numbers in the [0, 1] interval, such that all the elements in the same row or in the same column add up to 1, for every row and column. The use of bi-stochastic matrices in multidimensional inequality analysis is traceable to Kolm (1977).

7 S-convexity refers to a property of functions called Schur-convexity. If a function f: $\mathbf{R}^D \rightarrow \mathbf{R}$ is Schur-convex then $f(y) \geq f(x)$ for all D-dimensional vectors x and y such that x is majorized by y. If x is majorized by y then it is the case that

$$\sum_{i=1}^{k} y_i \geq \sum_{i=1}^{k} x_i \forall k = 1,2,\cdots,D \text{ and } \sum_{i=1}^{D} y_i \geq \sum_{i=1}^{D} x_i$$

where y_i and x_i are the elements of y and x ordered in decreasing order, respectively. It is easy to show that if x is obtained from y by either a Pigou-Dalton transfer or as the product of y and a bi-stochastic matrix, then x is majorized by y. For more details the reader is referred to Marshall *et al.* (2010).

8 However, as Alkire and Foster (2011) show other members of their family of measures do fulfil weak transfers. But these are only suitable for continuous variables.

9 Note that, as stressed by an anonymous referee, the traditional approach based on the expected utility theory (Tsui (1995), Atkinson (2003), Bourguignon and Chakravarty (2003) and the other articles cited in the chapter) adopts axioms based on rearrangements leaving unchanged the marginal distributions of the attributes. Aaberge and Peluso (2012) noticed however that in the deprivation analysis based on the counting approach this assumption can be weakened by imposing the less restrictive condition that rearrangements leave equal means.

10 In summarizing this theorem as well as a few others derived by Yaari (1987) we purposely simplify the exposition to give as much as possible an intuitive presentation of the results. For a rigorous presentation, see Yaari's (1987) original article.

References

Aaberge, R. and E. Peluso (2012) "A counting approach for measuring multidimensional deprivation", Discussion paper No. 700, Research Department, Statistics Norway.

Alkire, S. and J. Foster (2011) "Counting and multidimensional poverty measurement", *Journal of Public Economics* 95(7–8): 476–487.

Alkire, S. and M. E. Santos (2010) "Acute multidimensional poverty: a new index for developing countries", Human Development Research Paper 2010/11.

Atkinson, A. B. (2003) "Multidimensional deprivation: contrasting social welfare and counting approaches", *Journal of Economic Inequality* 1(1): 51–65.

Boland, P. and F. Proschan (1988) "Multivariate arrangement increasing functions with application in probability and statistics", *Journal of Multivariate Analysis* 25(2): 286–298.

Bossert, W., S. R. Chakravarty and C. D'Ambrosio (2012) "Multidimensional poverty and material deprivation with discrete data", *Review of Income and Wealth* 59(1): 29–43.

Bourguignon, F. and S. R. Chakravarty (2003) "The measurement of multidimensional poverty", *Journal of Economic Inequality* 1(1): 25–49.

Chakravarty, S. and C. D'Ambrosio (2006) "The measurement of social exclusion", *Review of Income and Wealth* 52(3): 377–398.

Decancq, K. and M. Lugo (2013) "Weights in multidimensional indices of well-being: an overview", *Econometric Reviews* 32(1): 7–34.

Donaldson, D. and J. A. Weymark (1980) "A single parameter generalization of the Gini indices of inequality", *Journal of Economic Theory* 22(1): 67–88.

Duclos, J. Y., D. Sahn and S. Younger (2006) "Robust multidimensional poverty comparisons", *Economic Journal* 116(514): 943–968.

Duclos, J. Y., D. Sahn and S. Younger (2007) "Robust multidimensional poverty comparisons with discrete indicators of wellbeing", in Jenkins, S. and J. Micklewright (eds), *Inequality and Poverty Re-examined*, 185–208, Oxford University Press.

Kolm, S. (1977) "Multidimensional egalitarianisms", *Quarterly Journal of Economics* 91(1): 1–13.

Lasso de la Vega, C. (2010) "Counting poverty orderings and deprivation curves", in Bishop, J. (ed.) *Research on Economic Inequality*, 18(7): 153–172. Emerald.

Marshall, A., I. Olkin and B. Arnold (2010) *Inequalities: Theory of Majorization and its Applications*, Springer Series in Statistics, 2nd edn.

Merz, J. and T. Rathjen (2012) "Intensity of time and income interdependent multidimensional poverty: well-being gap and minimum 2dgap – German evidence", Paper presented at the 32 IARIW Conference, Boston.

Nussbaumer, P., M. Bazilian, V. Modi and K. Yumkella (2011) "Measuring energy poverty: focusing on what matters", OPHI Working Paper 42.

OECD/EU (2008) *Handbook on Constructing Composite Indicators*, OECD Publishing.

Ravallion, M. (2011) "On multidimensional indices of poverty", *Journal of Economic Inequality* 9(2): 235–248.

Rippin, N. (2010) "Poverty severity in a multidimensional framework: the issue of inequality between dimensions", Courant Research Center, Discussion paper no. 47, University of Göttingen.

Saltelli, A., M. Ratto, T. Andres, F. Campolongo, J. Cariboni, D. Gatelli, M. Saisana and S. Tarantola (2008) *Global Sensitivity Analysis: The Primer*, Wiley.

Sen, A. (1979) "Issues in the measurement of poverty", *Scandinavian Journal of Economics* 81(2): 285–307.

Thorbecke, E. (2007) "Multidimensional poverty: conceptual and measurement issues", in Kakwani, N. and J. Silber (eds), *The Many Dimensions of Poverty*, Chapter 1, Palgrave Macmillan.

Townsend, P. (1979), *Poverty in the United Kingdom: A Survey of Household Resources and Standards of Living*, Penguin.

Tsui, K. Y. (1995) "Multidimensional generalizations of the relative and absolute indices: the Atkinson–Kolm–Sen approach", *Journal of Economic Theory* 67(1): 251–265.

UNDP (2010) *Human Development Report 2010. The Real Wealth of Nations: Pathways to Human Development*, 20th Anniversary Edition: Palgrave.

Yaari, M. E. (1987) "The dual theory of choice under risk", *Econometrica* 55(1): 95–115.

Yaari, M. E. (1988) "A controversial proposal concerning inequality measurement", *Journal of Economic Theory* 44(2): 381–397.

Yalonetzky, G. (2012) "Conditions for the most robust multidimensional poverty comparisons using counting measures and ordinal variables", ECINEQ Working Paper, 2012–2257.

3 Poverty and the dimensionality of welfare

Walter Sosa-Escudero,[1] *German Caruso and Marcela Svarc*

3.1 Introduction

The influential work by Sen (1985) and its related literature convincingly argued about the multidimensional nature of welfare, that is that well-being should be appropriately captured by more than one coordinate or dimension. This multidimensionality ultimately characterizes features of the distribution of welfare, like poverty rates, inequality or polarization.

In this multidimensional context poverty is understood as deprivation of *capabilities*. Sen (1992) defines a *function* as an activity or state that can affect the welfare of an individual, like eating, being calm or having a good job. Functions do not provide information about the possibilities a person faces before obtaining an outcome, but describe what a person is able to do. When a person is working, he is performing the functioning of working.

Capabilities are the set of functions that an individual can perform. Sen (1992) describes freedom as the possibility of an individual to pursue a combination of functions. In this sense, freedom is related to the number of elements of the capabilities set. This concept captures the multidimensional essence of welfare and the main reason why a single variable evaluating a final outcome (like income) is not an adequate measure of well-being.

In spite of this view, most of the applied literature on poverty or welfare analysis is still based on a strictly unidimensional notion, based on current income or consumption. This inconsistency, between the acknowledged multidimensional nature of welfare at a conceptual level, and the unidimensional approaches favoured in practice, is a reflection of both the operational and conceptual difficulties practitioners face when moving to a multidimensional framework. Kakwani and Silber (2008a) is a recent collection of articles on this issue.

The practical implementation of multidimensional poverty measurement requires to solve two fundamental issues. The first one is to decide which variables should be measured to capture multidimensional welfare. The second one refers to how to produce methodologically coherent assessments of poverty or inequality, based on this multiplicity of measures. This second problem has received relatively more attention in the literature, and most of the research on multidimensional poverty refers to how to define, identify and aggregate the poor in such context; see Deutsch and Silber (2005) for a comparative review, and Kakwani and Silber (2008b) for a collection of recent methods.

This literature assumes that empirical researchers have already agreed on a set of variables that are the input for multidimensional analysis. One possible route to find such a set is to agree at a conceptual and, very likely, at a multidisciplinary level, on a list of notions that jointly represent welfare. Alkire (2008) is a detailed investigation on this approach, balancing conceptual as well as operational aspect towards the goal of defining a set of variables that represent welfare multidimensionally.

Hence, at the practical level, the problem proceeds by finding a set of measurable variables that jointly represent these concepts. In such context, the *dimensionality* of welfare is the minimum number of coordinates or variables that must be quantified in order to represent welfare. Once such a list is available at a conceptual level, it is natural to proceed by finding, for each dimension, an appropriate empirical measure.

Nevertheless, the most common situation empirical researchers face is that there is available a large set of variables, usually from a household survey, that is believed to jointly represent welfare. In this framework the problem of dimensionality of welfare refers to: (1) which is the dimensionality of welfare, i.e., the minimum number of variables that can represent welfare (2) starting from an initially large set of variables, how to construct a reduced, hopefully small set of variables that can be used to assess welfare, poverty or inequality in a multidimensional context.

In this chapter we review recent strategies to estimate the dimensionality of welfare and to produce a reduced set of variables to measure multidimensional poverty. The subject is relevant from both a conceptual and an operative perspective. From a conceptual point of view, finding the minimum number of variables that represent welfare contributes to the ongoing discussion on the relevant dimensions of well-being, that are later used to measure poverty or inequality in a multidimensional framework. From a practical point of view, it helps researchers select and/or collect new data more efficiently to study multidimensional deprivation, by focusing on a conveniently small set of variables that can be sampled and interpreted easily.

The chapter is organized as follows. Section 3.2 links the problem of dimensionality of welfare to some recent literature on feature and variable selection, common in the field of statistical learning. Section 3.3 focuses on the notion of the poor as a cluster. Section 3.4 describes strategies based on reducing dimensionality first using factor analytic methods, as a previous stage to finding the poor using cluster analysis. Section 3.5 discusses alternative strategies that invert the order of the procedure, starting by finding the multidimensionally poor and then proceeding to reduce dimensionality through a variable selection procedure via cluster analysis. Section 3.6 concludes. The chapter is largely based on Caruso, Sosa-Escudero and Svarc (2012) to which we will refer for further details.

3.2 The dimensions of welfare and poverty

The practical predominance of welfare analysis based on income or consumption solely, should not be interpreted naively as ignoring the convincing efforts of Sen's capabilities approach, but as a pragmatic reaction to the methodological, conceptual and operational difficulties in adopting a fully multidimensional strategy. How much can a single dimension, like income, appropriately serve the purpose of capturing welfare, is indeed a relevant empirical question aimed at measuring the costs of abandoning the unidimensional approach. But the massive body of recent work on multidimensional deprivation clearly suggests that the cost of ignoring relevant dimensions is substantially high, leading to biased characterizations of the severity of poverty – in different regions and periods – and, even worse, complicating the assessment of policy measures aimed at poverty alleviation.

The mere fact that welfare is recognized as being multidimensional simply states that one dimension is not enough, without clear indications about how many coordinates or variables are necessary or sufficient to capture welfare properly. To complicate matters, the capabilities approach calls for a truly multidisciplinary perspective in the determination of welfare. That is, traditional economic perceptions of welfare, as captured by income or consumption, must be complemented with sociological, psychological, institutional, biological or anthropological aspects. The collection of articles in Kakwani and Silber (2008a, b) are a clear example of the complexity implicit in defining conceptually and practically a welfare space that can be properly measured to quantify multidimensional deprivation.

Sen (2004) recognized this complexity and warned applied researchers about the dangers of relying on simplified theories to produce a conclusive, fixed list of capabilities. More concretely Sen (2004, p. 77) stresses that

> The problem is not with listing important capabilities, but with insisting on one predetermined canonical list of capabilities, chosen by theorists without any general social discussion or public reasoning. To have such a fixed list, emanating entirely from pure theory, is to deny the possibility of fruitful public participation on what should be included and why.

For example, Nussbaum (2000) produced a detailed list of ten of these 'canonical' capabilities. The advantages of producing such lists are related to providing concrete directions for applied researchers when trying to move away from purely unidimensional visions of well-being. Its disadvantages are related to Sen's warnings about overlooking the complexity of welfare by focusing on an invariant list.

Consequently, the problem of deciding on a set of variables should be seen as a dynamic and idiosyncratic one, in light of the goals of particular research initiatives that call for measuring poverty multidimensionally. As a matter of fact, Sen himself attacked the problem of deciding on a narrowed down list for specific situations, as in Dreze and Sen (1989). In any case, Sen's reluctance to provide a definitive list should be interpreted as suggesting that the problem of listing capabilities is a delicate and complex one that requires substantial analytical and conceptual study.

Alkire (2008) explores these difficulties with detail. Her article reviews standard practices on how dimensions are chosen, and suggests that this choice is usually led by pragmatic (data availability, standard practices) and/or conceptual (theoretical assumptions of different disciplines) reasons. She concludes that even though, as Sen clearly stresses, it is impossible and eventually futile to agree on a definitive, time, context and regionally invariant list of dimensions, the practical need to restrict empirical analysis to a certain list must be accompanied by a detailed analysis (empirical as well as conceptual) of why and how a final set of variables is chosen to study multidimensional deprivation.

The problem of reducing dimensions is ubiquitous in the statistical learning literature (see Hastie *et al.* (2009) for a recent review of this literature). In terms of our concerns, there is a vector x of p random variables, each of them representing a feature of multidimensional welfare. Let $F(x)$ be its (multivariate) cumulative distribution function (CDF). The final goal is to produce a poverty measurement, g, which in abstract terms can be seen as a specific functional of the CDF, i.e. $g = T(F(x))$. That is, there is an initial set of variables that is used to produce a poverty measure through its CDF.

The problem of *feature extraction* consists in creating a new set of artificial variables, usually in a space of smaller dimension, and then proceeding to measure poverty on this new set. Classical principal component and factor analysis are examples of this strategy. Alternatively, the problem of *feature selection* proceeds by choosing a subset of the initial set of variables, and then measuring poverty on this smaller subset.

The methods proposed to handle these problems can be categorized into two families, *filters* and *wrappers*. *Filter* methods evaluate the relevance of each feature or subset of features regardless the relevance of the function $T(F)$, that is, in our case independently of the second stage where poverty is measured. On the other hand, *wrapper* methods assess the relevance of the produced features or the selected variables, in light of the performance of $T(F)$, that is, in our case, based on its ability to measure poverty importance based on the grouping ability of the clustering method. More concretely, in the case of poverty measurement, filters work directly with the initial space of variables measuring welfare and proceed to reduce it either by creating a new but smaller set of variables (feature extraction) or by selecting a subset of the initial variables (feature selection), but without paying attention at the problem of finding the poor. Wrappers solve the dimensionality problem while evaluating its performance in light of the ability of the reduced set to identify the poor.

The active literature on statistical learning pays particular attention to the case where $T(F)$ is handled via cluster analysis, that is the poor understood as a coherent group, statistically and conceptually different from its complement, the 'non-poor'. Consequently, we will discuss two alternative approaches. The first one proceeds by first solving a feature extraction problem (via factor analysis) and then by finding the poor based on this new reduced set of artificial variables. The second approach inverts the process by first solving the problem of finding the multidimensionally poor (via cluster analysis), and then proceed as a 'wrapper', that is, using the outcome of this first stage to select a subset of the variables in the original space that can reproduce as accurately as possible the poverty characterization found with the original set of variables.

3.3 Multidimensional poverty via cluster analysis

In light of the previous discussion, the problem of estimating and reducing the dimensionality of welfare is not entirely independent of that of defining poverty in a multidimensional sense. A relevant stream of literature is based on establishing poverty thresholds for different dimensions (Bourguignon and Chakravarty (2003) and, more recently, Alkire and Foster (2011)).

A complementary notion to measure poverty multidimensionally is based on the idea of the poor as a cohesive group. A cluster is by construction the solution to a within/between similarity trade-off, that is, individual or families classified as poor are similar among themselves, but different from the non-poor. The idea of social groups as clusters is central to the recent approaches in the polarization literature (Duclos *et al.*, 2004). Even though cluster methods have long been available in practice, recent advances in data mining and computer intensive techniques have driven considerable attention on such strategies, see Cherkassky and Mulier (2007) or Bishop (2006) for a recent overview. Cluster methods allow handling multiple dimensions easily and, as presented in the next section, may assist in defining relevant dimensions of welfare.

The input for cluster methods is an $N \times p$ matrix X, where rows correspond to N individuals or families, and columns to the p variables together representing (multidimensional) welfare. Each row can be seen as a point of dimension p. A *cluster* is a collection of these points. If there are K clusters, a clustering mechanism can be characterized by an 'encoder' function $C(i)$: $(1,\ldots, N) \rightarrow (1,\ldots, K)$ that assigns each point to only one group.

Let $d(x_j, x_i)$ be a any distance function for any two p dimensional points x_j and x_i. Let

$$W(C) = \frac{1}{2}\sum_{k=1}^{K}\left[\sum_{i,\frac{j}{C(i)}=C(j)=k} d(x_i, x_j)\right],$$

be a loss function. Cluster algorithms find the encoding that minimizes this penalty function, which measures aggregate discrepancies within each cluster. The choice of $d(.)$ to be the square of the euclidean distance $d(x_j, x_i) = \|x_i - x_j\|^2$, leads to the *k-means* algorithm (MacQueen (1967)). Alternative choices for distance functions and centres lead to different solutions, like the *k-medians* algorithm that replaces the mean by the median.

This characterization of the poor as a cluster leads to a natural comparison with available methods to find the poor in a multidimensional framework, like the recent proposal by Alkire and Foster (2011). Standard methods to identify the poor multidimensionally usually start by defining deprivation along each dimension, that is by defining a threshold along each welfare dimension, below which a person is considered poor. 'Union' methods define as multidimensionally poor a person who is deprived in *at least* one dimension, while 'intersection' methods require that the person is deprived in all dimensions. Alkire and Foster (2011) propose a 'counting', intermediate method that lies between one and all dimensions.

Cluster-based poverty can be seen as an alternative, intermediate strategy. First, the methods discussed above depend on exogenous cutoffs. In the cluster approach, the cutoffs are determined endogenously, as a solution to the dissimilarity optimization problem that simultaneously defines clustering. Second, the clustering algorithm discussed above leads to a specific partition of the welfare space that separates the poor from the non-poor. More concretely, suppose there are two groups. The optimal solution defines two centres, labelled as c_1 and c_2, so in the final step of the clustering algorithm all points belong to the cluster that leaves them the closest to either center. Consequently, the separation in two groups implicitly defines a partition, the so-called *Voronoi tesselation* whose frontier is a hyperplane that separates, in our case, the poor from the non-poor, defined implicitly by the inequality

$$d(z,c_1)^2 < d(z,c_2)^2,$$

where z is a point in $\mathbf{R}^2$. Union and intersection methods imply particular partitions. In the case of the union method, it is defined by the inequalities $z_j<p_j$ for at least one j in $(1,\ldots, p)$, where t_j is the poverty threshold for dimension j. For the intersection method, the separating hyperplane is defined by the inequalities $z_j<p_j$ for all j. Duclos *et al.* (2011), present recent advances on this 'frontier' approach to multidimensional poverty and inequality.

Understanding the scale, normalization and distance that will be considered in a clustering study is a key step to interpret the results of the procedure. Standard data sets for poverty analysis usually consist of mixed data, containing categorical (mostly binary) and continuous variables. A simple standardization is to divide variables by its range, i.e. for the observation x_{ij}, consider y_{ij}, the standardized observation,

$$y_{ij} = \frac{x_{ij}}{\max\limits_i(x_{ij}) - \min\limits_i(x_{ij})} \text{ for } i=1,\ldots,n \text{ and } j=1,\ldots,p.$$

Highly skewed (positive) variables, such as the income, are better standardized by taking the natural logarithm of the variable.

The *k-means* algorithm is sensitive to the choice of an appropriate distance. The L_1-norm is a natural additive distance to handle a mixed data set. The distance between two observations y_i and y_j is given by

$$d_{ij} = \sum_{l=1}^{p} | y_{il} - y_{jl} |,$$

so it can be seen as being the standard L_1-norm for continuous or ordinal variables, and in the case of binary variables, as the number of points where the observations take different values, that is, the same information as in the standard Jaccard index (see Hand *et al*, 2001), one of the most well-known measures of similarity for binary variables.

The iterative nature of k-means is also sensitive to the choice of initial conditions, that is, to the position of the initial centroids used to start the algorithm. Several proposals have been made to handle this effect (see Steinly and Brusco, 2007).

The k-means algorithm needs as an input the number of clusters. Classical methods to estimate the number of clusters assume the existence of more than one group. The *gap statistic* introduced by Tibshirani *et al.* (2001) can indicate if more than one group exists and if so how many of them are relevant. The statistic by Calinsky and Harabasz (1974) is another useful alternative to determine the number of clusters.

In the following sections we explore two alternative strategies to measure dimensionality and find a reduced space to measure multidimensional poverty. The first approach starts by first solving the problem of estimating and reducing the dimensionality first, and then apply cluster methods to find the multidimensionally poor. The second approach inverts these stages, by first finding the poor using cluster methods, and then using these results to reduce dimensionality. We will compare advantages and disadvantages of both approaches.

3.4 Factor analytic approaches

A standard approach to dimension reduction is to think that the vector of p observed variables x arises from a much smaller vector of latent variables f, where f is vector of m random variables. In practice x represents variables observed, for example, in a household survey, and f are unobserved variables ('factors') each of them representing a different dimension of welfare, with m usually much smaller than p. A simple linear structure linking these variables is

$$x = Lf + \varepsilon,$$

where L is a $p \times m$ matrix of unknown coefficients or 'loadings' and ε is a $p \times 1$ vector of unobserved random factors. Standard assumptions are that $E(f)=0$ (factors are normalized to have zero mean), $V(F)=I_m$ (factors are orthogonal and normalized to have unit variance), $E(\varepsilon)=0$, $V(\varepsilon)\equiv\Psi=diag(\psi_1, \psi_2,\ldots, \psi_p)$ (unobserved errors are uncorrelated but may have different variances) and $Cov(\varepsilon, f)=0$ (factors are uncorrelated with the error vector). The linear structure and the previous assumptions conform the *linear orthogonal factor* model. See Hardle and Simar (2012) for a detailed description of the factor model.

In the context of multidimensional welfare, f represents a minimum and orthogonal set of variables that jointly represent welfare. Naturally, m is the dimensionality of welfare, understood as the dimension of the (orthogonal) basis of the space of multidimensional welfare. Clearly, orthogonality is a restrictive constrain since capabilities are not necessarily unrelated (see Thorbecke (2008) for a detailed conceptual discussion on the interaction between capabilities). We will discuss how stringent this assumption is later on.

Given a sample of n observations for this model, the $n\times p$ matrix X contains observations for each of the p variables, for all n observations. Given X, the empirical problem consists in (a) determining m (the dimensionalilty of welfare), (b) estimating f (the latent factors that represent welfare).

Estimation proceeds by noting that

$$V(x)\equiv\Sigma=LL'+\Psi,$$

hence from a strictly parametric perspective, the unknown parameters of the model are the elements of L and Ψ.

The matrix Σ can be consistently estimated by its sample covariance matrix $S_n=n^{-1}X'x$. A standard strategy to estimate L and Ψ is based on the principal components approximation

$$\Sigma\simeq\sum_{i=1}^{m}\lambda_i a_i a_i'$$

where λ_i, $i=1,\ldots, m$ are the largest m eigenvalues of Σ and a_i the corresponding eigenvectors.

The choice of m is a crucial step in our problem, since it provides an estimate of the dimensionality of the welfare state. An heuristic rule for selecting the number of factors is to compute the proportion of total variance of the first m^{th} eigenvalues in respect to the total variance of the sample. Each consecutive factor provides less variability than the previous one. Then the proposal is to stop adding factors either when the rate of variance is beyond a threshold fixed by the user, for instance 0.7, or when the reduction in the variability proportion supplied by a factor is very small compared to the previous one. Other proposals can be based on the model selection ideas, that is, estimate the factor model for different number of factors and then compute a goodness-of-fit statistic for each model considering m factors versus a no factor structure. The most well-known model selection criteria are the Akaike Criteria and Bayesian Information Criteria (BIC).

The problem of estimating f first requires estimating L, in this context such estimator is given by

$$\hat{L}_{pc} = \left[\sqrt{\hat{\lambda}_1}\hat{a}_1 \sqrt{\hat{\lambda}_2}\hat{a}_2 \cdots \sqrt{\hat{\lambda}_m}\hat{a}_m \right],$$

where $\hat{\lambda}_i$ and $\hat{a}_i$ are, respectively, the first m eigenvalues and corresponding eigenvectors of S_n. When ε and f are assumed to be jointly normal,

$$E(f \mid x) = L'\Sigma^{-1}X,$$

suggesting that f can be estimated as

$$\hat{f} = \hat{L}S_n^{-1}x,$$

where $\hat{L}$ is any consistent estimator of L. It is immediate to observe that if $\hat{L} = \hat{L}_{pc}$, then $\hat{f}$ coincides with the first m principal components of S_n.

The matrix L provides useful information about how the latent variables f are weighted to form each of the observed variables x. In our case, this is indicative of how the orthogonal capabilities f are combined to produce the observed measures x. At this stage it is relevant to remark that the matrix L is not properly identified. In particular, for any orthogonal $m \times m$ matrix T, $Lf=LTT'f$, so given X, L is identified up to an orthogonal transformation. This fact plays an important role in the interpretation of factor models. A standard practice to facilitate interpretation is to 'rotate' L so that the components of x are conformed by positive and high weights on some factors and zero on the remaining ones. The usual *varimax* rotation maximizes the variance of the loadings in each factors.

Ferro Luzzi *et al.* (2008) and Gasparini *et al.* (2011) are recent references on the use of factor analytic approaches. See also Lelli (2001) for a comparison with fuzzy set methods. Ferro Luzzi *et al.* (2008) and Gasparini *et al.* (2011) start with a set of variables that is initially thought as being jointly representative of multidimensional welfare. Then, they use the factor methods described above to obtain a set of relevant factors. The fact that the number of retained factors is more than one is an indication of the multidimensionality of welfare, and the fact that it is considerably smaller than the original numbers of variables suggests that the original space can be reduced substantially and appropriately captured by a few variables. After reducing the dimensionality of the problem, they proceed to find the poor based on this reduced set of factors. Gasparini *et al.* (2011) identify the poor along each of the relevant dimensions, whereas Ferro Luzzi *et al.* (2008) applied cluster methods to the reduced set of factors found in their previous stage.

Gasparini *et al.* (2011) base their analysis on the Gallup World Poll, that in spite of its popularity, has seldom been used for multidimensional poverty analysis. Their initial data set contains 15 variables, including income, and other monetary and non-monetary measures of welfare, as well as some indicators related to subjective welfare. The application of standard factor analytic tools suggests that their initial space can be reasonably represented by only *three* factors. Interestingly, the one that contributes the most to the aggregate variability of welfare is income, suggesting that in spite of its limitations, it plays an important role in accounting for multidimensional welfare. The second factor includes variables related to subjective perceptions of welfare. The final factor can be linked to the standard 'basic needs' measures, like water access.

As mentioned previously, the poor are then identified along each dimension. For income, standard US$1 and 2 daily poverty lines are used. Factors are by construction unit free, so for the remaining two factors, they use a relative approach, that is, they set a poverty line in the index of non-monetary welfare and subjective welfare that implies a share of the population below that threshold equal to the income poverty headcount ratio with the US$2 line. They report that the rank correlation between income and subjective poverty is positive and significant, suggesting that subjective-based poverty is related to its objective counterpart. But, on the other hand, the correlation is far from high, suggesting that income represents only part of a more complex, multidimensional structure behind welfare. Finally, they find that the US$1 line appears to be a reasonable cut-off value to measure food deprivation.

Ferro-Luzzi *et al.* (2008) also start by applying factor techniques, on an initial set of 32 variables from the Swiss Household Panel. They are able to summarize the initial set using *four* latent factors that they relate to financial, health, neighbourhood and social exclusion dimensions. After finding the factors that resume the dimensions of welfare, they find the poor group using cluster methods based on the reduced set of factors. Consequently, the poverty thresholds are set implicitly by cluster algorithm.

To summarize, both papers find evidence that the original welfare space, composed of many relevant measures, can be drastically reduced to a few factors, and that more than one variable is needed to adequately represent it, even when income (in the case of Gasparini *et al.*, 2011) or variables closely related to it (the financial ones in the case of Ferro-Luzzi *et al.*, 2008) are included in their data sets. In spite of being strongly associated to a relevant factor, both studies clearly point towards the inadequacy of income solely to capture the multidimensional nature of welfare.

3.5 Cluster/variable selection approaches

There are several methodological concerns related to the use of the previous approach, which basically consists in a first stage where the dimensionality of the original welfare space is reduced using factor methods, and then the poor are found based on this reduced set. First, covariance methods like factors or principal components are subject to delicate identification issues which harm their direct interpretation. Basically, and as mentioned before, factors are linear combinations of the original variables, identified up to orthogonal transformations (see Elffers *et al*., 1978, for a detailed overview of these problems). The standard practice, and the one adopted in both Gasparini *et al.* (2011) and Ferro Luzzi *et al.* (2008), is to rely on 'rotations' or other algebraic transformations to produce interpretable results. Second, factors are not directly observable, and hence for practical reasons, new information must be constructed by sampling the whole set of initial variables. For example, suppose that the analysis must be repeated for a different period or region, then all the initial variables must be measured in order to construct the factors, even under the assumption that the underlying latent structure remains unchanged. Third, reducing the dimensionality first may unnecessarily complicate the identification of a coherent group (the poor) that can be safely distinguished from its complement (the non-poor). This is particularly relevant when most variables in the welfare space consist of categorical (in most cases, binary) variables. The aggregation process implicit in the factor analytic approach may smooth out relevant differences contained in the original welfare space. For example, standard income-based poverty lines have serious trouble distinguishing the poor from the non-poor when the distribution of income is densely populated around the poverty line. Other categorical indicators may actually help separating the poor from the non-poor.

An alternative strategy that may help overcome these issues is the one adopted by Caruso *et al.* (2012). They suggest inverting the order and start by applying clustering methods to the initial data set, which produces a partition of the observations into poor/non-poor groups. This classification is later used to assist a variable selection method to select a few variables of the initial data set, that are able to reproduce as accurately as possible the initial classification of poor/non-poor. The main advantage of this approach is that, by construction, in a feature selection/wrapper fashion, the resulting variables are directly interpretable since they are originally in the data set used as a starting point to represent the welfare space.

Caruso *et al.* (2012) adopt the *blinding* strategy in Fraiman *et al.* (2008). Fraiman *et al.*'s (2008) procedure selects relevant variables after a satisfactory clustering procedure has been implemented. Their approach is based on the idea of blinding unnecessary non-informative or redundant variables. Even though a detailed description of the method is beyond the scope of this chapter (and we refer to the references mentioned before for further details), the main idea is simple to summarize. Suppose there are only two variables in the original space, X and Y. Given an appropriate clusterization based on X solely, Y is redundant if (a) it is strongly related to X, so given X it adds little information to the clusterization, (b) it is independent of X and non-informative about any clusterization (it only adds 'noise'). In these cases, the clusterization remains relatively unaltered if Y is replaced by its best prediction based on X, its conditional expectation $E(Y|X)$. In the extreme versions of the previous cases, Y will be replaced by X (X strongly related to Y or by a constant (Y just adding noise)). Consequently, the goal is to find the smallest group of original variables that can reproduce the original clusterization as accurately as possible, by replacing redundant variables by their expectations conditional on this reduced subset.

More concretely, let $x=(X_1,\ldots, X_p)$ be a random vector with distribution P and consider any partition of the space $\mathbf{R}^p$, like that produced by cluster methods. When p is large, there might be dependences among several variables of x and/or some of them are not relevant. Then, if the information of the noisy or dependent variables is removed, the cluster allocation should not change, meaning that the data should be kept in the original partition.

Fraiman *et al.* (2008) propose to look for a subset of indices $I\subset\{1,\ldots, p\}$ for which the original partition rule applied to a new 'less informative' vector $Z^I\in \mathbf{R}^p$ built up from x occurs with minimal loss, that is, the goal is to find the smallest subset of variables that achieves a certain reallocation rate with respect to the original partition based on all variables.

The variables whose indices are in I are the same as those in x, i.e. $Z_i=x_i$ while the rest of the variables are replaced by their conditional expectation, $Z_i=E(x_i|x[I])$, where $X[I]$ denotes the set of variables whose indices are in I. This is the *blinding* procedure. In the case of noisy variables $E(x_i)=E(x_i|x[I])$. For a fixed integer $d<p$, the target population is the set $I\subset\{1,\ldots, p\}$, $\#I=d$, for which the *population objective function*, given by

$$h(I)=\sum_{k=1}^{K} P(f(x)=k, f(Y^I)=k),$$

attains its maximum. The empirical implementation of the procedure is based on the following steps.

1 Given $X_1,\ldots, X_n\in \mathbf{R}^p$ i.i.d data, consider the partitioning procedure f_n: $\mathbf{R}^p\rightarrow\{1,\ldots, K\}$.
2 For a fixed value of $d<p$, given a subset of indices $I\subset\{1,\ldots, p\}$, with $\#I=d$, fix an integer value r (the number of nearest neighbour to be used). For each $j=1,\ldots, n$, find the set of indices C_j of the r-nearest neighbour's of $X_j[I]$ among $\{X_1[I],\ldots, X_n[I]\}$, where $X_j[I]=\{X_j[i]: i\in I\}$. And define,

$$Z_j^*[i]=\begin{cases} X_j[i] & \text{if } i\in I \\ \frac{1}{r}\sum_{m\in C_j} X_m[i] & \text{otherwise} \end{cases}$$

where $X[I]$ stands for the i-coordinate of the vector X.
3 Calculate the empirical objective function

$$h_n(I)=\frac{1}{n}\sum_{k=1}^{K}\sum_{j=1}^{n} J_{\{f_n(x_j)=k\}} J_{\{f_n(x_j^*)=k\}'}$$

where J_A stands for the indicator function of set A. The empirical objective function measures the proportion of observations that are reallocated on the same group as in the original partition after blinding the variables.

4 Look for a subset $I_{d,n}=I_n$, with $\#I_n=d$, that maximizes the empirical objective function h_n.

The procedure is computationally expensive, but Fraiman *et al.* (2008) suggest a forward–backward algorithm in order to find a subset of variables with the desired properties.

Naturally, and as in the case of factor methods, cluster ideas to represent the poor and to solve the dimensionality problem are not free from conceptual disadvantages. As discussed above, deciding the number of groups is not a trivial issue, specially from a conceptual perspective. Due to its characterization, cluster algorithms cannot guarantee that the number of groups will necessarily be two, and even if two groups are found, not necessarily one of them will contain the poor. For example, it might be the case that one group is the rich, so its complement includes the poor and likely the middle class. Hence, it is important to study carefully that one of the groups indeed contains the poor, this is usually done by confirmatory analysis, studying the characteristics of each group.

Caruso *et al.* (2012) apply this variable selection/cluster analysis method to the original variable set in Gasparini *et al.* (2011). The results using the Calinsky/Harabasz index and the Gap statistics suggest that the optimal number of clusters is indeed two. In order to safely label one of these groups as containing the poor, they computed the means of the three optimal factors obtained by Gasparini *et al.* (2011), interpreted by these authors as representing monetary, subjective and non-monetary aspects of welfare. Group one contains 73.52 per cent and group two the remaining 26.48 per cent of the individuals in the sample. Group two presents substantially lower values for the three dimensions of welfare, suggesting that this group contains those individuals with low levels of welfare. Consequently, they refer to this group as the 'deprived cluster', because they see group two as a statistically and economically different entity with respect to its complement, in the sense that it contains individuals with significantly lower levels of welfare indicators.

After having found an acceptable clusterization, they solved the dimensionality problem by finding a reduced set of variables, initially in the welfare space, that can reproduce the initial clusterization as accurately as possible. Interestingly, the blinding strategy shows that the initial classification of poor–non-poor can be adequately reproduced with only three variables in the original data set: *monthly household income*, *not having had enough money to buy food over the last year in at least three opportunities* and *having a computer at your home or the place you live*. The correct cluster reallocation rate is always between 90 and 92 per cent, that is, almost all individuals classified as poor with the initial set of 15 variables are correctly classified as poor based on this much smaller set of three variables.

The fact that the reduced welfare space needs more than one variable to adequately reproduce the original welfare space is an indication of its multidimensionality. Nevertheless, income turns out to be one of the variables chosen in the reduced set. This result is consistent with the previous literature, like Ferro Luzzi *et al.* (2008) and Gasparini *et al.* (2011) that suggest that, though important, income is not enough to capture all the dimensions of welfare. As a matter of fact, when the reduced set of variables is forced to keep only income, only 60 per cent of the observations are reallocated on the correct cluster.

It is interesting to compare the results of the cluster-based multidimensional approach, with a standard one based on income solely. Out of those identified as poor by our cluster approach, only 45 per cent are labelled as poor by a poverty line set at US$2 a day, and only 25 per cent when the line is lowered to US$1 a day. Though monotonically increasing with income, this result clearly speaks about the severe discrepancies between a multidimensional notion of poverty (as implicit in our cluster analysis) and that based on income solely.

3.6 Final remarks

The problem of choosing and measuring relevant variables for multidimensional welfare analysis is a delicate one, where researchers must weigh conceptual and operational restrictions. From a conceptual perspective, the complexity of the problem is mirrored by Sen's reluctance to produce a definitive and invariant list of canonical variables, which only speaks about the need to clearly specify the conceptual criteria used to select variables for concrete cases. From an operative perspective, researchers usually face the problem of the availability of a large number of variables, that in the best scenario it is comprehensive enough to quantify welfare multidimensionally, hence a relevant problem is whether for conceptual and operative purposes it can be adequately summarized into a few indicators that facilitate computations and lead to simple interpretations.

In light of this discussion, any mechanical attempt to solve the dimensionality problem must necessarily be seen as a complement (and not a substitute) of a general, conceptual and methodological strategy that balances the complexities of the issue with the pragmatic needs of specific regions or periods where multidimensional poverty is measured. The methods discussed in this chapter are shown to be able to summarize an initially large list of variables into a few new variables (as in factor analytic methods) or a subset of the original ones (as in feature selection/cluster methods), that can serve the purpose of characterizing the poor. These methods can assist the conceptual search for relevant dimensions of welfare, or provide confirmatory analysis of alternative, very likely multidisciplinary studies aimed at isolating relevant factors for poverty analysis.

Note

1 We thank Adriana Conconi for her help with the background literature. All errors and omissions are our responsibility.

References

Alkire, S. (2008) 'Choosing dimensions: the capability approach and multidimensional poverty', in Kakwani, N. and Silber, J. *Quantitative Approaches to Multidimensional Poverty Measurement*, Palgrave Macmillan, Basingstoke.

Alkire, S. and Foster, J. (2011) 'Counting and multidimensional poverty measurement', *Journal of Public Economics*, 95(7–8), 476–487.

Bishop, C. M. (2006) *Pattern Recognition and Machine Learning*, Springer, New York.

Bourguignon, F. and Chakravarty, S. R. (2003) 'The measurement of multidimensional poverty', *Journal of Economic inequality*, 1(1), 25–49.

Calinsky, R. B. and Harabasz, J. (1974) 'A dendrite method for cluster analysis', *Communications in Statistics*, 3(1), 1–27.

Caruso, G., Sosa-Escudero, W. and Svarc, M. (2012) 'Deprivation and the dimensionality of welfare: a variable-selection cluster-analysis approach', Mimeo.

Cherkassky, V. and Mulier, F. M. (2007) *Learning from Data: Concepts, Theory and Methods*, 2nd edn, Wiley, New York.

Deutsch, J. and Silber, J. (2005) 'Measuring multidimensional poverty. an empirical comparison of various approaches', *Review of Income and Wealth*, 51(1), 145–174.

Duclos, J. Y., Esteban, J. and Ray, D. (2004) 'Polarization: concepts, measurement, estimation' *Econometrica*, 72(6), 1737–1772.

Duclos, J. Y., Sahn, D. E. and Younger, S. D. (2011) 'Partial multidimensional inequality orderings', *Journal of Public Economics*, 95(3), 225–238.

Dreze, J. and Sen, A. (1989) *Hunger and Public Action*, Oxford University Press, Oxford.

Elffers, H., Bethlehem, J. and Gill, R. (1978) 'Indeterminacy problems and the interpretation of factor analysis results', *Statistica Neerlandica*, 32(4), 181–199.

Ferro Luzzi, G., Fluckiger, Y. and Weber, S. (2008) 'A cluster analysis of multidimensional poverty in Switzerland', in Kakwani, N. and Silber, J. *Quantitative Approaches to Multidimensional Poverty Measurement*, Palgrave Macmillan, Basingstoke.

Fraiman, R., Justel, A. and Svarc, M. (2008) 'Selection of variables for cluster analysis and classification rules', *Journal of American Statistical Association*, 103(483), 1294–1303.

Gasparini, L., Sosa Escudero, W., Marchionni, M. and Olivieri, S. (2011) 'Multidimensional poverty in Latin America and the Caribbean: new evidence from the Gallup World Poll', *Journal of Economic Inequality*, October, 1–20.

Hastie, T. J., Tibshirani, R. J. and Friedman, J. H. (2009) *The Elements of Statistical Learning: Data Mining, Inference, and Prediction*, Springer, New York.

Hand, D., Mannila, H. and Smyth, P. (2001) *Principles of Data Mining*, MIT Press, Cambridge, MA.

Hardle, W. and Simar, L. (2003) *Applied Multivariate Statistical Analysis*, Springer, New York.

Kakwani, N. and Silber, J. (2008a) *The Many Dimensions of Poverty*, Palgrave Macmillan, Basingstoke.

Kakwani, N. and Silber, J. (2008b) *Quantitative Approaches to Multidimensional Poverty Measurement*, Palgrave Macmillan, Basingstoke.

Lelli, S. (2001) 'Factor analysis vs. fuzzy sets theory: assessing the influence of different techniques on Sen's functioning approach', *Center of Economic Studies Discussion Paper*, KU Leuven.

MacQueen, J. B. (1967) 'Some methods for classification and analysis of multivariate observations', in *Proceedings of 5th Berkeley Symposium on Mathematical Statistics and Probability*, University of California Press, Berkeley, CA.

Nussbaum, M. C. (2000) *Women and Human Development: The Capabilities Approach*, Cambridge University Press, Cambridge.

Sen, A. (1985) *Commodities and Capabilities*, Oxford University Press, Oxford.

Sen, A. (1992) *Inequality Reexamined*, Harvard University Press, Cambridge, MA.

Sen, A. (2004) 'Capabilities, lists, and public reason: continuing the conversation', *Feminist Economics*, 10(3), 77–80.

Steinly, D. and Brusco, M. J. (2007) 'Initializing k-means batch clustering: a critical evaluation of several techniques', *Journal of Classification*, 24(1), 99–121.

Thorbecke, E. (2008) 'Multidimensional Poverty: Conceptual and Measurement Issues' in: Kakwani, N., Silber, J., *The Many Dimensions of Poverty*. Palgrave Macmillan, Basingstoke.

Tibshirani, R., Walther, G. and Hastie, T. (2001) 'Estimating the number of clusters in a data set via the gap statistic', *Journal of the Royal Statistical Society*, Serie B (Statistical Metodology), 63(2), 411–423.

4 Income, material deprivation and social exclusion in Israel

Naama Haron

4.1 Introduction

There is a great similarity and even an overlapping in the definitions of economic poverty, material deprivation and social exclusion; however, the terminological multiplicity is part of an expanding process of measuring poverty in a multi-dimensional manner rather than a solely economic one, and studies have shown that there is a limited overlapping between the segments of the population identified as being poor according to each method (Bradshaw *et al.*, 2000; Burchardt *et al.*, 1999; Perry, 2002; Saunders and Adelman, 2004; Tsakloglou and Papadopoulos, 2002; Whelan *et al.*, 2001). Therefore, the question to be asked is, who are the poor? This chapter is an attempt at identifying and comparing the population segments in Israel that suffer from economic poverty, material deprivation and social exclusion.

Measuring economic poverty in Israel is a known and accepted procedure, but no single method has been established for measuring deprivation and social exclusion. Part of the difficulty inherent in measuring these issues in Israel, and around the world as well, stems from the vagueness of their definition. Today no agreement has yet been reached concerning the definition of material deprivation and its scope. What are the fields in which a shortage thereof is considered a deprivation that causes physical hardship to the individual? Social exclusion is an even more vague term and also includes social and emotional deprivations that do not possess an immediate material expression. The significance of the term 'social exclusion' is unclear, leaving various theoreticians and empirical studies to grant it differing content. The primary ambiguity surrounds the relationship between social exclusion and poverty. Are the two terms synonymous for the same situation? Does social exclusion expand the definition of poverty? Perhaps these are two totally different phenomena (Atkinson and Hills, 1998; Burchardt *et al.*, 2002; Hayes *et al.*, 2008; Larsen, 2004; Levitas *et al.*, 2007; Morgan *et al.*, 2007; Omtzigt, 2009; Silver, 1994; Washington and Paylor, 1998)?

In spite of these difficulties, it is common practice to identify deprivation as a tool for measuring poverty, using an index of material deprivation stemming from a shortage in resources (Burchardt *et al.*, 2002; Larsen, 2004). Social exclusion is defined as a state of individuals and populations precluded from participation in various aspects of life in their surrounding society (Bossert *et al.*, 2005). This approach includes populations excluded from society for reasons such as racism, discrimination and poverty.

The aim of this chapter is to illustrate the state of economic poverty, material deprivation and social exclusion in Israel. We will examine the correlation between these phenomena and attempt to identify and characterize the afflicted population. The research presented herein is based upon Israeli societal survey data of 2007. The Central Bureau of Statistics has carried out this survey since 2002 and it enables the development of an index uniquely adapted to these needs. In order to maintain clarity and create as clear a distinction as possible between economic poverty, material deprivation and social exclusion in this study, the definitions of the paradigm of social quality will be followed (Berman and Phillips, 2000). This paradigm serves as a superstructure for understanding and organizing the various levels of an individual's life in the society in which he/she resides, including the absence of economic security and social exclusion.

For the purpose of identifying the population segments afflicted by these various phenomena, we will examine the known risk factors for poverty in Israel and examine their relationship, if any, to deprivation and exclusion. These risk factors include unemployment and an inadequate participation in the labor market, insufficient education, advanced age, large families and single-parent families, as well as an affiliation to the Arab and Ultra-Orthodox sectors (Endweld *et al.*, 2010).

The remainder of the chapter is constructed as follows. First, we will present some of the challenges that occur when measuring material deprivation and social exclusion and then continue by formulating the definitions of deprivation and exclusion used in this study. In the methods section, we will propose a method for measuring deprivation and exclusion that refers to the intensity of the phenomena in the population. In the findings section, in the first stage a comparison will be made between the population segments identified as afflicted by each of the phenomena, and, during the second stage, we will implement a cluster analysis in an effort to identify whether or not there is a unifying trend that draws together all those suffering these negative social phenomena. We will conclude with some remarks concerning the significance of the findings.

4.1.1 Measuring deprivation and social exclusion

The exclusion approach has a close relationship with that of deprivation. Both approaches measure poverty directly by examining and observing the individuals' living conditions and comparing the findings to a certain standard (Saunders and Adelman, 2004). These two approaches represent the concept of poverty that focuses on the poor person's reality – his living conditions, his material and social hardships, and does not focus solely on economic resources, as commonly accepted in classic methods of measuring poverty. However in contrast to social exclusion that examines socially significant aspects, the deprivation approach was developed primarily as an instrument for measuring poverty by using an index of material deprivation stemming from an absence of resources (Burchardt *et al*., 2002; Larsen, 2004).

Measuring deprivation is conducted by gathering information on the state of individuals and households, the resources and commodities they possess, consumer products and their ability to participate in various social activities in accordance with the resources in their possession. Accordingly, the method is characterized by implementation of deprivation indexes made up of a varying quantity of indicators. The deprivation indicators are selected by experts and researchers in the field or by the public that is asked to determine which of the items and activities are critical to normal life in today's society. A situation in which there are multiple deprivations in various dimensions points to a high probability of poverty (Townsend, 1979). In many studies, measuring social exclusion relies on the tradition of indicators of deprivation studies (Burchardt *et al*., 2002).

Identifying the various dimensions of deprivation and social exclusion, and selecting the most suitable indicators for each, poses a challenge for the field's researchers. At present, deprivation studies commonly contend with the issue of selecting the deprivation indicators by surveying the public as to what it considers as deprivation. However, this approach is not without its problems (McKay, 2004) since various population sectors (e.g. young people versus the elderly, individuals as opposed to families) identify different items as critically deprived needs which they were unsuccessful at attaining, as opposed to noncritical items.

Indicator selection for examining social exclusion is varied and even more controversial than the term itself (Larsen, 2004; Tsakloglou and Papadopoulos, 2002). Various studies operationalized the term exclusion differently, while including various dimensions under the title exclusion, as well as selecting different indicators for each field. For example, contrary to the perception that does not view unemployment as representing social exclusion (Larsen, 2004), some of the studies included exclusion from the labor market as a form of social exclusion and used unemployment as one of their indicators (Bradshaw *et al*., 2000; Burchardt *et al*., 1999).

Furthermore, while a number of researchers pooled the indicators into a single social exclusion index (Larsen, 2004; Saunders and Adelman, 2004; Tsakloglou and Papadopoulos, 2002), others chose not to unify the various exclusion dimensions (Bradshaw *et al*., 2000; Burchardt *et al*., 1999); the differences stem from the perception that views multi-dimensionalism as a meaningful part of the definition of exclusion as opposed to the perception that views social exclusion even when it exists in a single dimension (Scutella *et al*., 2009). An additional reason is the differing needs for which the study was conducted, i.e., whether to identify specific needs or a general policy.

Both the deprivation and poverty indices to be presented in the method section herein are not exempt from criticism; however, an attempt was made to reduce the arbitrariness by which indicators were selected.

4.1.2 The Social Quality Paradigm and deprivation and social exclusion definitions

In this study, the theoretical framework for defining material deprivation and social exclusion is the Social Quality Paradigm. This paradigm was developed as an overall framework for the theoretical understanding and measuring of human life in society and is promoted by the European Foundation on Social Quality. Beck (quoted in Berman and Keizer, 2006) defines social quality as 'the extent in which citizens are capable of participating in the socio-economic life of their communities under conditions that improve their general situation and personal potential'.

The paradigm proposes a theoretical foundation from which social quality components and the indicators for their review stem, and includes four components found on a continuum: social-economic security/insecurity, social inclusion/exclusion, social cohesion/anomie and empowerment/disempowerment. In this study we will focus on examining the relationship of the first two components in the model: socio-economic security (i.e., the level of poverty and material deprivation) and social inclusion (exclusion).

Social–economic security/insecurity. This component examines the extent to which people have sufficient resources (material and non-material) for a dignified existence, in the context of social relations. This refers to the manner in which the various welfare systems meet its citizens' needs. In a state of socio-economic security, the citizen will be protected from *poverty*, unemployment, illness and various forms of *material deprivation*. In other words, the deprivation component in the social–economic security component concerns physical existence and the quantity of resources.

Social inclusion/exclusion. This component examines the extent of accessibility people have to social institutions and relationships. It relates to the principles of social equality and equity, and addresses the reasons inherent within the society for their absence. In a state of social inclusion, the structural factors that cause exclusion are reduced. Berman and Phillips (2000) expanded the definition of social exclusion and divided it into two levels. In the first level, social exclusion is examined at the state level and exclusion identified as preventing people from fulfilling the universal rights and civil services for which they are eligible. At its second level, exclusion is identified on the community level, not just in the physical element but also from the emotional and subjective area of identification with the surrounding society and satisfying social relationships.

4.2 Methodology

The research is a study in a survey design, based upon the Social Survey 2007 of the Central Bureau of Statistics (CBS) (Shaprinski and Dror-Cohen, 2009) designed particularly to supply unique social information pertaining to residents of the State of Israel. The sixth cycle of the study, conducted in 2007, dealt with two issues. The study measured the population's welfare (consumption and waiving in various areas of life as a consequence of the economic situation, attitudes toward the state of the economy and their emotional state) and satisfaction from various governmental services (a range of services, such as health and law, finishing with fields such as education and recreation).

The research population and sampling: the 2007 Social Survey sampling included over 9000 individuals age 20 and older from all areas of the country, representing the permanent population of the State of Israel, approximately 4.5 million individuals within the relevant age group. Of these persons, 7391 (82 percent) responded to the survey. The research population was distributed into 52 percent males and 48 percent females, primary households composed of couples with one to three children (40 percent). Most of those surveyed were non-Ultra-Orthodox Jews (74 percent), a minority (6 percent) was composed of Ultra-Orthodox Jews, while the most significant non-Jewish minority segment was Muslim (12 percent). The majority of the subjects (60 percent) had completed more than 12 years of study and most were employed (59 percent).

4.2.1 Research tools: indices of economic poverty, material deprivation and social exclusion

As previously noted, in this study we chose to maximize the differences between poverty, deprivation and social exclusion as much as possible in order to facilitate a clear study of their relationships.

Poverty was identified as it is measured in the State of Israel, using income levels. Material deprivation was identified as tangible living conditions (consumer products or services) affected as a consequence of a shortage in economic resource. The third index is the social exclusion index comprising direct indicators pertaining to man's social–emotional existence, and his accessibility to various civil rights. The study indicators were selected by the researcher, relying on the Social Quality theory, and are limited to the questions existing in the survey.

Economic Poverty Index

An 11-point scale serves for identifying the income level of each household evaluated in the survey. The scale ranges from zero (net per household) income to 24,000 NIS and more. The income level was divided into the standard number of persons[1] per household in order to enable a comparison between the income of the survey's subjects and their deprivation and exclusion scores.

Material Deprivation Index

The Material Deprivation Index was developed in order to identify the level of deprivation suffered by each of the survey subjects. Five indicators that are valid for the entire population were found to be suitable for inclusion in the deprivation index (α=67), the value in the parentheses contains the weight of each indicator in the population for which the calculation will be explained herein.

A The lack of medical insurance because of economic difficulties (0.87).
B Sacrificing home heating/cooling because of economic difficulties (0.65).
C Cutting off telephone/electricity because of economic difficulties (0.86).
D Sacrificing food because of economic difficulties (0.79).
E The ratio between the number of rooms in the home to the number of individuals in the household (0.68).[2]

Following the selection of the indicators deemed suitable for the index, the weight was calculated for each indicator. The aforementioned weight was determined according to a division of the number of respondents that did not suffer from a deprivation in society by the number of overall respondents to the same indicator, i.e. identifying the percentage of the population that was not deprived. Each score in the indicator series that was examined was multiplied by the indicator's weight. After multiplying each score in its weight, all of the scores of all the indicators were added together. The index is a normalized summation index comprising the weights of all of the scores of all of the deprivation indicators. The scores range from 0 for full material welfare, to 1 for maximum material deprivation. This is the method proposed in the British Ministry of Labour Report for Developing Deprivation Indices (Willitts, 2006) and relies on the work of Desai and Shah (1988).

The utilization of indicator scores, as it is, grants an identical weight to each indicator (consequently cutting off the phone service is identical in its significance as is sacrificing food). This problem is overcome by granting additional weight to indicators representing fields in which most of the population does not encounter any difficulty. The high weight values given to the indicators in the deprivation index testify to the relatively high ability of these indicators to identify individuals who suffer significant deprivation in comparison to the rest of the country's population.

Social Exclusion Index

As a means for examining *social exclusion* an index was developed to measure the level of exclusion of each survey respondent. Similar to the deprivation index, this is a normalized summation index comprising the weights of all of the scores of all of the exclusion indicators. The scores range between 0 representing full inclusion to 1 as a score for full exclusion.

A set of 19 indicators in the questionnaire was found to be suitable for an examination of exclusion and compiled into a comprehensive social exclusion index, as presented in Table 4.1. The 19 indicators were compiled into three sub-indices: environmental exclusion, personal security exclusion and social contacts exclusion. The division into sub-indices is solely for theoretical purposes. In this study, emphasis was placed upon exclusion as an expression of the *accumulation* of various indicators for exclusion and there is no preliminary expectation that a respondent excluded in a specific field will also be excluded in other fields as well. From this theoretical point of view, although the results of Alpha-Cronbach

Table 4.1 Social Exclusion Index

Environmental exclusion α=0.75	Are you satisfied with the area in which you reside? (0.83) Are you satisfied with the quantity of green spaces/public gardens/ parks in the area where you reside? (0.55) Are you satisfied with the cleanliness of the area where you reside? (0.53) Are you satisfied with the cleanliness of the air in the area where you reside? (0.57) Are you satisfied with the quantity of waste containers stationed in the area where you reside? (0.71) Are you satisfied with road conditions in the area where you reside? (0.53)
Personal security exclusion α=0.63	During the past year, has your apartment been burglarized/broken into? (0.92) During the past year, have you been physically harmed? (0.99) During the past year, has something been stolen without the use of force? (0.94) During the past year, have you been sexually harassed? (0.98) Do you feel safe walking alone in the dark in the area where you reside? (0.73)
Social contacts exclusion α=0.54	Are you satisfied with the relationship you have with your neighbors? (0.85) Have you used a computer? (0.62) Is there a situation in which you feel lonely? (0.69) During a crisis or distress, are there people whose help you can count on? (0.90) Are you satisfied with the relationship you have with your friends? (0.91) How often do you meet or speak on the telephone with friends? (0.86) Are you satisfied with the relationship you have with your family? (0.95) Have you served in the IDF/done Civilian National Service? (0.53)

Credibility Test are represented here, they are of limited significance. The primary objective of the sub-indices is to describe and illustrate the subjects contained within the exclusion index (nonetheless, the level of credibility represented here for the environmental and personal security exclusion sub-indices is commonly accepted as sufficient when utilizing data taken from existing surveys and not those from designated indices (Berthoud *et al.*, 2004)). Following the selection of the indicators deemed suitable for the index, the weight was calculated for each indicator.

A “Poverty Line” determining the population suffering from economic poverty, material deprivation and social exclusion will not be proposed; however, subsequent to an analysis of the relationships between these phenomena, we will present a cluster analysis enabling maximum distinction and differentiation between population groups based upon their various characteristics.

The following section will present statistical analysis results that examine the relationships between the various economic poverty scores, material deprivation scores and the level of social exclusion of each of the survey participants.

4.3 Results

The objective of the tests was to examine the relationship between economic poverty, material deprivation and social exclusion in Israel and compare the populations suffering from the aforementioned phenomena. Since no exclusion or poverty line was determined, we examined the correlation between the individual's income level to the exclusion that he suffered as well as the measure of overlapping between 25 percent of the population with the lowest income levels, 25 percent of the population with the highest exclusion levels and 25 percent of the population with the highest level of deprivation. Table 4.2 presents the (Pearson) correlations between the various income, deprivation and exclusion levels as well as corrections with additional significant variables.

When examining the measure of overlap between the 25 percent of those surveyed with the lowest income level, the 25 percent with the highest level of exclusion and the 25 percent of those surveyed with the highest level of deprivation, we find a very small overlap. The percentages noted in Figure 4.1 represent the entire sample population. Referring to the lower quartiles as a population unto itself (100 percent), we can see that the greatest overlap is between those with the lowest income levels and those surveyed individuals with the highest level of deprivations – approximately 57 percent of those with low income (economically poor) also suffer from a high level of deprivation and vice versa. Approximately only 41 percent of the population suffering a high level of deprivation also suffers from a high level of exclusion and vice versa. Between those with low income and those socially excluded the overlap is lowest, less than 39 percent. There is an overlapping of less than 25 percent between the economically poor, the excluded and the deprived population.

In order to characterize the populations suffering from economic poverty, material deprivations and social exclusion, a relationship was examined between these phenomena and various socio-demographic characteristics known as risk factors for poverty in Israel. A series of T tests on independent samples and one-way ANOVA tests were conducted with subsequent Scheffé type analysis to examine the source of the differences found. The various risk factors include sector designations (Jewish, Non-Jewish, Ultra-Orthodox (*Haredi*), Muslim, Christian, Druze and Atheist), household types (Single, Childless couple, Small

Table 4.2 Simple correlations between exclusion, deprivations and income

	Exclusion	*Deprivations*	*Income*	*Muslim*	*Years of study*
Deprivations	0.391(**)				
Income	–0.283(**)	–0.472(**)			
Muslim	0.322(**)	0.411(**)	–0.286(**)		
Years of study	–0.215(**)	–0.342(**)	0.346(**)	–0.275(**)	
Work	–0.146(**)	–0.249(**)	0.334(**)	–0.137(**)	0.304(**)

Note
** $p<0.01$.

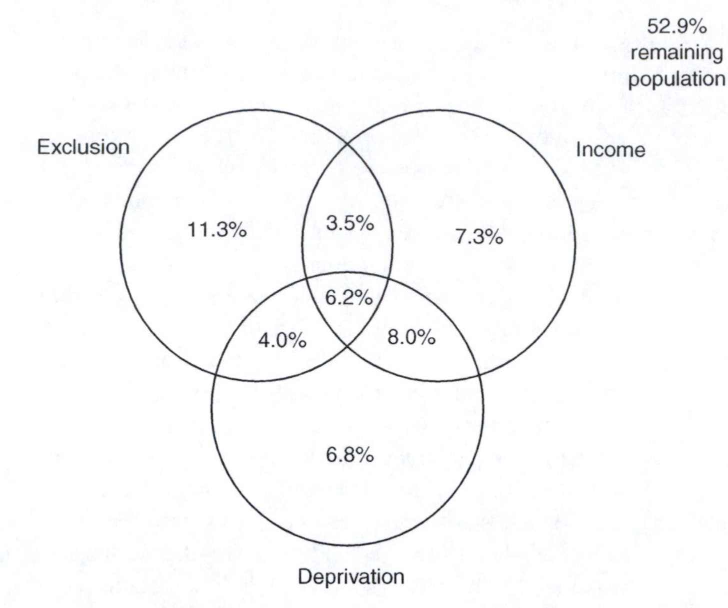

Figure 4.1 The measure of overlapping between the lower quartiles of income, exclusion and deprivation.

family up to three children, Large family of four children or more, Single parent, New immigrant (immigrated during the ten years prior to the survey), a household with only elderly people (males over the age of 65 and females age 60 and older)), years of study and work (worked for wages in the last week as opposed to did not work/unemployed), health and gender.

Table 4.3 is a summation of the results of the relationships found between the various factors to income levels, material deprivation and the level of social exclusion.

Figure 4.1 suggests that almost all of the known risk factors for poverty examined here are related to low income, a high level of deprivation and a high exclusion level. A relationship was found between sector affiliation to deprivation and social exclusion levels – Jewish sector affiliation assures lower levels of deprivation and exclusion in comparison to other sectors, primarily the Muslim sector, but this is also true in the case of the Druze and Christian sectors. From an analysis of the sub-indices of exclusion a similar picture also emerges: Muslims are the most excluded sector in the area of environmental exclusion, as opposed to the Jews who are the most inclusive. An interesting finding is that the environmental exclusion score of Ultra-Orthodox Jews was similar to the exclusion score of non-Orthodox Jews. In other words, Jews (Ultra-Orthodox and non-Orthodox) are the most satisfied as far as their physical environment is concerned. No significant difference was found between the sectors concerning their level of exclusion from personal security. Jews indicated the significantly lowest level of exclusion from social contacts, in comparison to the other sectors, while Muslims indicated the highest level of exclusion.

The findings suggest that Ultra-Orthodox Jews suffer a significantly higher level of exclusion from social contacts than the non-Orthodox brethren. However, since the Ultra-Orthodox group is considered as having a number of unique characteristics, among them a strong sense of community, a number of additional tests was conducted on this population. It was found that by removing the question concerning military/national service from the sub-index, there was no longer any significant difference concerning the level of exclusion from social contacts between Orthodox and non-Orthodox Jews. When the question pertaining to computer use is also removed from the index, the findings are inverted and the Ultra-Orthodox become the sector with the lowest exclusionary score in the area of social contacts ($M=0.077$, $SD=0.118$), ($F(6,5461)=27.90$, $p<0.001$), and even significantly lower than that of non-Orthodox Jews ($M=0.110$, $SD=0.147$). On the other hand, even without the inclusion of military/national service and computer use, Muslims remain the most excluded sector in terms of their social contacts level ($M=0.181$, $SD=0.162$). In this context, it is worth noting that the Druze population as well possesses a level of exclusion from social contacts which is not significantly different ($M=0.149$, $SD=0.154$) from secular Jews, Christians or Muslims when examined without questions pertaining to computer use and military service. A discussion on the meaning of selecting various indicators as part of examining the social exclusion will be presented herein.

Anticipated relationships were also found between deprivation, exclusion and affiliation to certain households, education, participating in the labor market, health and gender. The affiliation to single, childless couples or small family households assured better outcomes in comparison to large family or single parent households. As the level of education increases (number of school years) there is a corresponding decrease in deprivation and exclusion levels. Immigrants tend to suffer from a higher level of deprivation in comparison to veteran Israelis and also from greater social exclusion, primarily as a consequence of their exclusion from social contacts. Working individuals enjoy lower levels of deprivation and exclusion than persons not participating in the labor market. Men are less exposed than women to deprivation and exclusion and have a better state of health, so their level of deprivation and exclusion is decreased.

Advanced age is an uncommon risk factor. No difference was found between the average income and exclusion level of elderly households and non-elderly households, and their deprivation level is in fact even lower than that experienced by the remaining population.

Subsequent to this presentation of the basic relationships between risk factors leading to poverty and income, deprivation and exclusion, the relative effect of each risk factor on these phenomena was examined by implementing regression analysis. For this purpose, a series of stepwise regression analyses was conducted. The first regression presented examines the relative impact of the sociodemographic risk factors on the respondent's level of income. Table 4.4 presents the first through fifth and final steps in the regression, out of the 17 steps

Table 4.3 Average scores of income level, material deprivation and social exclusion tests (overall, environmental, personal security and social contacts)

	Average income level	*Average deprivation level*	*Average overall exclusion level*	*Average environmental exclusion level*	*Average personal security exclusion level*	*Average social contacts exclusion level*
	Religion					
Christian	2302	0.247	0.203	0.335	0.081	0.146
Jewish – Non-Ultra-Orthodox	3060	0.136	0.168	0.560	0.083	0.148
Ultra-Orthodox (*Haredi*)	1382	0.243	0.186	0.415	0.081	0.146
Muslims	1253	0.442	0.286	0.469	0.059	0.146
Druze	1712	0.302	0.229	0.359	0.066	0.137
Atheist	2860	0.189	0.194	0.334	0.075	0.127
Others	3249	0.081	0.172	0.331	0.083	0.119
ANOVA	$p<0.001$	$p<0.001$	$p<0.001$	$p<0.001$	$p>0.05$	$p<0.001$
	Household					
Single	2753	0.156	0.192	0.341	0.010	0.157
Childless couple	3501	0.124	0.174	0.321	0.0.72	0.149
Small family	2737	0.176	0.181	0.376	0.074	0.149
Large family	1427	0.339	0.226	0.441	0.079	0.161
Single parent	1962	0.300	0.231	0.384	0.105	0.173
ANOVA	p<0.001	$p<0.001$	$p<0.001$	$p<0.001$	$p<0.001$	$p<0.001$
Non-elderly	2644	0.194	0.185	0.376	0.081	0.152
Elderly	2757	0.147	0.194	0.311	0.083	0.149
T test	p>0.05	$p<0.001$	$p>0.05$	$p<0.001$	$p>0.05$	$p<0.001$
Veteran Israelis	2729	0.184	0.183	0.374	0.082	0.155
Immigrants	2355	0.205	0.201	0.344	0.079	0.207
T test	p<0.001	$p<0.01$	$p<0.001$	$p<0.001$	$p>0.05$	$p<0.001$

	Education					
1–4 years of study	1473	0.303	0.243	0.368	0.078	0.154
5–8 years of study	1531	0.369	0.265	0.433	0.089	0.145
9–10 years of study	1734	0.327	0.250	0.438	0.075	0.158
11–12 years of study	2226	0.209	0.185	0.366	0.086	0.152
13–15 years of study	2752	0.155	0.180	0.361	0.087	0.142
16 or more years of study	3528	0.108	0.164	0.342	0.072	0.135
ANOVA	$p<0.001$	$p<0.001$	$p<0.001$	$p<0.001$	$p<0.001$	$p<0.001$
	Work					
Not working (including unemployed)	1872	0.258	0.209	0.376	0.088	0.162
Working	3191	0.140	0.173	0.364	0.077	0.140
T test	$p<0.001$	$p<0.001$	$p<0.001$	$p<0.05$	$p<0.001$	$p<0.001$
	Health					
Health problem	2462	0.230	0.211	0.386	0.086	0.165
Healthy	2778	0.165	0.174	0.359	0.078	0.141
T test	$p<0.001$	$p<0.001$	$p<0.001$	$p<0.001$	$p<0.001$	$p<0.001$
	Gender					
Females	2485	0.200	0.201	0.376	0.103	0.154
Males	2841	0.176	0.170	0.361	0.059	0.149
T test	$p<0.001$	$p<0.001$	$p<0.001$	$p<0.05$	$P<0.001$	$p<0.001$

Table 4.4 Standardized coefficients (α) for stepwise regression to account for the variance of income

Step/index	*1*	*2*	*3*	*4*	*5*	*17*
Years of study	0.346***	0.296***	0.235***	0.229***	0.229***	0.252***
Jewish		0.286***	0.266***	0.233***	0.180***	0.068***
Work			0.216***	0.245***	0.256***	0.224***
Childless couple				0.207***	0.153***	0.070***
Number of persons in household					–0.159***	–0.143***
Immigrant						–0.151***
Ultra-Orthodox Jew						–0.118***
Age						0.220***
Health						–0.117***
Muslim						–0.074***
Single parent						–0.099***
Elderly						–0.091***
Large family						–0.099***
Small family						–0.097***
Single person						–0.054**
Atheist						0.040**
Gender						0.027*
R^2	0.12***	0.20***	0.24***	0.28***	0.30***	0.36*
ΔR^2		0.08***	0.04***	0.04 ***	0.02***	0.06*

Notes
* $p<0.05$.
** $p<0.01$.
*** $p<0.001$.

conducted. In the analysis it was found that 35.9 percent of the variance of income could be accounted for by the various risk factors when the primary effecting factors are education, age and participation in the labor market.

The regression conducted in order to examine the contribution of all of the variables to account for the variance of deprivation showed that it is possible to account for 35.8 percent of the deprivation variances. Table 4.5 presents the first through fifth and final steps in the regression, out of the 16 steps conducted. In order to conduct a deprivation (as well as a social exclusion) regression, an adjustment was made in this variable. Since the deprivation index ranges between a score of zero and one, the range is unsuitable for use as a dependent variable in linear regression, the variable's natural logarithm ($\ln_e$) was implemented. In other words, instead of using Deprivation Index E, the following was used as an index:

$$\ln \frac{E}{1-E}$$

It can be seen that the variables with the greatest effect on deprivation levels are years of study, state of health as well as the number of individuals in the

Table 4.5 Standardized coefficients (α) for stepwise regression in order to account for the variance of material deprivation

Step/index	*1*	*2*	*3*	*4*	*5*	*16*
Jew	–0.383***	–0.337***	–0.258***	–0.254***	–0.236***	–0.131***
Years of study		–0.270***	–0.272***	–0.198***	–0.206***	–0.188***
Number of persons in household			0.218***	0.256***	0.229***	0.167***
Health				0.222***	0.285***	0.229***
Age					–0.137***	–0.144***
Single parent						0.116***
Muslim						0.148***
Immigrant						0.110***
Working						–0.098***
Small family						0.091***
Large family						0.091***
Atheist						–0.031***
Single person						0.31***
Gender						–0.025*
Ultra-Orthodox Jew						0.032*
Elderly						–0.029*
R^2	0.15***	0.22***	0.26***	0.30***	0.31***	0.36*
ΔR^2		0.07***	0.04***	0.04***	0.01***	0.05*

Notes
* $p<0.05$.
** $p<0.01$.
*** $p<0.001$.

household. The better the respondent's state of health is and his age increases, the lower the deprivation level that can be expected. Table 4.6 presents the first to fifth, and final steps in the regression for social exclusion dependent variables out of the 13 steps conducted. The analysis showed that 17.1 percent of the exclusion variances could be accounted for.

It was found that sector affiliation had the greatest effect on the level of exclusion – affiliation to the Muslim, Ultra-Orthodox and Druze sectors all showed an increased risk of social exclusion and these variables led to the removal of the Jewish sector variable from the regression. A deteriorating state of health is one of the most significant factors in exposure to social exclusion. A low number of years in school and affiliation to the female gender also pose risk factors for exclusion.

In order to understand better the relationship between economic poverty, social exclusion and deprivation in Israeli society, cluster analysis was conducted, according to these variables. We chose to use the two-step clustering method, first presented by Zhang *et al.* (1996). This method is intended to enable cluster analysis with a large number of cases. In conducting the cluster analysis based upon the three parameters, income, deprivation and social exclusion,

Table 4.6 Standardized coefficients (α) for stepwise regression to account for the variance of social exclusion

Step/index	*1*	*2*	*3*	*4*	*5*	*13*
Muslim	0.291***	0.274***	0.279***	0.204***	0.186***	0.275***
Health		0.225***	0.220***	0.220***	0.203***	0.195***
Gender			–0.121***	–0.123***	–0.126***	–0.121***
Jew				–0.114***	–0.113***	
Years of study					–0.071***	–0.071***
Single parent						0.042**
Large family						0.040**
Immigrant						0.062***
Druze						0.072***
Single person						0.042**
Ultra-Orthodox Jew						0.061***
Elderly						–0.029*
R^2	0.09***	0.14***	0.15***	0.16	0.16***	0.17**
ΔR^2		0.05***	0.01***	0.01	0.00***	0.01**

Notes
* $p<0.05$,
** $p<0.01$,
*** $p<0.001$.

it was found that the cases divide into only two clusters, as presented in Figure 4.2. These clusters differ significantly from one another from the standpoint of income, deprivation and social exclusion. In the first cluster income is low; exclusion and deprivation are high, in contrast to the second cluster characterized by high income, and low levels of exclusion and deprivation. Table 4.7 presents the clusters' characteristics.

An observation into the standard deviation (distribution shape) shows that in the second cluster the standard deviation of income is more than double the standard deviation of income in the first cluster. A possible reason for this is that while high income assures the avoidance of deprivation and exclusion, it is not inevitable that a low income person will necessarily suffer from deprivation and exclusion. Therefore, in the cluster containing low levels of deprivation and exclusion one can expect to see high income levels, but also low income levels. We will see, however, small standard deviations of exclusion and deprivation that point to a homogeneity of the group in the second cluster, since low deprivation is related to low exclusion and vice versa. In the first cluster containing low income with high levels of deprivation and exclusion, one can see a low standard deviation of income, while on the contrary, relatively large standard deviations for deprivation and exclusion as low income does not mean high levels of deprivation or exclusion.

The first cluster (low income, high exclusion and high deprivation) expresses a variety of risk factors for poverty, e.g. more people in a household, younger ages and a poor state of health, less workers and more women and new

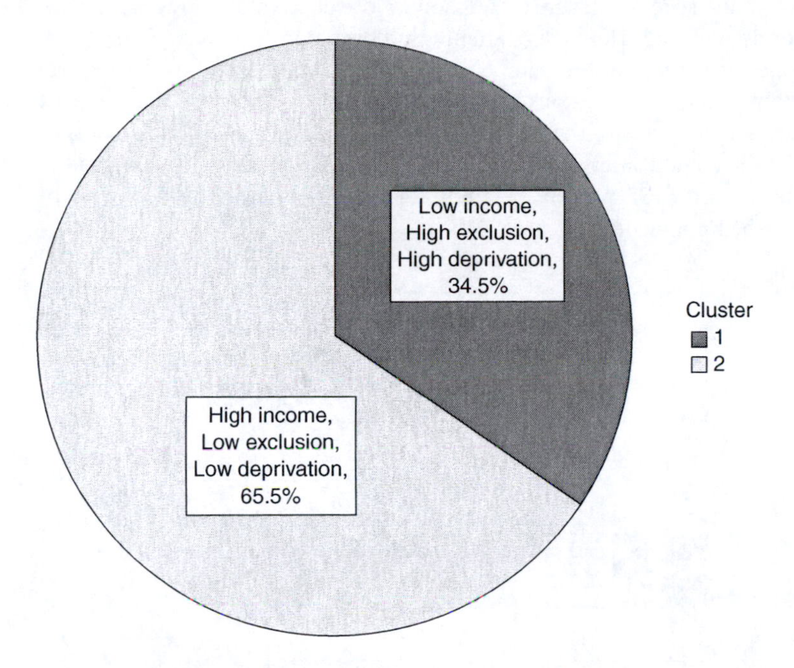

Figure 4.2 Population distribution into the income, deprivation and exclusion clusters – CBS social survey.

Table 4.7 Personal characteristics of the clusters' population – income, deprivation and social exclusion

Cluster no./variable	*1*		*2*		*T-test score for independent samples (t)*	*Significance*
	M	*SD*	*M*	*SD*		
Income level	1538	938	3572	1963	t(4253)=–45.93	$p<0.001$
Deprivation level	0.427	0.203	0.041	0.056	t(4253)=72.72	$p<0.001$
Exclusion level	0.259	0.119	0.152	0.096	t(4253)=31.13	$p<0.001$
Number of persons in household	4.41	2.36	3.39	1.54	t(4253)=17.01	$p<0.001$
Age	42.80	16.23	43.60	15.99	t(4253) =–1.55	$p>0.05$
Health (1, very good; 4, terrible)	2.09	0.90	1.60	0.73	t(4253)=17.91	$p<0.001$
	N	*%*	*N*	*%*		
Working	725	47.7	2018	73.8	$\chi^2(2)=288.86$	$p<0.001$
New immigrants	347	22.8	436	15.9	$\chi^2(2)=30.87$	$p<0.001$
Elderly households	141	9.3	305	11.2	$\chi^2(2)=3.64$	$p>0.05$
Gender (number of women)	841	55.3	1309	47.9	$\chi^2(2)=21.80$	$p<0.001$

immigrants. In the second cluster (high income, low deprivation, low exclusion), the situation is reversed. Herein is a graphic presentation (Figures 4.3–4.5) that illustrates the differences in the number of years in school, sector affiliation and household types in the various clusters.

In the low income cluster there is a high concentration (above 53 percent) of individuals with an education level of up to 12 years of schooling, as opposed to a high concentration (73 percent) of individuals with more than 13 years of schooling in the high income cluster.

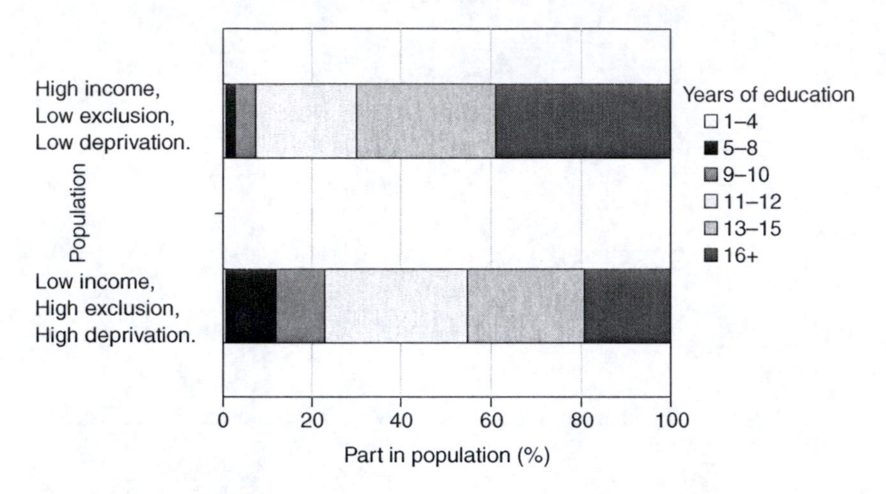

Figure 4.3 Education distribution in the income, deprivation and exclusion clusters.

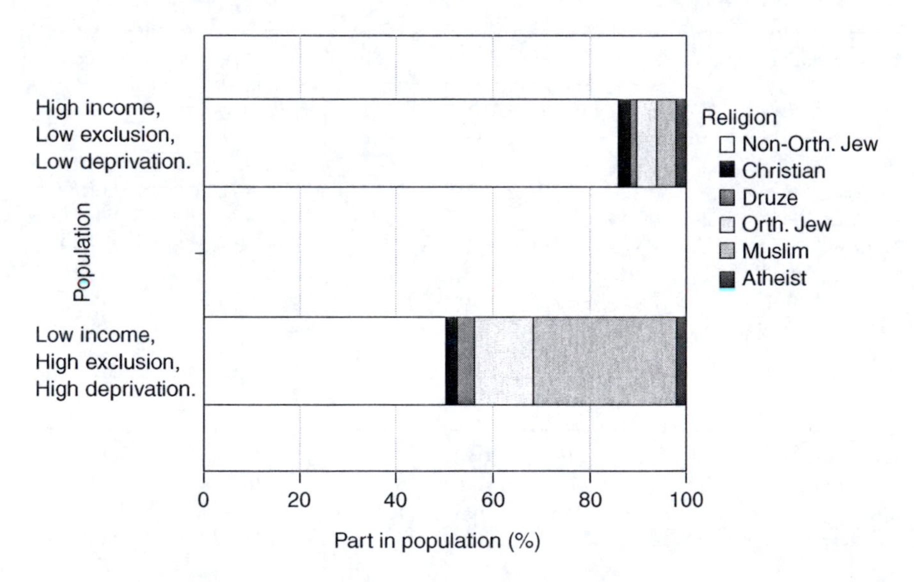

Figure 4.4 Sector affiliation distribution in the income, deprivation and exclusion clusters.

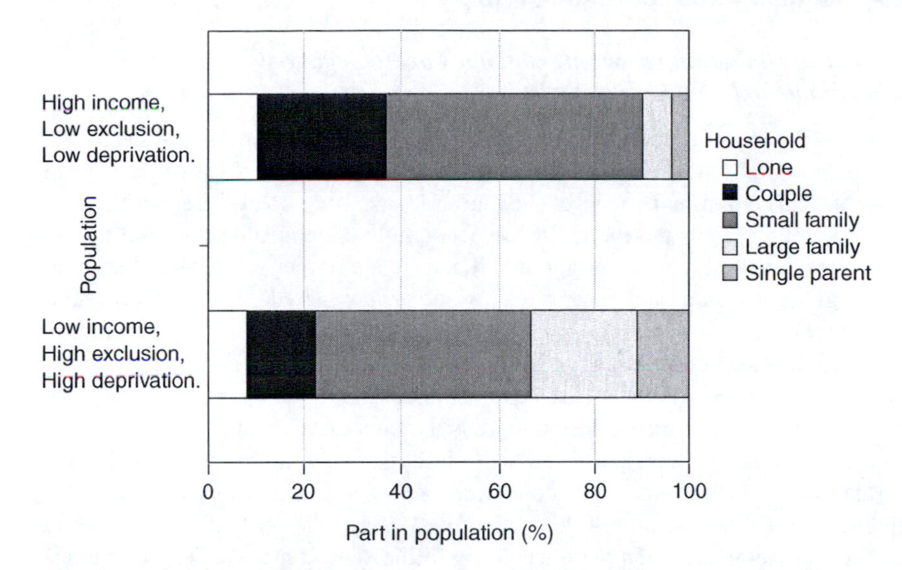

Figure 4.5 A distribution of household types in the income, deprivation and exclusion clusters – CBS social survey.

Although they only include 14 percent of the sample's population, Muslims comprise 31.5 percent of the low income, high deprivation and exclusion cluster. On the other hand, Jews, who comprise 74 percent of the sample's population, make up only 52 percent of the low income population cluster. Ultra-Orthodox Jews comprise 9 percent of the low income cluster, even though they make up only approximately 5 percent of the sample's population.

Large and single parent families are over-represented in the low income cluster as opposed to their representation in the population. While large families make up little more than approximately 12 percent of the sample's population, they comprise 22 percent of the low income cluster population, and single parent families comprise 11 percent of the same cluster, while comprising only 6 percent of the sample's population.

To summarize, when comparing the two clusters under analysis, it can be seen that in the second cluster the state of the respondents is better from the standpoint of their state of income, deprivation and exclusion. Correspondingly, this cluster is composed primarily of individuals from the Jewish sector, with a minority of Muslims, Druze and Ultra-Orthodox Jews relative to the first cluster. The average years of study in the second cluster is higher. The number of people per household is lower and there are fewer large and single parent families compared to the first cluster. In the second cluster, the respondents' state of health is better, age is higher, a greater number of respondents participate in the labor force and there are fewer immigrants and women. In the next section we will discuss the aforementioned findings.

4.4 Summary and concluding remarks

4.4.1 The relationship between economic poverty, material deprivation and social exclusion in Israel

The findings point to a connection between economic poverty, social exclusion and deprivation. Those afflicted by economic poverty are faced with an enhanced risk of deprivation and social exclusion and vice versa. Nevertheless, the risk level is not identical and the existence of a single phenomenon does not necessarily lead to the existence of another. There is a stronger relationship between material deprivation and social exclusion as compared to the relationship between economic poverty and social exclusion. The greatest correlation between the research variables is the one between income and deprivation levels, and the lowest correlation is between exclusion and income, while between the levels of deprivation and exclusion there is a moderate correlation in comparison to the two other correlations. The study's findings suggest that although there is a relationship between poverty, deprivation and social exclusion, these are different phenomena and not all those identified as poor according to one method are also identified as being poor according to the second method. These findings are consistent with previous findings published in the research literature comparing these phenomena (Bradshaw and Finch, 2003; Bradshaw *et al.*, 2000; Burchardt *et al.*, 1999; Devicienti and Poggi, 2007; Hallerod and Larsson, 2008; Nolan and Whelan, 2009; Saunders and Adelman, 2004; Whelan *et al.*, 2001).

4.4.2 Concerning the populations exposed to economic poverty, material deprivation and social exclusion

Examining risk factors for economic poverty, deprivation and exclusion it was found that all low income risk factors are connected significantly to deprivation and social exclusion (except in the case of the elderly, who were not found to suffer any enhanced risk concerning social exclusion). However, an observation of the study's overall findings presents a different profile of risk to income, deprivation and exclusion. Income is affected greatly by education, participation in the labor market, sector, age and type of household, including immigrants and the elderly. On the other hand, the primary risk factors for exclusion include sector affiliation, state of health and gender. The primary risk factors for deprivation comprise a more mixed profile including education, sector affiliation as well as age, state of health and household type and size.

First we will review the influence of the level of education as it is a known risk factor for low income. Education has a strong relationship to all the parameters of income, deprivation and exclusion levels. As the level of education increases, so does income, and correspondingly the levels of deprivation and exclusion decrease. The regression analyses showed that education was the strongest protective factor against low income and a significant defensive factor against material deprivation. On the other hand, in the regression analysis conducted for the social exclusion variable, education proved to be a relatively marginal protective factor in relation to other factors.

The findings show that sector affiliation has a large and significant impact concerning income, deprivation and exclusion. It was found that the Muslim population suffers significantly from lower income, higher deprivation and higher exclusion when compared to non-Orthodox Jews. Muslim sector affiliation comprises the primary risk factor for both social exclusion and for material deprivation according to the data presented in the CBS Social Survey. The income of the Ultra-Orthodox sector is not significantly different from that of the Muslim sector; and it also suffers from a significantly lower level of income and a significantly higher level of deprivation in comparison to the non-Orthodox Jewish sector. On the other hand, this sector is not afflicted by as high a level of deprivation as is the Muslim population and no significant difference was found between the overall level of social exclusion of non-Orthodox and Ultra-Orthodox Jews. The regression analyses conducted suggest that sector affiliation impact was greater on deprivation and exclusion levels in comparison to its impact on income levels.

The type of household and its size had a significant impact on income, deprivation and exclusion levels. The findings show that those households exposed to deprivation and exclusion are those of large families and single-parent families that also suffer the lowest levels of income. Small families had the lowest exclusion levels although their income was not the highest. The highest income was reserved for childless couples that also enjoyed the lowest levels of deprivation and a level of exclusion not significantly different from that of small families. The regression analyses showed greater influence for household size and type variables on the levels of income and deprivation as opposed to the impact these variables had on the level of social exclusion.

An examination of the elderly population suggests that while this sector is known as a group at risk of poverty, it does not suffer from economic poverty to an extent greater than the entire population and the level of deprivation they suffer is in fact lower than that suffered by the remaining sectors of the population. One of the reasons for this is that in Israel, the elderly population is granted greater defence against poverty in the form of an old-age pension, in comparison to defence against poverty supplied by other allowances, e.g. income support benefits (Endweld *et al*., 2010). The overall level of exclusion of the elderly is not higher than the remaining population sectors, although it was found that their level of exclusion from social contacts is significantly higher than the remaining population. The regression analyses point to elderly sector affiliation as a risk factor for low income, but not for social exclusion and deprivation. Increased age was found to be a defensive factor against low income and material deprivation, but not introduced as an influential variable into the regression analysis of social exclusion. It should be remembered that the correlation between advanced age, increased income and a decrease in the level of exclusion is limited and 'discontinued' with the departure from the labor market and aging, represented in the elderly variable.

The immigrant population suffers from a lower income level and a greater overall level of deprivation and exclusion than that experienced by veteran Israelis. Nevertheless, they do suffer less from environmental exclusion and there is no difference between them and the remaining population with respect to personal security and interest in current affairs. The regression analyses testify to immigrant sector affiliation as a greater risk factor for low income and deprivation, but not for high social exclusion.

Participation in the labor market assures a higher income, less deprivation and a lower level of exclusion. Nevertheless, regression analysis points to the fact that while work bears a great impact on the risk to low income as well as an impact on deprivation levels; when controlling for the other socio-demographic variables (sector affiliation, household type, education, etc.) participating in the labor market has no impact on the level of social exclusion. Participation in the labor market affects deprivation levels but does not influence income levels to the same extent.

The individual's state of health has a tremendous and significant impact on his levels of income, deprivation and exclusion. Persons suffering from health-related disabilities suffer from lower levels of income and higher levels of deprivation and exclusion. Regression analyses suggest that a negative state of health poses a more significant risk factor for exclusion and deprivation when compared to its measure of negative impact on one's income potential.

In comparison to men, women have lower income and suffer from higher levels of exclusion and deprivation. Women's level of exclusion from personal security is nearly twice as high as that of men's. Regression analysis results point to the fact that female gender affiliation raises the risk of social exclusion more than the risk of low income and deprivation.

The different profile of risk factors also strengthens the need to address poverty, deprivation and exclusion as different phenomena that various segments of the population are exposed to in various amounts. In the cluster analysis conducted, the population clustered into only two clusters. One cluster included levels of high income, and low deprivation and exclusion levels (65 percent of the population), while the other cluster was characterized by low income and high levels of deprivation and exclusion. When observing the nature of distribution of the cluster variables one can identify a wider distribution of income versus a narrower distribution of deprivation and exclusion in the 'positive' cluster (i.e., high income, low deprivation and exclusion), while there is a relatively wide distribution of deprivation and exclusion versus only a narrow distribution of income in the 'negative' cluster. This suggests that low levels of deprivation and exclusion do not require a high income but rather enable a wider range of income. On the other hand, low income does not require high levels of deprivation and exclusion; rather it enables a wider range of deprivation and exclusion. Consequently, the cluster analysis findings support the position that distinguishes and separates between the phenomena of economic poverty, social exclusion and material deprivation.

These findings pose a number of possible implications on policies for fighting poverty and social exclusion. One possible implication involves the question of identifying and measuring social exclusion. The research supports the position that implementing alternative indices for poverty other than income can lead to a change in identifying poor populations from the standpoint of its socio-demographic characteristics (Whelan, 2007) and hence the importance of a multi-dimensional poverty measure.

Emphasizing social exclusion requires addressing poverty as a social phenomenon and not solely as an economic issue. Social policies must be designed to cope with the phenomena of poverty, deprivation and exclusion while weighing their social basis (Hickey and du Toit, 2007). The Income Level Index has proven to be limited in identifying populations that also suffer from material deprivation and social exclusion, and it is reasonable to believe that it will not facilitate any improvement in the state of those afflicted by deprivation and social exclusion (Devicienti and Poggi, 2007). Reviewing these findings, one might wonder whether policy based solely upon poverty data according to income level, such as those policies designed to encourage entrance into the labor market, can in fact aid those suffering from social exclusion. Since the relationship between income and social exclusion is extremely partial, it would seem that any attempt to raise the level of income in order to reduce the level of exclusion is doomed to be less effective than focusing attention on social exclusion itself and its causes.

Notes

1 The standard number of individuals scale accepted in Israel and published by the Central Bureau of Statistics was used here.

2 For 68 percent of the population there is a ratio of 1/1 between the number of individuals in the home and the number of rooms. A ratio lower than this, is considered as deprivation.

References

Atkinson, A. B. and Hills, J. (Eds). (1998). Exclusion, employment and opportunity. Centre for Analysis of Social Exclusion, LSE.

Berman, Y. and Keizer, M. (2006). Indicators of social-economic security based on the social quality paradigm. *Social Security, 71*, 109–128 (in Hebrew).

Berman, Y. and Phillips, D. (2000). Indicators of social quality and social exclusion at national and community level. *Social Indicators Research, 50*(3), 329.

Berthoud, R., Bryan, M. and Bardasi, E. (2004). The dynamics of deprivation: the relationship between income and material deprivation over time. Department for Work and Pensions, London.

Bossert, W., D'Ambrosio, C. and Peragine, V. (2005). Deprivation and social exclusion. *Economica, 74*(296), 777–803.

Bradshaw, J. and Finch, N. (2003). Overlaps in dimensions of poverty. *Journal of Social Policy, 32*(4), 513–525.

Bradshaw, J., Williams, J., Levitas, R., Pantazis, C., Patsios, D., Townsend, P. and Middleton, S. (2000). The relationship between poverty and social exclusion in Britain. Paper presented at the International Association for Research in Income and Wealth Conference Cracow, Poland.

Burchardt, T., Le Grand, J. and Piachaud, D. (1999). Social exclusion in Britain 1991–1995. *Social Policy and Administration, 33*(3), 227–244.

Burchardt, T., Le Grand, J. and Piachaud, D. (2002). Introduction. In J. Hills, J. Le-Grand and D. Piachaud (eds), *Understanding Social Exclusion*. Oxford; New York: Oxford University Press.

Desai, M. and Shah, A. (1988). An econometric approach to the measurement of poverty. *Oxford Economic Papers, 40*(3), 505–522.

Devicienti, F. and Poggi, A. (2007). Poverty and social exclusion: two sides of the same coin or dynamically interrelated processes? Torino: LABOR-Centre for Employment Studies.

Endweld, M., Fruman, A. and Barkali, N. (2010). Poverty and social gaps: poverty and income distribution. In M. Endweld and Gotlieb, D. (eds), *Annual Survey: 2009*. Jerusalem: National Insurance Institute – Research and Planning Administration (in Hebrew).

Hallerod, B. and Larsson, D. (2008). Poverty, welfare problems and social exclusion. *International Journal of Social Welfare, 17*(1), 15–25.

Hayes, A., Gray, M. and Edwards, B. (2008). Social inclusion: origins, concepts and key themes. Report prepared by Australian Institute of Family Studies for Social Inclusion Unit, Department of Prime Minister and Cabinet, Canberra.

Hickey, S. and du Toit, A. (2007). Adverse incorporation, social exclusion and chronic poverty. Working Paper 81. Manchester: Chronic Poverty Research Centre.

Larsen, J. E. (2004). Social inclusion and exclusion: conceptual issues and measurement of inclusion and exclusion in Denmark 1976 to 2000. Paper presented at the RC 19 annual conference: Welfare state restructuring: processes and social outcomes, Paris, France.

Levitas, R., Pantazis, C., Fahmy, E., Gordon, D., Lloyd, E. and Patsios, D. (2007). The multi-dimensional analysis of social exclusion. Department of Sociology and School for Social Policy, University of Bristol, Bristol.

McKay, S. (2004). Poverty or preference: what do 'consensual deprivation indicators' really mean? *Fiscal Studies, 25*(2), 201–223.

Morgan, C., Burns, T., Fitzpatrick, R., Pinfold, V. and Priebe, S. (2007). Social exclusion and mental health: conceptual and methodological review. *British Journal of Psychiatry, 191*, 477–483.

Nolan, B. and Whelan, C. (2009). Using non-monetary deprivation indicators to analyse poverty and social exclusion in rich countries: Lessons from Europe? UCD School of Applied Social Science, Dublin, Ireland.

Omtzigt, D.-J. (2009). Survey on social inclusion: theory and policy. Oxford Institute for Global Economic Development, Oxford University.

Perry, B. (2002). The mismatch between income measures and direct outcome measures of poverty. *Social Policy Journal of New Zealand, 19*, 101–127.

Saunders, P. and Adelman, L. (2004). Resources, deprivation and exclusion approaches to measuring well-being: a comparative study of Australia and Britain. Paper presented at the General Conference of the International Association for Research in Income and Wealth, Cork, Ireland.

Scutella, R., Wilkins, R. and Horn, M. (2009). Measuring poverty and social exclusion in Australia: a proposed multidimensional framework for identifying socio-economic disadvantage. Melbourne Institute of Applied Economic and Social Research, University of Melbourne.

Shaprinski, M. and Dror-Cohen, S. (2009) *Annual Society Survey for 2007*. Jerusalem: Central Bureau of Statistics (in Hebrew)

Silver, H. (1994). Social exclusion and social solidarity: three paradigms. *International Labour Review, 133*(5–6), 531–579.

Townsend, P. (1979). *Poverty in the United Kingdom: A Survey of Household Resources and Standards of Living*. Harmondsworth, England: Penguin Books.

Tsakloglou, P. and Papadopoulos, F. (2002). Aggregate level and determining factors of social exclusion in twelve European countries. *Journal of European Social Policy, 12*(3), 211–225.

Washington, J. and Paylor, I. (1998). The role of social work in social exclusion. *Social Work in Europe, 4*(3), 14–19.

Whelan, C. T. (2007). Understanding the implications of choice of deprivation index for measuring consistent poverty in Ireland. *Economic and Social Studies, 38*(2), 211.

Whelan, C. T., Layte, R., Maitre, B. and Nolan, B. (2001). Income, deprivation, and economic strain: an analysis of the European community household panel. *European Sociological Review, 17*(4), 357–372.

Willitts, M. (2006). Measuring child poverty using material deprivation: possible approaches. Working Paper no. 28, Department for Work and Pensions, UK.

Zhang, T., Ramakrishnan, R. and Livny, M. (1996). BIRCH: an efficient data clustering method for very large databases. *ACM, 25*, 103–114.

5 Multidimensional and fuzzy measures of poverty and inequality at national and regional level in Mozambique

Gianni Betti, Francesca Gagliardi and Vincenzo Salvucci

5.1 Introduction

This study provides a step-by-step account of how fuzzy measures of non-monetary deprivation and monetary poverty may be constructed based on the Mozambican Household Budget Survey 2008–09 (IOF08). To our knowledge, this is the first attempt to apply Fuzzy Set Theory to poverty measurement in Mozambique.

The dataset we use is the most recent budget survey available for Mozambique, which is representative of the national, regional (North, Centre, South), provincial, and urban/rural level.

In order to construct a Fuzzy Set index of poverty, monetary as well as non-monetary indicators are considered, and two different measures of deprivation are subsequently constructed: Fuzzy Monetary (FM) and Fuzzy Supplementary (FS).

Including non-monetary dimensions the analysis of poverty in Mozambique is important and informative as the majority of Mozambicans live close to the poverty line, meaning that small changes in their income levels are likely to produce perceptible changes in the Head Count statistics.

The chapter proceeds as follows: in Section 5.2 we present previous studies and official statistics concerning poverty in Mozambique; Section 5.3 illustrates the concept of multidimensional poverty, as well as the Fuzzy Set technique and its application to poverty estimation. In Section 5.4 we introduce the dataset that is used throughout the study, while in Section 5.5 we set out the empirical analysis and the resulting poverty estimates. Section 5.6 concludes.

5.2 Poverty in Mozambique

Mozambique is among the poorest countries in the world, with a per capita income level of approximately $428, ranking 197 out of 210 countries (World Bank, 2010). After the end of the civil war in 1992, Mozambique underwent a process of sustained growth and poverty reduction, leading the country to be considered as a success story by the World Bank and international donors (World Bank, 2008).

Nevertheless, poverty levels remain very high and poor living conditions are widespread throughout the country. The process of poverty reduction has been closely monitored and analysed by three official national assessments (MPF,

1998, 2004; MPD-DNEAP, 2010) and several other studies by both Mozambican and international analysts (Hanlon, 2007; Castel-Branco, 2010; Ossemane, 2010; Van den Boom, 2011).

What emerges from the three main household surveys conducted in the 1996–2008 period, and from other field-specific surveys, is that the situation of Mozambican citizens has improved substantially with respect to some non-monetary dimensions: access to education and health services, household asset ownership, and quality of housing. On the other hand, monetary poverty remained fairly stable between 2002 and 2008: the Head Count Ratio increased slightly from a value of 54.1 per cent in 2002–03 to 54.7 per cent in 2008–09. However, it is important to note that this stabilization followed a sharp fall from its previous levels (69 per cent) in 1996–97.

The Third National Poverty Assessment (MPD-DNEAP, 2010) provides an analysis of both monetary and non-monetary poverty. The Mozambican Government and international donors have invested considerably in reducing non-monetary poverty. In particular, education and health are considered key areas of intervention, and progressively more people have been granted access to schools and health facilities in urban as well as rural areas (Chao and Kostermans, 2002; Government of Mozambique, 2005; Republic of Mozambique, 2006). Nonetheless, monetary poverty did not decrease between 2002 and 2008. At provincial level, the Southern provinces and some of the rural areas in the North experienced a sharp fall in their Head Count Ratio, while Central regions witnessed an increase. Nationwide, rural poverty increased from 55.3 per cent in 2002–03 to 56.9 per cent in 2008–09, whereas urban poverty decreased from 51.5 to 49.6 per cent in the same period.

Regarding non-monetary dimensions, each of the three dimensions considered in 2008–09 (housing conditions, ownership of durable goods, and access to public goods and services) is compared separately with the same dimension six years before, but without computing a general composite welfare indicator. The results indicate that, on average, housing conditions improved between 2002 and 2008, though differences at sub-national level remain high. Ownership of durable goods also improved: the percentage of households owning a radio, a TV, a fridge, a mobile, a telephone, a car, and a bike or motorbike increased by 5.7 points. Turning to access to public goods and services, it emerges that access to education peaked, to the extent that in 2008–09 more than 76 per cent of all children aged six to 13 were attending school, reflecting a large jump compared to the figure of 66.8 per cent in 2002–03. Moreover, geographic inequality in access to education decreased over time and access to health facilities improved. At the same time, other non-monetary dimensions of deprivation did not improve in a substantial way: access to safe water and chronic malnutrition, for example, remained more or less stable.[1]

5.3 Multidimensional poverty and Fuzzy Set Theory

In order to understand poverty and social exclusion, it is necessary to consider deprivation in different terms (such as low income, as well as different non-monetary aspects of deprivation) simultaneously. The need to adopt a multidimensional approach has been noted, among others, by Kolm (1977), Atkinson and Bourguignon (1982), Maasoumi (1986), Tsui (1995), and Sen (1999). Moreover, the multidimensional nature of poverty is widely recognized, not only by the international scientific community, but also by many official statistical agencies (e.g. Eurostat, Istat) as well as by international institutions (United Nations, World Bank).

In the present work we go beyond the conventional study of poverty, based simply on the poor/non-poor dichotomy defined in relation to a chosen poverty line. Instead, poverty and multidimensional deprivation are treated as matters of degree, based on the individual's position in the distribution of income and other aspects of living conditions. The state of deprivation is thus seen in the form of 'fuzzy sets' to which all members of the population belong to varying degrees. This fact brings with it a more complete and realistic view of the phenomenon, but also increased complexity at both the conceptual and analytical levels.

A number of authors have applied the concepts of fuzzy sets to the analysis of poverty and living conditions (Chiappero Martinetti, 1994; Vero and Werquin, 1997, *inter alia*). Our application is based on the specific methodology developed by Cerioli and Zani (1990), Cheli (1995), Cheli and Lemmi (1995), Betti and Verma (1999), Betti *et al.* (2006).

Under the so-called traditional approach, poverty is characterized by a simple dichotomization of the population into poor and non-poor, defined in relation to a chosen poverty line, z. This approach presents two main limitations: first, it is unidimensional (i.e. it refers to only one proxy of poverty, namely low income or consumption expenditure), and second it reduces the population to a simple dichotomy. However, poverty is a much more complex phenomenon, which is not formed solely of its monetary dimension, but must also take account of non-monetary indicators of living conditions. Moreover it is not an attribute that characterizes an individual as being either present or absent, but rather a difficult-to-define predicate that manifests itself in different shades and degrees.

The fuzzy approach considers poverty as a matter of degree rather than an attribute that is simply present or absent in the individuals of a population. In this case, two additional aspects have to be introduced:

i the choice of membership functions (m.f.), i.e. quantitative specifications of individuals' or households' degrees of poverty and deprivation;
ii the choice of rules for the manipulation of the resulting fuzzy sets.

The traditional approach can be seen as a special case of the fuzzy approach, in which the membership function may be seen as $\mu_i^H=1$ if $y_i<z$, $\mu_i^H=0$ if $y_i \geq z$, where y_i is the income of the individual i and z is the poverty line.

An early attempt to incorporate the concept of poverty as a matter of degree at a methodological level was made by Cerioli and Zani (1990), who drew inspiration from the theory of *Fuzzy Sets* initiated by Zadeh (1965). Subsequently, Cheli and Lemmi (1995) proposed the so-called Totally Fuzzy and Relative

(TFR) approach, in which the m.f. is defined as the distribution function $F(y_i)$ of income, normalized (linearly transformed) so as to equal one for the poorest and zero for the richest person in the population.

5.3.1 Income poverty: the Fuzzy Monetary (FM) measure

In the present study we make use of a fuzzy monetary indicator, as found in Betti *et al.* (2009). The proposed FM indicator is defined as a combination of the $(1-F_{(M),i})$ indicator, namely the proportion of individuals richer than individual i (Cheli and Lemmi, 1995), and the $(1-L_{(M),i})$ indicator, namely the share of the total income received by all individuals richer than individual i (Betti and Verma, 1999). Formally:

$$\mu_i = FM_i = \left(1-F_{(M),i}\right)^{\alpha-1}\left(1-L_{(M),i}\right)$$
$$= \left(\frac{\sum_{\gamma=i+1}^{n} w_\gamma \mid y_\gamma > y_i}{\sum_{\gamma=2}^{n} w_\gamma \mid y_\gamma > y_1}\right)^{\alpha-1}\left(\frac{\sum_{\gamma=i+1}^{n} w_\gamma y_\gamma \mid y_\gamma > y_i}{\sum_{\gamma=2}^{n} w_\gamma y_\gamma \mid y_\gamma > y_1}\right) \tag{5.1}$$

where y_γ is the income, $F_{(M),i}$ is the income distribution function, w_γ is the sample weight of the individual of rank γ ($\gamma=1,\ldots, n$) in the ascending income distribution, and $L_{(M),i}$ represents the value of the Lorenz curve of income for individual i. The parameter α is estimated so that the overall FM indicator (which is calculated simply as the weighted mean of the individual FM_i) is equal to the Head Count Ratio computed for the official poverty line.

5.3.2 Non-monetary poverty: the Fuzzy Supplementary (FS) measure

In addition to the level of monetary income, the standard of living of households and individuals can be described by a host of indicators, such as housing conditions, possession of durable goods, health conditions, education, perception of hardship. Several steps are necessary in order to quantify and gather together diverse indicators of deprivation. In particular, decisions are required to assign numerical values to the ordered categories, weighting the scores to construct composite indicators, choosing their appropriate distributional form, and scaling the resulting measures in a meaningful way.

First, from the large set which may be available, a selection has to be made of indicators which are substantively meaningful and worthwhile in relation to the analysis in question. Second, it is useful to group different indicators into statistical components (or dimensions) in order to reduce dimensionality. Whelan *et al.* (2001) suggest systematically examining the range of deprivation items as a first step in an analysis of life-style deprivation, to see whether the items cluster into distinct groups. Factor analysis can be used to identify such clusters of interrelated variables. To quantify and gather together diverse indicators several steps are necessary:

1. identification of items;
2. transformation of the items into the [0, 1] interval;
3. exploratory and confirmatory factor analysis;
4. calculation of weights within each dimension (each group);
5. calculation of scores for each dimension;
6. calculation of an overall score and the parameter α;
7. construction of the fuzzy deprivation measure in each dimension (and overall).

Aggregation over a group of items in a particular dimension h (h=1, 2,..., m) is given by a weighted mean taken over j items:

$$s_{hi} = \frac{\sum w_{hj} \cdot s_{hj,i}}{\sum w_{hj}}$$

where w_{hj} is the weight of the j-th deprivation variable in the h-th dimension. An overall score for the i-th individual is calculated as the unweighted mean:

$$s_i = \frac{\sum_{h=1}^{m} s_{hi}}{m} \tag{5.2}$$

Then, we calculate the FS indicator for the i-th individual over all dimensions as:

$$FS_i = (1-F_{(S),i})^{\alpha-1}(1-L_{(S),i}) \tag{5.3}$$

As for the FM indicator, the parameter α is determined so as to make the overall non-monetary deprivation rate (which is calculated simply as the weighted mean of the individual FS_i) numerically identical to the Head Count Ratio computed for the official poverty line. The parameter α estimated is then used to calculate the FS indicator for each dimension of deprivation separately. The FS indicator for the h-th deprivation dimension and for the i-th individual is defined as a combination of the $(1-F_{(S),hi})$ indicator and the $(1-L_{(S),hi})$ indicator.

$$\mu_{hi} = FS_{hi} = \left(1-F_{(S),hi}\right)^{\alpha-1}\left(1-L_{(S),hi}\right) =$$

$$\left[\frac{\sum_{\gamma=i+1}^{n} w_{h\gamma} \mid s_{h\gamma} > s_{hi}}{\sum_{\gamma=2}^{n} w_{h\gamma} \mid s_{h\gamma} > s_{h1}}\right]^{\alpha-1} \left[\frac{\sum_{\gamma=i+1}^{n} w_{h\gamma} s_{h\gamma} \mid s_{h\gamma} > s_{hi}}{\sum_{\gamma=2}^{n} w_{h\gamma} s_{h\gamma} \mid s_{h\gamma} > s_{h1}}\right], \tag{5.4}$$

$$h = 1,2,...,m; i = 1,2,...,n; \mu_{hn} = 0$$

The $(1-F_{(S),hi})$ indicator for the *i*-th individual is the proportion of individuals who are less deprived, in the *h*-th dimension, than the individual concerned. $F_{(S),hi}$ is the value of the score distribution function evaluated for individual *i* in dimension *h*, and $w_{h\gamma}$ is the sample weight of the *i*-th individual of rank γ in the ascending score distribution in the *h*-th dimension.

The $(1-L_{(S),hi})$ indicator is the share of the total lack of deprivation score assigned to all individuals less deprived than the person concerned. $L_{(S),hi}$ is the value of the Lorenz curve of the score in the *h*-th dimension for the *i*-th individual.

As for the Fuzzy Monetary and the Fuzzy Supplementary indicators, the overall index corresponding to each dimension FS_h is calculated simply as the weighted mean of the individual FS_{hi}. Here it is interesting to note that the overall ranking of the *FS* indicator cannot be obtained directly from the rankings in each dimension; however, the ranking obtained with FS_i is consistent with the ranking obtained from FS_{hi}.[2]

5.4 Data

The dataset used in the study is the Mozambican Household Budget Survey 2008–09 (IOF08) (*Inquerito aos Agregados Familiares sobre Orçamento Familiar 2008–09*), a nationally representative household survey conducted by the National Institute of Statistics (INE). The IOF08 was conducted from August 2008 to September 2009. The survey has a stratified structure with three steps of selection: (i) selection of the primary sampling units (PSUs), (ii) selection of the enumeration areas[3] within the PSUs, and (iii) selection of the households within the enumeration areas. Twenty-one strata were constructed, one for each urban/rural sample of the 11 provinces in Mozambique (the province of Maputo City does not have a rural area). The IOF08 has a sample size of 10,832 households and it is representative at the national, regional (North, Centre, South), provincial, and urban/rural level. The survey includes information on the general characteristics of individuals and households concerning daily, monthly and durable goods in relation to final consumption expenditures, own consumption, transfers, and gifts. Supplementary information for the IOF08 can be found in (INE, 2010; MPD-DNEAP, 2010).

Concerning socioeconomic status, we use data on (real) per capita daily consumption available from the IOF08. This variable is used by the Government for official analyses of poverty, which makes our results immediately comparable to existing ones. This measure of income also considers the inflation that occurred during the implementation of the surveys, the different values of the *Metical* (the Mozambican currency) in different periods of the year, and spatial differences in price levels among different provinces and rural/urban areas.[4]

In order to compute a measure for non-monetary poverty we use information on ownership of durable goods, housing quality, health status, and education level.

5.4.1 Problems with the data

The IOF08 is a very rich and detailed dataset. It has been carefully designed and implemented to provide reliable information and statistical results. Nonetheless, a few problems with the data were encountered while conducting the analysis on multidimensional poverty. In particular, it was found that sampling weights were not calibrated at the household level following a non-response or other problems occurring in the surveying process. Moreover, such weights ranged from 54.6 to 93,452.2, and, as a result of this, a few households with very high weights significantly influenced the statistical results.

In terms of household real consumption – the variable used to assess socioeconomic status – we conformed to official analyses, which divide it by the number of household members, and estimated poverty rates on the basis of this variable. However, this overlooks issues of intra-household allocation of resources and economies of scale, which might matter considerably when dealing with poverty estimates in a country whose average household size is of approximately six members. Indeed, the Head Count Ratio computed by dividing household consumption per adult equivalent produces very different poverty estimates, in which the percentage of poor is 36.8 per cent vs 54.69 per cent.

5.5 Empirical analysis and results

In this section we describe the steps involved in the measurement of multidimensional poverty in Mozambique at national, provincial, and urban/rural level, as outlined in the previous sections. This is followed by an analysis of the results regarding Mozambique's monetary poverty and the different dimensions of non-monetary deprivation.

As introduced in Section 5.4, the Fuzzy Monetary measures, FM, are based on a household's real consumption divided by its size. Real consumption is obtained by taking into account regional differences in price levels, inflation, and seasonal fluctuations. In order to obtain FM, we needed to take into account both the proportion of households richer than each particular household and the cumulative share of consumption that such richer households receive. Finally, the resulting distribution was transformed so that its mean was equal to the Head Count Ratio: this ensured comparability between the two measures and the two approaches, namely the traditional and the multidimensional one.

As for the Fuzzy Supplementary measures, we used information about 32 basic items, as described in Section 5.4. The deprivation dimensions were initially determined using an exploratory factor analysis: this procedure allowed us to describe the variability among the variables observed – our basic items – in terms of a lower number of unobserved, uncorrelated variables, called factors. In the exploratory factor analysis the variables observed were expressed as a linear combination of the underlying factors, without any a priori assumption about the factor structure.

The results of the exploratory factor analysis were then calibrated according to the literature and the experience acquired during the fieldwork in Mozambique.

Finally, a confirmatory factor analysis was performed by imposing a priori assumptions on the underlying factor structure. This allowed us to test whether the proposed calibration of initial items into a lower number of dimensions made statistical sense (Table 5.2).

After these preliminary steps, 32 basic indicators were grouped into six dimensions, roughly corresponding to: (i) housing conditions; (ii) more widespread and affordable durable goods; (iii) less common, more expensive durable goods; (iv) housing quality; (v) income-related deprivation; (vi) health and education. The complete list of the indicators selected and the resulting dimensions are reported in Table 5.1, while the results of the confirmatory factor analysis are summarized in Table 5.2.

Concerning the aggregation of different indicators in each individual dimension, a weighting procedure was carried out, as described in Section 5.3. Depending on the distribution of each indicator in the population and its correlation with other indicators in the same dimension, we constructed item-specific composite weights with equal value for all households in the population. The item-specific weights, W_j, are composed of two parts: Wa_j, which is an inverse function of the percentage of people deprived in item j, and Wb_j, an inverse function of the correlation between item j and all the other items in the same dimension. For each dimension we have that $W_j = Wa_j \ldots Wb_j$.

Intuitively, the first component of the weights, Wa_j, takes into account that if a high percentage of people possess j, then the few who do not possess j are very deprived; the second component, Wb_j, tries to achieve parsimony by assigning a lower weight to items that are highly correlated in the same dimension (e.g. high-quality walls and high-quality roof in the 'housing conditions' dimension).

The result is the identification of six different fuzzy supplementary measures, one for each dimension: FS1, FS2, FS3, FS4, FS5, FS6. Subsequently, we aggregated the different non-monetary dimensions into a single composite Fuzzy Supplementary poverty indicator, FS. This was done by assigning equal weights to each supplementary dimension, based on the assumption that all dimensions are equally important in determining supplementary deprivation. The resulting FS distribution was also scaled so that its mean was equal to the Head Count Ratio, as we did for the monetary poverty indicator, FM. The rescaling ensured that the traditional and the fuzzy indicators were comparable.

Standard errors were computed using Jackknife Repeated Replication (JRR). For a general description of JRR and other practical variance estimation methods in large-scale surveys, see Verma (1993). For a comparative analysis between JRR and Taylor linearization methods see Verma and Betti (2011).

5.5.1 Poverty estimates at national level

As outlined above, the overall Fuzzy Monetary (FM) and Fuzzy Supplementary (FS) dimensions are constructed in a way that their mean is equal to the official

Table 5.1 Dimensions and indicators of non-monetary deprivation

1	*Housing conditions*	*2*	*More widespread durable goods*	*3*	*Less common, expensive durable goods*	*4*	*Housing quality*	*5*	*Income-related deprivation*	*6*	*Health and education*
d1	Bed	d8	Telephone or mobile phone	d14	Tools	d21	Energy source for cooking	d26	A bank account	d29	Ability to read and write
d2	Proper house	d9	TV	d15	Electric or coal iron	d22	Energy source for in-house lighting	d27	A formal or informal job	d30	Educational level
d3	High-quality walls	d10	Bike	d15	Fridge or freezer	d23	Hotplate or gas ring	d28	More than two meals per day	d31	Chronic illness
d4	High-quality roof	d11	Radio	d17	New or second-hand car	d24	Has access to safe water	–	–	d32	Child malnutrition
d5	High-quality floor	d12	Watch	d18	Computer	d25	Fan or air conditioner	–	–	–	–
d6	Has WC or latrine	d13	Motorbike	d19	Printer	–	–	–	–	–	–
d7	Number of rooms/(household members)^2	–	–	d20	Sewing machines	–	–	–	–	–	–

Table 5.2 Confirmatory factor analysis results

Goodness of fit (GFI)[a]	0.834
Adjusted GFI[b]	0.8086
Parsimonious GFI[c]	0.77
Root Mean Square Residual[d]	0.0807
RMSEA[e]	0.0748

Notes

a Based on the ratio of the sum of squared discrepancies to the observed variances; it ranges from 0 to 1 with higher values indicating a good fit.

b GFI adjusted for degrees of freedom of the model, that is the number of the fixed parameters. It can be interpreted in the same manner.

c Adjusts GFI for the number of estimated parameters in the model and the number of data points.

d Fit is considered really good if RMR is equal or below 0.06.

e Root Mean Squared Error of Approximation (RMSEA) is based on the analysis of residuals, with small values indicating a good fit.

Head Count Ratio, so they do not convey additional information to our analysis at national level (Head Count Ratio = FM = FS = 54.69 per cent). Hence, in this subsection we focus only on the values of the supplementary dimensions FS1–FS6.

From Table 5.3, it can be seen that the factor with highest level of deprivation is FS3, corresponding to less common, expensive durable goods. Most Mozambicans do not possess any of the items included in this dimension and a level of deprivation of about 0.75 is thus reasonable. Conversely, the deprivation value for less expensive durable goods (FS2) is lower, showing that some durable goods – especially mobile phones and bikes – are becoming more common in the country.

The level of deprivation for housing conditions (FS1) is also very high (0.53), and reflects the fact that many households lack basic facilities in their dwellings. Even so, the proportion of households lacking decent household quality (FS4) is significantly lower (0.31). Income-related deprivation (FS5) appears to be relatively low: this result is probably influenced by the inclusion of a dummy for whether someone in the household had a job (formal or informal) or not. Since most of the households interviewed (about 98 per cent) had at least one member with a formal or informal job, the entire dimension was pushed towards low levels of deprivation (0.12). When this variable is eliminated from the FS5 dimension, the average deprivation rises significantly (0.64). This was then taken into account in the following analyses.

Table 5.3 Deprivation by dimension, national level

	HCR	*FM*	*FS*	*FS1*	*FS2*	*FS3*	*FS4*	*FS5*	*FS6*
Mean	0.5469	0.5469	0.5469	0.5456	0.5043	0.7457	0.3079	0.1158	0.3224
SE	0.0119	0.0078	0.0056	0.0065	0.0070	0.0076	0.0103	0.0020	0.0063

Note

Mozambique (n = 10,831).

Finally, the result for health and education (FS6) shows that education and health conditions in Mozambique are improving. However, it should be pointed out that the relatively low average value of deprivation for this dimension (0.32) is likely to be affected by the low level of deprivation characteristic of chronic illnesses and ability to read and write. Indeed, the level of child malnutrition in Mozambique is still among the highest in the world (WHO, 2011).

5.5.2 Poverty estimates at provincial level

When fuzzy set poverty analysis is carried out at sub-national level it becomes evident how the inclusion of multiple dimensions substantially increases the amount of information available.

Mozambique is divided into 11 provinces.[5] These territories are quite heterogeneous with regard to economic development, culture, and ethnic and linguistic composition. Consequently, huge differences in poverty rates exist between different zones and provinces in Mozambique. Even though some insights emerged from the official Head Count reports, the multidimensional analysis of poverty we undertook using Fuzzy Set Theory allowed us to highlight important characteristics that would otherwise go unnoticed in a traditional poverty assessment.

In particular, one of the striking aspects of the Fuzzy Monetary (FM) and Fuzzy Supplementary (FS) statistics presented in Table 5.4 is that some of the provinces with low rates of monetary poverty are also much more deprived in other dimensions, and the converse is also true. The Northern provinces (Niassa, Cabo Delgado, Nampula) and the Central province of Tete all have much higher FS averages with respect to their FM ones. The other Central provinces (Zambezia, Manica, Sofala) have similar statistics in both the FM and FS dimensions,

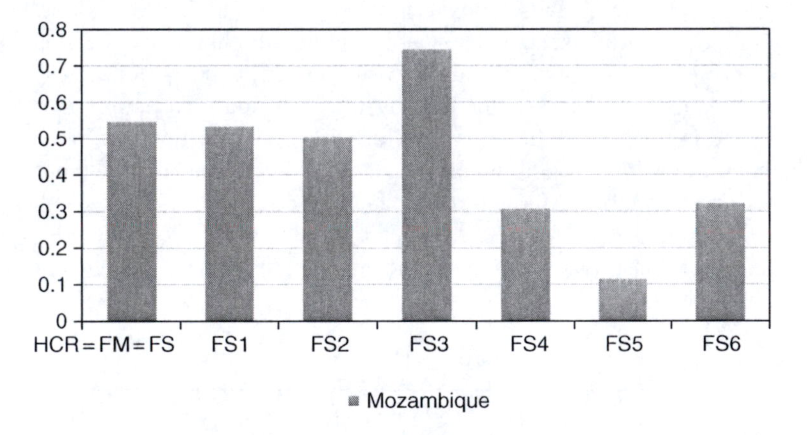

Figure 5.1 Deprivation by dimension, national level.

Note

HCR=Head Count Ratio; FM=Fuzzy Monetary; FS=Fuzzy Supplementary; FS1=housing conditions; FS2=more widespread and affordable durable goods; FS3=less common, more expensive durable goods; FS4=housing quality; FS5=income-related deprivation; FS6=health and education.

Table 5.4 Deprivation by dimension, provincial level

	Niassa	*Cabo Delgado*	*Nampula*	*Zambezia*	*Tete*	*Manica*	*Sofala*	*Gaza*	*Inhambane*	*Maputo Pr.*	*Maputo Cid.*
Head Count Ratio	0.3194	0.3740	0.5468	0.7054	0.4203	0.5509	0.5803	0.5795	0.6254	0.6746	0.3615
SE	0.0322	0.0353	0.0291	0.0339	0.0444	0.0427	0.0484	0.0475	0.0320	0.0245	0.0239
Mean FM	0.4068	0.4386	0.5459	0.6423	0.4903	0.5605	0.5690	0.5504	0.6132	0.6134	0.4054
SE	0.0237	0.0217	0.0186	0.0218	0.0273	0.0249	0.0377	0.0329	0.0233	0.0165	0.0182
Mean FS	0.5649	0.5829	0.6431	0.6406	0.9915	0.5313	0.5265	0.4603	0.4385	0.2970	0.1672
SE	0.0244	0.0171	0.0143	0.0123	0.0192	0.0208	0.0255	0.0225	0.0249	0.0159	0.0066
Mean FS1	0.4941	0.4229	0.6353	0.7319	0.7200	0.6101	0.6179	0.4021	0.3076	0.1233	0.0323
SE	0.0284	0.0190	0.0148	0.0150	0.0273	0.0284	0.0231	0.0290	0.0298	0.0159	0.0029
Mean FS2	0.3615	0.5175	0.5969	0.4948	0.5132	0.3345	0.4360	0.5578	0.5554	0.5305	0.5031
SE	0.0213	0.0211	0.017	0.0141	0.0285	0.0190	0.0408	0.0236	0.0222	0.0231	0.0137
Mean FS3	0.8316	0.8319	0.8710	0.8349	0.8796	0.7410	0.7511	0.5181	0.6439	0.4253	0.2768
SE	0.0200	0.0209	0.0139	0.0280	0.0179	0.0245	0.0236	0.0291	0.0288	0.0240	0.0159
Mean FS4	0.4048	0.3763	0.3269	0.3471	0.3945	0.3545	0.2756	0.2048	0.2187	0.2355	0.0327
SE	0.0396	0.0342	0.0285	0.0245	0.0358	0.0395	0.0367	0.0278	0.0438	0.0368	0.0042
Mean FS5	0.1186	0.1131	0.1293	0.1172	0.1150	0.0910	0.0957	0.1316	0.1355	0.1060	0.0979
SE	0.0065	0.0057	0.0042	0.0064	0.0067	0.0057	0.0075	0.0043	0.0044	0.0073	0.0069
Mean FS6	0.3489	0.3755	0.3652	0.3843	0.3445	0.3017	0.2775	0.3272	0.2936	0.1517	0.1234
SE	0.0146	0.0168	0.0125	0.0228	0.0147	0.0157	0.0242	0.0305	0.0162	0.0156	0.0087
n	814	780	1575	1523	768	804	851	803	814	900	1199

while the Southern provinces show lower FS averages than their respective FM averages. The estimated averages for the Head Count Ratio, the FM and FS dimensions, together with their standard errors, are presented in the first rows of Table 5.4 and in Figure 5.2.

The analysis of FS dimensions indicates that the South is generally less deprived than the Centre and the North, with Maputo City being much less deprived than all the other provinces. These characteristics remained hidden using the standard poverty Head Count analysis. This is probably due to various causes: first, consumption is highly dependent on temporary and/or seasonal fluctuations (e.g. a bad harvest in 2008), while other dimensions, such as those included in the computation of the FS statistics, are more robust to such changes. Indeed, buying an asset, a durable good or investing in education requires an evaluation of a household's economic status that is only partially related to the level of income/consumption in a given year. Moreover, a large part of the Mozambican population has consumption levels that are close to the poverty line, hence even small fluctuations can alter the poverty Head Count statistics in a substantial way. This is one of the main drawbacks of using a dichotomous index like the Head Count Ratio for the analysis of a complex phenomenon such as poverty. In fact, poverty Head Count analyses based on Mozambican Budget Surveys generally yield strange or non-robust results, with strong fluctuations in the Head Count Ratio and re-ranking of poor and rich provinces (Van den Boom, 2011: pp. 7–8).

A more in-depth investigation into supplementary factors yields additional results (Table 5.4). As for housing conditions (FS1), we can identify three distinct groups of provinces on the basis of their FS1 averages: the Central provinces (Zambezia, Tete, Manica, Sofala) and the province of Nampula are the most deprived in this dimension, with an average of about 0.60 for Nampula, Manica, and Sofala but roughly 0.70 for Zambezia and Tete. In the second group, with an average deprivation of about 0.40, we find two Northern

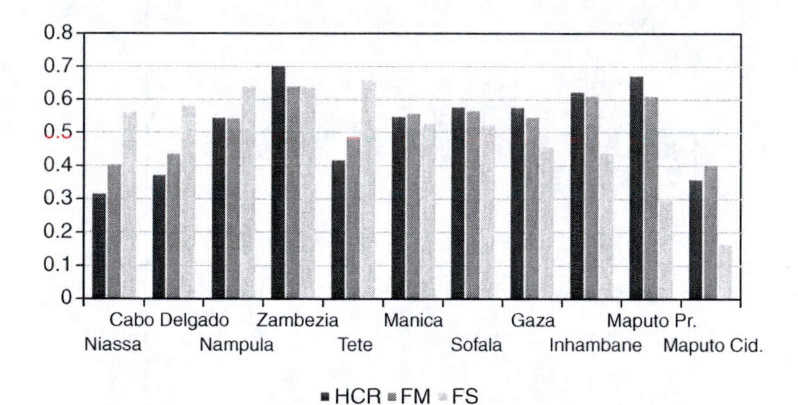

Figure 5.2 Head Count Ratio (HCR), Fuzzy Monetary (FM), and Fuzzy Supplementary (FS), by province.

provinces (Niassa, Cabo Delgado) and two Southern provinces (Gaza, Inhambane). Finally, the least deprived provinces are again Maputo Province and Maputo City, the latter with an average level of deprivation of 0.03.

In the FS2 dimension we grouped together some durable goods that are more widespread than others, such as mobile phones, bikes and motorbikes, radios, watches, and TVs. Indeed, most provinces show similar average levels of deprivation in this dimension, in the 0.44–0.55 range, while Nampula is the most and Niassa and Manica the least deprived, with scores of 0.60, 0.36, and 0.33 respectively.

The FS3 dimension, instead, consists of durable goods that are less affordable and thus less common among Mozambicans, such as cars, fridges or freezers, irons, computers, printers, other electronic tools, and sewing machines. As evidenced in the analysis at national level, this is the factor for which average levels of deprivation are highest. This is particularly true in the North and in the Centre, where five provinces (Niassa, Cabo Delgado, Nampula, Zambezia, Tete) have average values that exceed 0.80, in contrast to the other Central provinces (Manica, Sofala), which perform a little better with values around 0.75. Once again, the Southern provinces of Gaza, Maputo Province, and Maputo City have much lower deprivation levels, with scores of 0.52, 0.43, and 0.28 respectively, confirming the finding that the Southern provinces are less deprived than the Northern and Central provinces in various dimensions.

As for access to safe water, energy sources for cooking, in-house lighting and the like (included in the FS4 dimension), we find that the average level of deprivation is relatively low. For the Northern and Central provinces it ranges between 0.28 for Sofala and 0.41 for Niassa, while all the Southern provinces perform comparatively better.

As presented in Section 5.1, the FS5 dimension (income-related deprivation) is the one for which average levels of deprivation are lowest. In this case, there are no noticeable differences between provinces. However, when the variable 'formal or informal job' is taken out, it emerges that there is a group of provinces (including Manica, Sofala, Maputo, and Maputo City) with average deprivation values of between 0.40 and 0.50, whilst all the other provinces perform comparatively worse, with values of around 0.65–0.75.

Finally, the last supplementary dimension (FS6) takes into account education, measured as level of education and ability to read and write, and health, measured as child malnutrition and chronic illnesses. In this case, Maputo Province and Maputo City record an average level of 0.12–0.15, while the estimated values for the other provinces range between 0.28 for Sofala and 0.38 for Zambezia, which amounts to more than twice the level of deprivation of the two Southern-most provinces.

What this subsection makes clear is that the analysis of dimensions other than consumption substantially improves the mapping of provincial differences regarding poverty. In particular, the higher level of development of the Southern provinces clearly emerged in more than one dimension (FS, and particularly FS1, FS3, FS4, FS6). At the same time, understanding which factors have the greatest influence on deprivation yields deeper insight about which characteristics are more unequally distributed throughout the country.

The estimated averages and standard errors for the supplementary dimensions are found in Table 5.4, while a graphical analysis of the results is shown in Figures 5.2, 5.3, 5.4, and 5.5. In Figure 5.2 we present the Head Count Ratio, Fuzzy Monetary and Fuzzy Supplementary averages for all provinces, which highlights the differences that exist between monetary and overall non-monetary deprivation. In Figure 5.3, instead, all the different supplementary dimensions

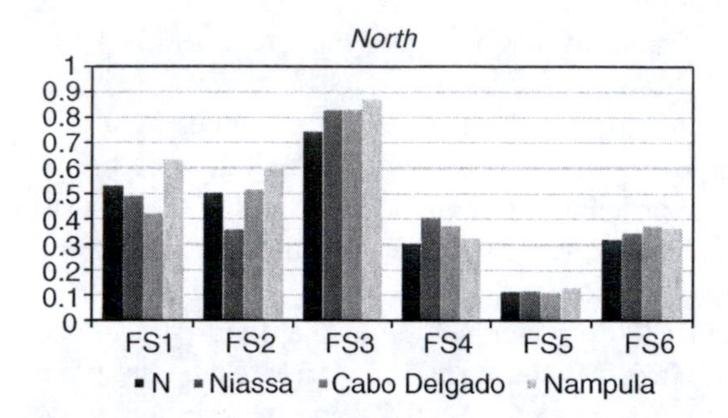

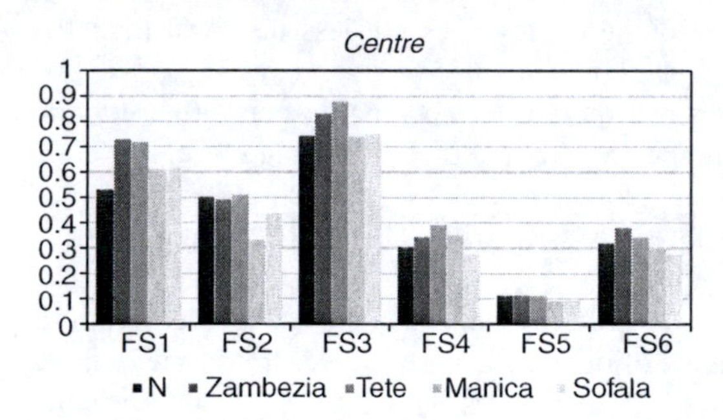

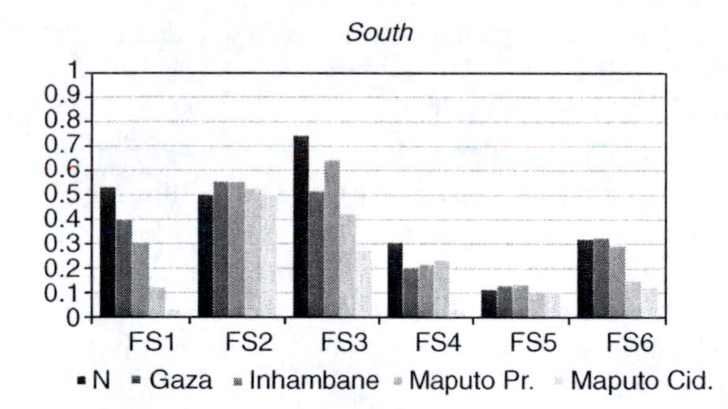

Figure 5.3 Fuzzy Supplementary dimensions (FS1–FS6), by region and province.

Note

FS1 = housing conditions; FS2 = more widespread and affordable durable goods; FS3 = less common, more expensive durable goods; FS4 = housing quality; FS5 = income-related deprivation; FS6 = health and education.

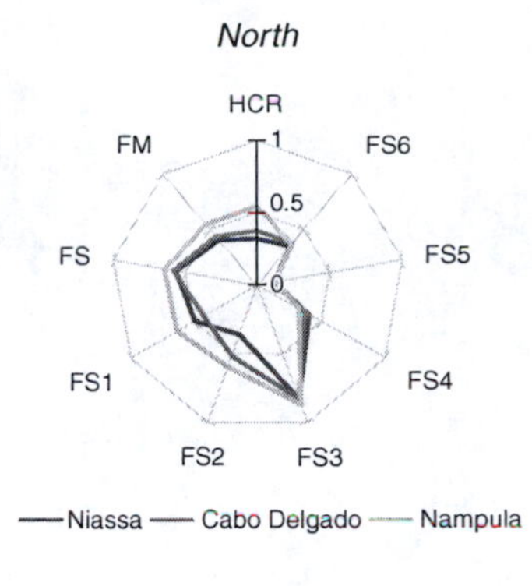

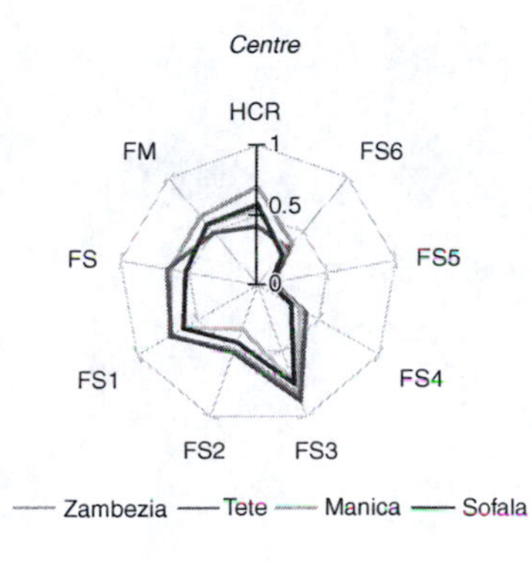

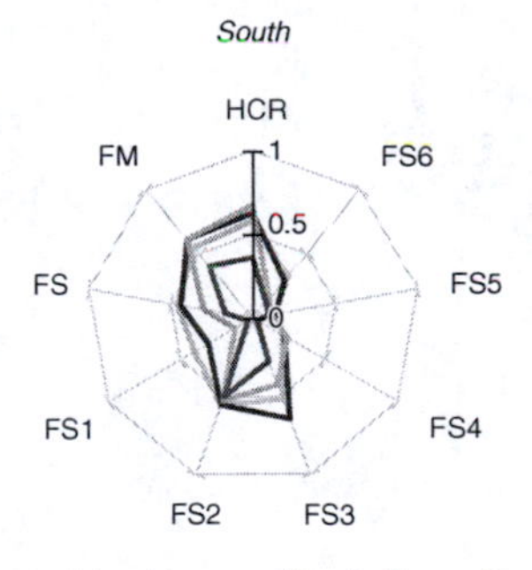

Figure 5.4 Deprivation by dimension and province.

Note

HCR = Head Count Ratio; FM = Fuzzy Monetary; FS = Fuzzy Supplementary; FS1 = housing conditions; FS2 = more widespread and affordable durable goods; FS3 = less common, more expensive durable goods; FS4 = housing quality; FS5 = income-related deprivation; FS6 = health and education.

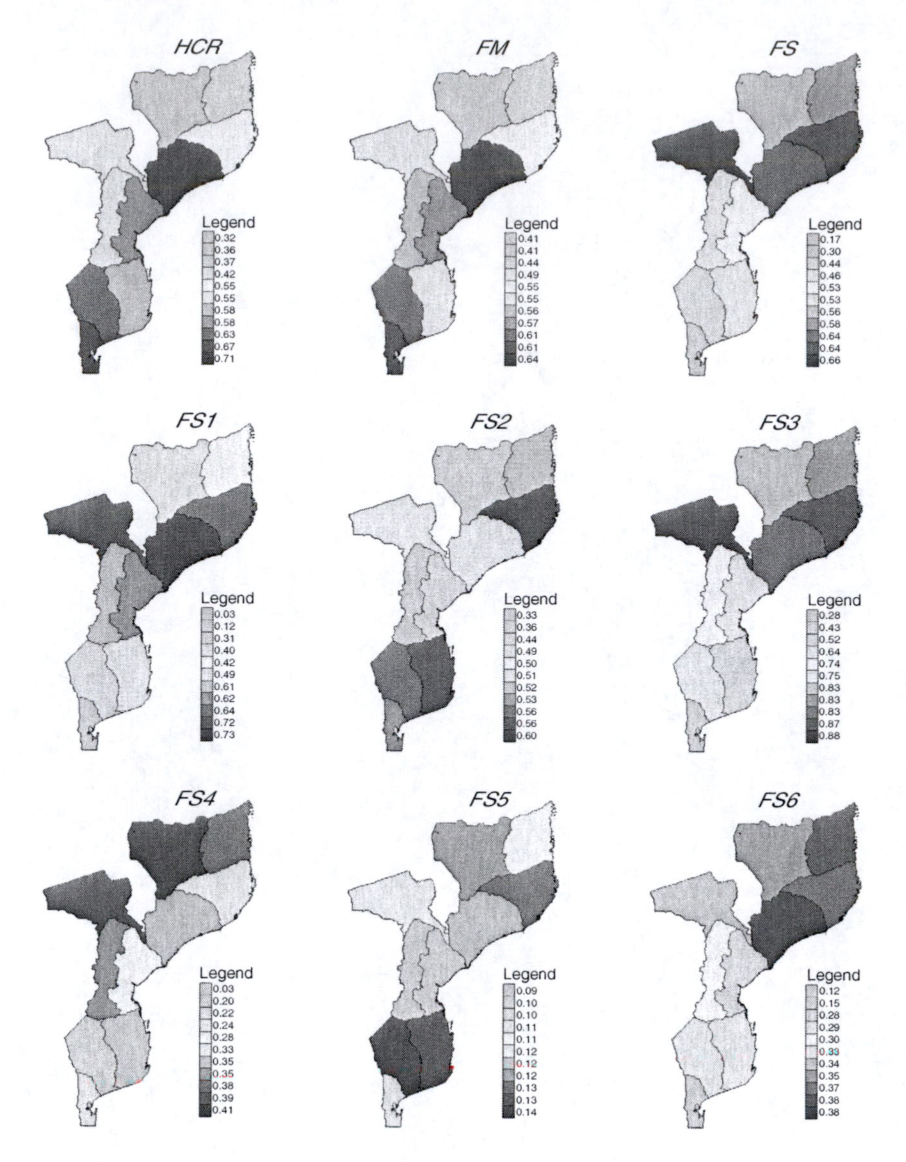

Figure 5.5 Maps of deprivation, by dimension.

Note

HCR = Head Count Ratio; FM = Fuzzy Monetary; FS = Fuzzy Supplementary; FS1 = housing conditions; FS2 = more widespread and affordable durable goods; FS3 = less common, more expensive durable goods; FS4 = housing quality; FS5 = income-related deprivation; FS6 = health and education.

are shown, divided by region and province. In Figure 5.4 both the monetary and individual non-monetary dimensions are shown for each region and province on a net graph. This kind of graph provides additional information about the overall condition of each province compared to the other provinces in the same region. Figure 5.5 is particularly informative since it allows a comparison of all the provinces in all dimensions, and clearly illustrates the gap between the Centre–North and the South regarding the supplementary dimensions of deprivation.

5.5.3 Multidimensional poverty estimates by province and area of residence (urban/rural)

In this section, we present multidimensional deprivation as estimated by province and by area of residence (urban/rural). The huge differences in poverty estimates that exist between urban and rural areas at both national and subnational level in Mozambique have already been introduced in Section 5.2. Nonetheless, unexpected results emerge from the analysis of supplementary dimensions of deprivation (Table 5.5): when these are introduced, the urban/rural deprivation gap widens substantially, contrasting with the official analyses based on consumption, which estimate a differential of about seven percentage points. Indeed, at national level the Head Count Ratio in rural and urban areas for 2008–09 is about 56.9 per cent and 49.6 per cent, respectively (MPD-DNEAP, 2010).

However, when supplementary dimensions of deprivation are considered, a different picture emerges. The aggregated FS deprivation level for urban areas is 0.34, whereas that for rural areas exceeds 0.63. This difference is due to the urban/rural gap seen in the underlying supplementary dimensions. In particular, housing conditions (FS1), possession of less common, more expensive durable goods (FS3), housing quality (FS4), and (to a lesser extent) health and education (FS6) all show very different deprivation levels for urban and rural areas (Figure 5.6).

Regarding housing conditions (FS1) the urban deprivation level is 0.26, while the rural level is as high as 0.65. For more expensive durable goods (FS3) they are equal to 0.52 and 0.84 respectively. The values for the housing quality dimension (FS4) are 0.13 for urban areas and 0.38 for rural areas, while those for the health and education dimension (FS6) are 0.21 for urban and 0.37 for rural areas. Much smaller differences exist in the more widespread durable goods (FS2) and income-related (FS5) dimensions of deprivation.[6] The wide gap in deprivation between urban and rural areas typical of most supplementary dimensions at national level is also reflected at the provincial level. Point estimates and standard errors are found in Table 5.5, divided by province and area of residence.

The central regions of Manica and Sofala exhibit the greatest differences between supplementary deprivation values in urban and rural areas. In the supplementary dimensions FS1 (housing conditions), FS3 (more expensive, less affordable durable goods) and FS4 (housing quality) this difference is conspicuous, ranging from 30 to 60 percentage points. While the urban areas of these two

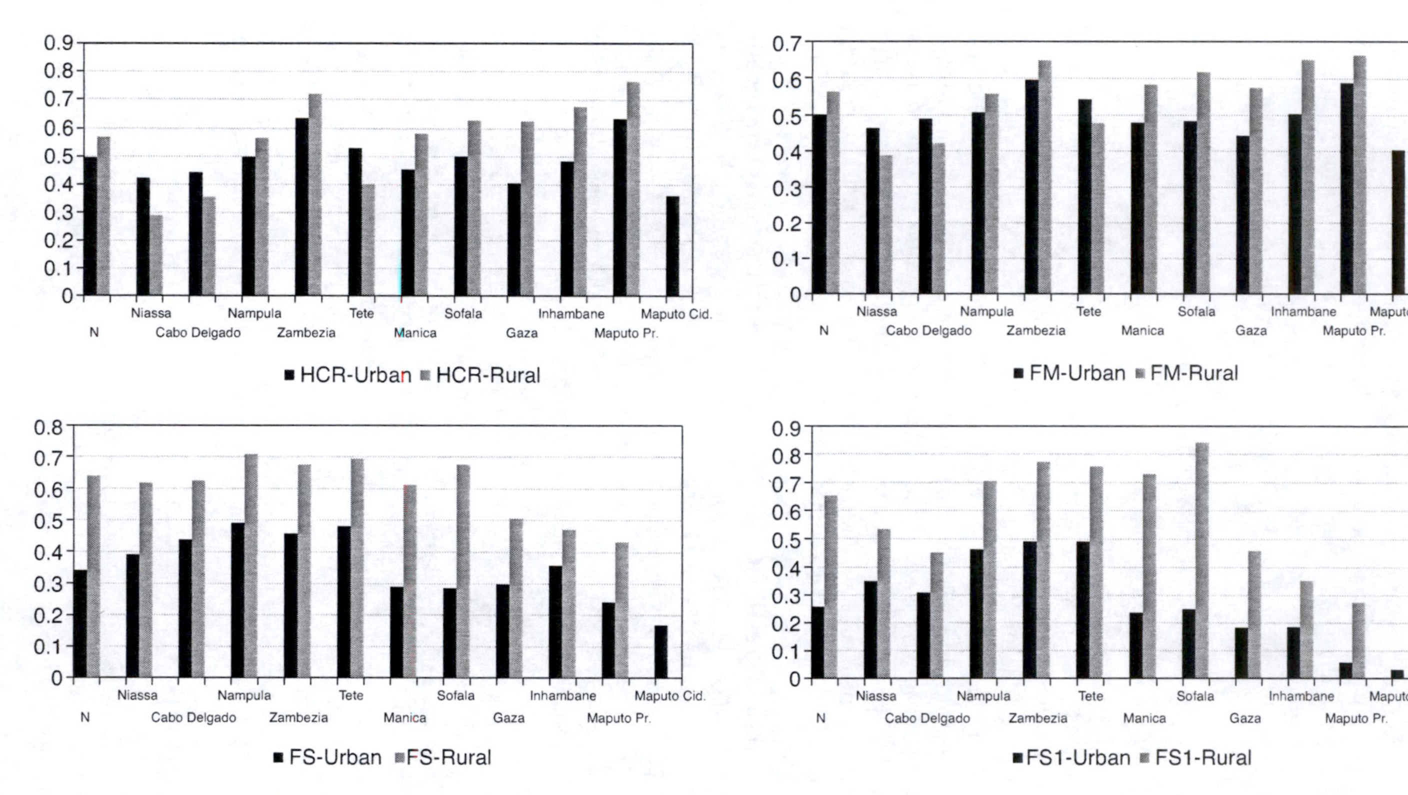

Figure 5.6 Deprivation by dimension, provincial and urban/rural level.

Note
HCR = Head Count Ratio; FM = Fuzzy Monetary; FS = Fuzzy Supplementary; FS1 = housing conditions; FS2 = more widespread and affordable durable goods; FS3 = less common, more expensive durable goods; FS4 = housing quality; FS5 = income-related deprivation; FS6 = health and education. The province of Maputo City does not have a rural area.

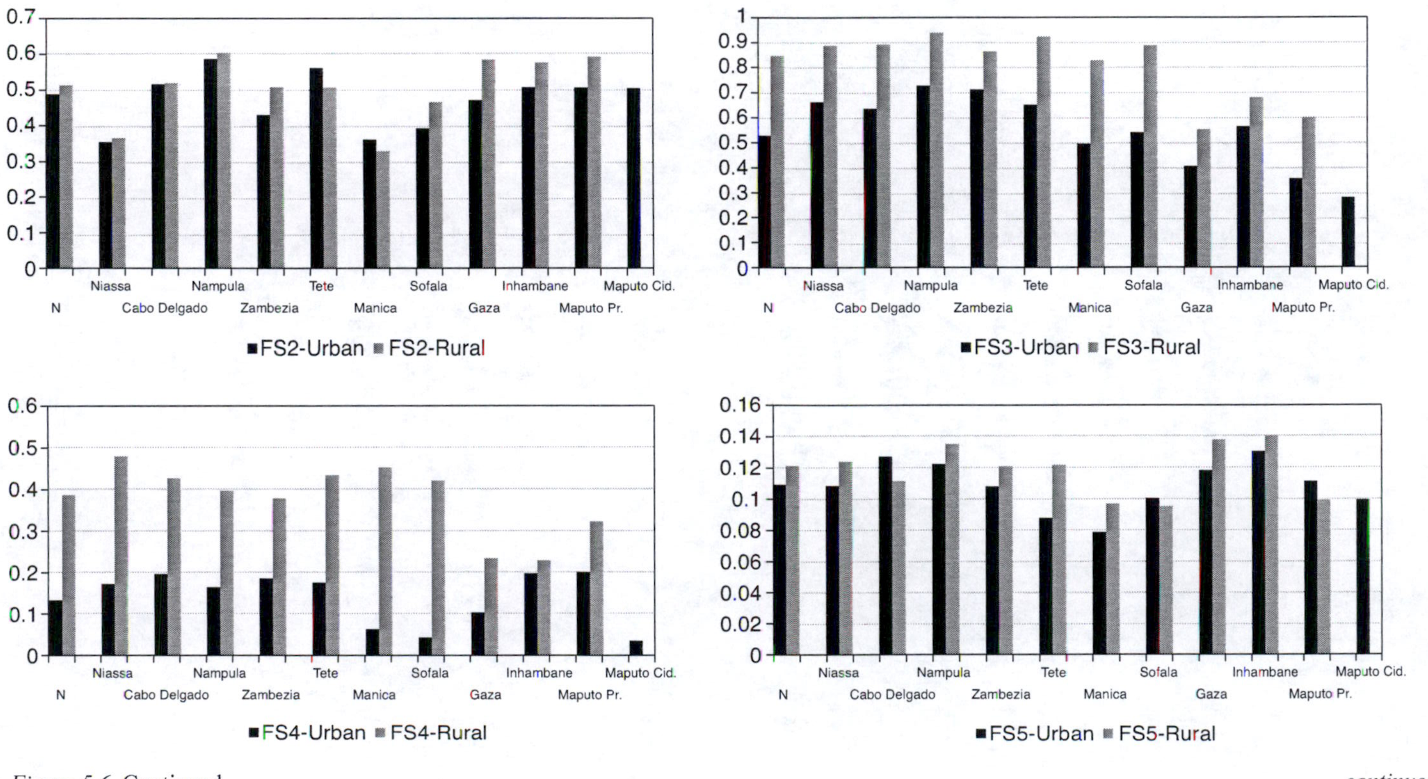

Figure 5.6 Continued

continued

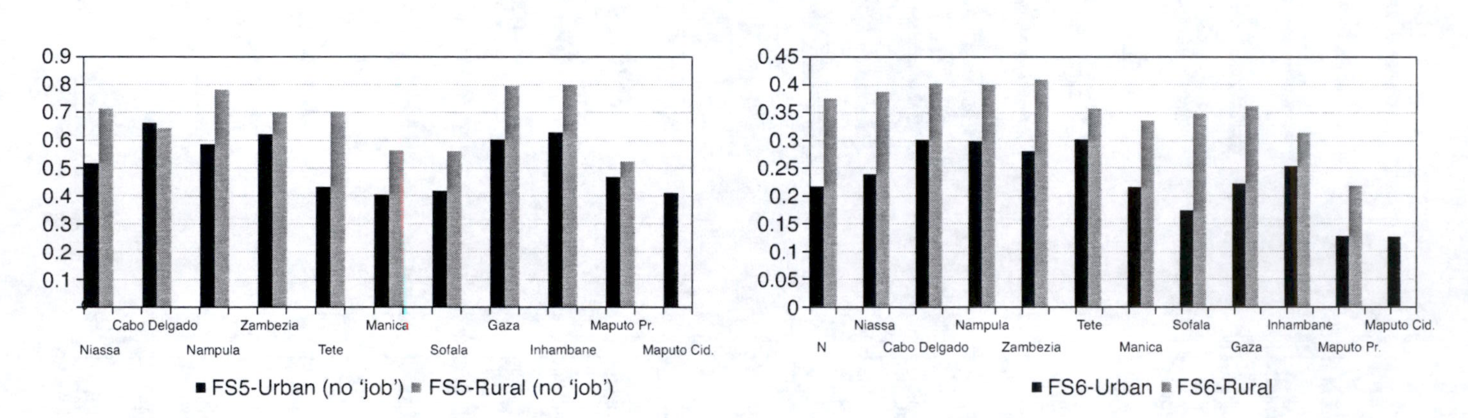

Figure 5.6 Continued

provinces are among the less deprived areas of Mozambique in all dimensions, the opposite is true for their rural counterparts. The urban/rural deprivation gap for the dimensions FS1, FS3, and FS4 is also substantial for other provinces, such as Niassa, Cabo Delgado, and especially Nampula, Zambezia, and Tete. Moreover, the Southern provinces of Gaza and Maputo Province also show significant differences between rural and urban areas.

Urban and rural deprivation levels are, instead, comparable for more widespread durable goods (FS2) and income-related (FS5) supplementary dimensions. Some of the rural areas score even better than their relative urban areas in FS2 (Tete, Manica). As pointed out in the previous paragraphs, excluding the variable of 'formal or informal job' from the FS5 dimension changes the results for this dimension substantially. When this variable is excluded, the difference between urban and rural areas increases noticeably for Niassa, Nampula, Tete, Sofala, Gaza, and Inhambane. Both figures, with and without the variable of 'formal or informal job', are presented in Figure 5.6.

Concerning FS6 (health- and education-related indicators), rural areas are systematically more deprived than urban areas. This is plausible, as healthcare facilities and schools are more widespread in urban areas. The average gap between areas of residence amounts to more than ten percentage points, notwithstanding the commitment of the Mozambican Government to increase the availability of health and education facilities in rural areas (Chao and Kostermans, 2002; Government of Mozambique, 2005; Republic of Mozambique, 2006).

As previously shown, both monetary deprivation dimensions – Head Count and Fuzzy Monetary – analysed at national level determine radically different results from non-monetary dimensions. This also holds true for the analysis at provincial and urban/rural level. Head Count Ratio and Fuzzy Monetary estimates show that the poorest region in Mozambique is the rural area of Maputo Province, while the rural areas of Niassa, Cabo Delgado, and Tete are richer than their urban counterparts and present the same low deprivation levels as Maputo City, the capital. In these monetary dimensions the urban/rural deprivation gap in Manica and Sofala is not as wide as that found in the supplementary dimensions, while the Head Count Ratio urban/rural gap in Gaza and Inhambane is significantly wider.

Introducing supplementary dimensions to the analysis of poverty in Mozambique substantially increases the amount and quality of information available, providing figures that often contrast with those derived from exclusively monetary poverty estimates.

In Figure 5.6 we present the average deprivation levels for each dimension and for all provinces, divided by area of residence. The province of Maputo City only shows one bar as it does not have a rural area.

5.6 Conclusions

In this study we have shown how it is possible to construct poverty measures regarding monetary and non-monetary dimensions using Fuzzy Set Theory.

Table 5.5 Deprivation by dimension, provincial and urban/rural level

	Mozambique		*Niassa*		*Cabo Delgado*		*Nampula*		*Zambezia*		*Tete*	
	Urban	*Rural*	*Urban*	*Rural*	*Urban*	*Rural*	*Urban*	*Rural*	*Urban*	*Rural*	*Urban*	*Rural*
Head Count Ratio	0.4962	0.5691	0.4224	0.2892	0.4429	0.3552	0.499	0.5667	0.6362	0.718	0.5302	0.4015
SE	0.0164	0.0170	0.0512	0.0388	0.0866	0.0382	0.0548	0.0343	0.0680	0.0381	0.0798	0.0502
Mean FM	0.5042	0.5656	0.4664	0.3893	0.4922	0.4239	0.5105	0.5607	0.5985	0.6503	0.5462	0.4808
SE	0.0114	0.0108	0.0221	0.0300	0.0588	0.0226	0.0400	0.0204	0.0485	0.0243	0.0478	0.0309
Mean FS	0.339	0.6378	0.3887	0.6165	0.4358	0.623	0.4901	0.7072	0.4557	0.6743	0.479	0.6926
SE	0.0103	0.0067	0.0313	0.0301	0.0387	0.0191	0.0312	0.0155	0.0457	0.0119	0.0709	0.0190
Mean FS1	0.2596	0.6545	0.3521	0.5357	0.3099	0.4538	0.4646	0.7068	0.4923	0.7756	0.4927	0.7588
SE	0.0113	0.0094	0.0269	0.0359	0.0232	0.0233	0.0325	0.0160	0.0543	0.0147	0.0958	0.0274
Mean FS2	0.487	0.5119	0.3531	0.3639	0.5145	0.5183	0.5847	0.602	0.4285	0.6068	0.5599	0.5053
SE	0.0117	0.0087	0.0232	0.0267	0.0527	0.0227	0.0437	0.0159	0.0278	0.0159	0.0497	0.0323
Mean FS3	0.5248	0.8421	0.6586	0.8822	0.6328	0.8862	0.7224	0.9333	0.7072	0.8581	0.6481	0.9191
SE	0.0133	0.0095	0.0334	0.0240	0.0675	0.0192	0.0385	0.0113	0.0484	0.0319	0.0779	0.0161
Mean FS4	0.1317	0.3848	0.1711	0.4779	0.1951	0.4258	0.1637	0.3953	0.184	0.3768	0.1743	0.432
SE	0.0119	0.0136	0.0475	0.0493	0.0434	0.0418	0.0454	0.0357	0.0401	0.0280	0.0500	0.0410
Mean FS5	0.1075	0.1195	0.1069	0.1221	0.1251	0.1099	0.1205	0.133	0.1064	0.1191	0.0866	0.1199
SE	0.0033	0.0024	0.0103	0.0079	0.0127	0.0064	0.0092	0.0046	0.0127	0.0072	0.0136	0.0075
Mean FS6	0.2135	0.3699	0.2355	0.3821	0.2972	0.3969	0.2951	0.3945	0.2769	0.4039	0.2978	0.3525
SE	0.2135	0.3699	0.0221	0.0178	0.0232	0.0204	0.0161	0.0164	0.0349	0.0262	0.0371	0.0160
n	5222	5609	384	430	240	540	570	1005	336	1187	192	576

	Manica		*Sofala*		*Gaza*		*Inhambane*		*Maputo Pr.*		*Maputo Cid.*
	Urban	*Rural*	*Urban*	*Rural*	*Urban*	*Rural*	*Urban*	*Rural*	*Urban*	*Rural*	*Urban*
Head Count Ratio	0.4537	0.5828	0.5005	0.6293	0.4049	0.6267	0.4836	0.6784	0.6366	0.7633	0.3615
SE	0.0505	0.0543	0.0721	0.0643	0.0429	0.0592	0.0612	0.0376	0.0276	0.0502	0.0239
Mean FM	0.4818	0.5864	0.4865	0.6197	0.4459	0.5786	0.5059	0.6533	0.5914	0.6649	0.4054
SE	0.0322	0.0314	0.0474	0.0535	0.0268	0.0412	0.0427	0.0277	0.0195	0.0308	0.0182
Mean FS	0.2888	0.6111	0.2849	0.6751	0.2969	0.5045	0.3566	0.469	0.2401	0.43	0.1672
SE	0.0255	0.0264	0.0318	0.0362	0.0269	0.0276	0.0545	0.0275	0.0134	0.0427	0.0066
Mean FS1	0.2387	0.7322	0.2507	0.8436	0.185	0.4609	0.1875	0.3525	0.0584	0.2748	0.0323
SE	0.0314	0.0363	0.0404	0.0279	0.0276	0.0361	0.0534	0.0357	0.0058	0.0514	0.0029
Mean FS2	0.3591	0.3263	0.3906	0.4639	0.4695	0.5817	0.5061	0.5738	0.5049	0.5902	0.5031
SE	0.0320	0.0230	0.0334	0.0626	0.0325	0.0287	0.0271	0.0288	0.0244	0.0520	0.0137
Mean FS3	0.4919	0.823	0.5364	0.8831	0.4025	0.5494	0.56	0.6752	0.3523	0.5958	0.2768
SE	0.0350	0.0305	0.0488	0.0235	0.0363	0.0357	0.0466	0.0355	0.0219	0.0616	0.0159
Mean FS4	0.0615	0.4509	0.0414	0.4196	0.1017	0.2327	0.1958	0.2272	0.1987	0.3213	0.0327
SE	0.0234	0.0519	0.0109	0.0588	0.0220	0.0348	0.0908	0.0496	0.0268	0.1058	0.0042
Mean FS5	0.0775	0.0955	0.0989	0.0937	0.1164	0.1357	0.1285	0.1381	0.1096	0.0976	0.0979
SE	0.0069	0.0072	0.0147	0.0080	0.0076	0.0050	0.0089	0.0050	0.0082	0.0151	0.0069
Mean FS6	0.2123	0.3311	0.1703	0.3435	0.2185	0.3567	0.2499	0.3099	0.1248	0.2145	0.1234
SE	0.0228	0.0195	0.0216	0.0367	0.0263	0.0382	0.0260	0.0200	0.0151	0.0380	0.0087
n	336	468	527	324	335	467	382	432	720	180	1199

We applied this technique to the Mozambican Household Budget Survey 2008–09 (IOF08) dataset, the most recent budget survey available for Mozambique.

Our main contribution to the analysis and measurement of poverty in Mozambique is twofold. On the one hand, we estimate a broader concept of poverty than monetary poverty, therefore involving supplementary dimensions. At the same time, we obtain reliable estimates of poverty rates at sub-national and urban/rural level, by using the Jackknife Repeated Replications method to compute standard errors.

To our knowledge, this is the first study to apply Fuzzy Set Theory to the measurement of poverty in Mozambique. As a result, the figures provided in the study substantially increase the amount and quality of information available about Mozambican household deprivation. Our estimates – especially those obtained for non-monetary dimensions – complement those derived solely from the Head Count Ratio. They also provide new evidence with respect to provincial and urban/rural deprivation levels.

With regard to monetary poverty, the Fuzzy Monetary estimates essentially confirm the official results obtained using the Head Count Ratio. In particular, the ranking of poorer and richer provinces remains unchanged, also when the analysis is carried out at the urban/rural level. This is due to both measures, the Head Count and Fuzzy Monetary, being based on consumption data.

Instead, innovative results come from the inclusion of six supplementary dimensions of deprivation in the analysis of poverty: housing conditions; more widespread and affordable durable goods; less common, more expensive durable goods; housing quality; income-related deprivation; health and education. When these dimensions are considered, some of the provinces showing relatively low Head Count Ratios are found to be among the most deprived with respect to the supplementary dimensions of deprivation, and vice versa. In particular, the Northern provinces and the Central province of Tete all show much higher Fuzzy Supplementary (FS) averages with respect to their Fuzzy Monetary (FM) averages. The remaining Central provinces have similar statistics in both the FM and FS dimensions, while the Southern provinces show lower FS averages than their respective FM averages.

The higher level of development of the Southern provinces becomes distinctly relevant to more than one supplementary dimension: housing conditions (FS1); less common, more expensive durable goods (FS3); housing quality (FS4); and, to a lesser extent, health and education (FS6).

Furthermore, in our analysis we point out that deprivation values found in urban and rural areas are very different. When we consider non-monetary dimensions of deprivation it emerges that the urban/rural gap is much wider than shown by Head Count Ratio or Fuzzy Monetary statistics. The aggregated Fuzzy Supplementary deprivation level for urban areas is estimated to be 0.34, whereas the level for rural areas exceeds 0.63. Moreover, while the ranking of some rural areas such as Cabo Delgado, Niassa, and Tete perceptibly worsens, it improves for the provinces of Zambezia and Maputo Province.

One partial explanation of this large difference between the results for monetary and non-monetary poverty is that some of the items included in the supplementary analysis are non-essential items, such as a fridge, car, or PC. For example, the highest average level of deprivation is found for FS3 (less common, expensive durable goods), as most Mozambicans do not possess any of the items included in this dimension, especially in the North and in the Centre. In fact, it might be objected that the inclusion of these items in the analysis of poverty is not entirely justified. However, the said difference between monetary and non-monetary poverty is also large for the supplementary dimensions such as housing conditions, housing quality, or health and education, which certainly denote a situation of deprivation.

Our results are particularly relevant since Mozambique is among the poorest countries in the world, given that its per capita income level is approximately $428 (World Bank, 2010), and several donor countries and international agencies are involved in poverty reduction plans. Accurate information and measurement of poverty at the local level is thus necessary and may be used to redirect funds.

Notes

1 Chronic malnutrition (stunting) is still suffered by 46.4 per cent of children under five, which is among the highest percentages in the world (WHO, 2011).

2 A possible alternative definition of the overall Fuzzy Supplementary indicator could be the simple average of the corresponding indicators. An advantage would be that the overall indicators would fulfil consistency properties with respect to decomposition (Chakravarty *et al.*, 1998; among others). A drawback would be that the weighted mean of the individual would not be equal to the Fuzzy Monetary and the Head Count Ratio indicators.

3 An enumeration area (EA) represents the area assigned to each enumerator for the distribution of questionnaires to households and is the smallest building block in the geographical framework of the Mozambican Household Budget Surveys.

4 This is the methodology used in official analyses of poverty.

5 The 11 provinces of Mozambique are grouped into three bigger zones: the North, which includes the provinces of Niassa, Cabo Delgado, and Nampula; the Centre, with the provinces of Zambezia, Tete, Manica, and Sofala; the South, containing the provinces of Gaza, Inhambane, Maputo Province, and Maputo City.

6 Again, the urban/rural difference increases for dimension FS5 when the variable 'formal or informal job' is not considered: in this case the average urban deprivation becomes 0.51, while rural deprivation becomes 0.70.

References

Atkinson, A., Bourguignon, F., 1982. The comparison of multi-dimensioned distributions of economic status. *Review of Economic Studies* 49(2), 183.

Betti, G., Cheli, B., Lemmi, A., Verma, V., 2006. Multidimensional and longitudinal poverty: an integrated fuzzy approach. In: Lemmi, A., Betti, G. (eds), *Fuzzy set approach to multidimensional poverty measurement*, Springer, New York.

Betti, G., Ferretti, C., Gagliardi, F., Lemmi, A., Verma, V., 2009. Proposal for new multidimensional and fuzzy measures of poverty and inequality at national and regional level. Working Paper no. 83, Dipartimento di Metodi Quantitativi, Università di Siena.

Betti, G., Verma, V., 1999. Measuring the degree of poverty in a dynamic and comparative context: a multi-dimensional approach using fuzzy set theory. Proceedings, ICCS-VI, Vol. 11, pp. 289–301, Lahore, Pakistan, 27–31 August 1999.

Castel-Branco, M., 2010. Intervention at the launch of the book 'Economia extractiva e desafios de industrializacao em Mocambique'. Instituto de Estudos Sociais e Economicos, Maputo, 21 October 2010.

Cerioli, A., Zani, S., 1990. A fuzzy approach to the measurement of poverty. In: Dagum, C., Zenga, M. (eds), *Income and wealth distribution, inequality and poverty*, Springer Verlag, Berlin.

Chakravarty, S., Mukherjee, D., Ranade, R., 1998. On the family of subgroup and factor decomposable measures of multidimensional poverty. *Research on Economic Inequality* 8, 175–194.

Chao, S., Kostermans, K., 2002. Improving health for the poor in Mozambique: the fight continues. Health, Nutrition and Population Discussion Paper. World Bank, Washington DC.

Cheli, B., 1995. Totally fuzzy and relative measures of poverty in dynamic context. *Metron* 53, 183–205.

Cheli, B., Betti, G., 1999. Fuzzy analysis of poverty dynamics on an Italian pseudo panel, 1985–1994. *Metron* 57, 83–104.

Cheli, B., Lemmi, A., 1995. A 'totally' fuzzy and relative approach to the multidimensional analysis of poverty. *Economic Notes* 24, 115–134.

Chiappero Martinetti, E., 1994. A new approach to evaluation of well-being and poverty by fuzzy set theory. *Giornale degli Economisti e Annali di Economia* 53, 367–388.

Government of Mozambique, 2005. Education sector strategy plan II 2005–2009.

Hanlon, J., 2007. Is poverty decreasing in Mozambique? IESE Conference Paper no. 14.

INE (2004) Inquérito Nacional aos Agregados Familiares Sobre Orçamento Familiar 2002/03, www.ine.gov.mz/inqueritos_dir/iaf/.

INE (2010) Inquérito Sobre Orçamento Familiar, 2008/09: Quadros Básicos, www.ine.gov.mz/inqueritos_dir/iaf/.

Kolm, S., 1977. Multidimensional egalitarianisms. *Quarterly Journal of Economics* 91, 1–13.

Maasoumi, E., 1986. The measurement and decomposition of multi-dimensional inequality. *Econometrica* 54, 771–779.

MPF, 1998. Poverty and well-being in Mozambique: the first national assessment (1996–1997). Ministry of Planning and Finance, Universidade Eduardo Mondlane, International Food Policy Research Institute.

MPF, 2004. Poverty and well-being in Mozambique: the second national assessment (2002–2003). National Directorate of Planning and Budget, Ministry of Planning and Finance, Economic Research Bureau, Ministry of Planning and Finance, International Food Policy Research Institute, Purdue University.

MPD-DNEAP, 2010. Poverty and wellbeing in Mozambique: third national poverty assessment. Maputo: Ministério da Planificação e Desenvolvimento – Direcção Nacional de Estudos e Análise de Políticas.

Ossemane, R., 2010. O dilema do crescimento empobrecedor. Tete, Mozambique, 26 October 2010.

Republic of Mozambique, 2006. Action plan for the reduction of absolute poverty 2006–2009.

Sen, A., 1999. *Development as freedom*. Oxford University Press, Oxford.

Tsui, K., 1995. Multidimensional generalizations of the relative and absolute inequality indices: the Atkinson–Kolm–Sen approach. *Journal of Economic Theory* 67(1), 251–265.

Van den Boom, B., 2011. Analysis of poverty in Mozambique, http://undp.org.mz/en/Publications/Other-Publications/Analysis-of-poverty-in-Mozambique.

Verma, V. (ed.), 1993. Sampling errors in household surveys: a technical study. United Nations Department for Economic and Social Information and Policy Analysis, Statistical Division, INT-92-P80-15E, New York.

Verma, V., Betti, G., 2011. Taylor linearization sampling errors and design effects for poverty measures and other complex statistics. *Journal of Applied Statistics* 38(8), 1549–1576.

Vero, J., Werquin, P., 1997. Reexamining the measurement of poverty: how do young people in the stage of being integrated in the labor force manage. *Economie et Statistique* 8(10), 143–156.

Whelan, C., Layte, R., Maitre, B., Nolan, B., 2001. Income, deprivation, and economic strain: an analysis of the European Community Household Panel. *European Sociological Review* 17(4), 357.

WHO, 2011. Nutritional status, www.childinfo.org/undernutrition_nutritional_status.php.

World Bank, 2008. *Beating the odds: sustaining inclusion in Mozambique's growing economy*, World Bank, Washington DC.

World Bank, 2010. *World development report 2010: development and climate change*, World Bank, Washington DC.

Zadeh, L., 1965. Fuzzy sets. *Information and control* 8(3), 338–353.

Part II

Longitudinal and chronic poverty

6 On assessing the time-dimension of poverty

Vijay Verma and Francesca Gagliardi

6.1 Introduction

The conceptualisation and assessment of poverty involves at least four dimensions: multiple aspects covering both monetary and non-monetary deprivation; diverse statistical measures in order to capture different facets of poverty; the time-dimension; and comparability over space and time.

This chapter, and the following four chapters in this Part, address various issues concerning the time-dimension of poverty. This involves two types of measures:

1 Measures of poverty trend over time at the aggregate level.
2 Measures of persistence or otherwise of poverty at the micro level.

The common grounds for both these types of measures are provided by the construction of cross-sectional measures, which define the state and other aspects of the individuals' poverty situation at each time.

This and one of the following chapters focus on (1), and the other chapters address specific technical issues in relation to (2).

The main measurement problem in the assessment of poverty trends is *the definition of the poverty threshold and its consistency over time.* The main measurement problem in the assessment of persistence of poverty concerns the effect of *random measurement errors on the consistency of the individuals' observed poverty situation at different times.*

The focus of the present chapter is on practical problems faced in the assessment of poverty trends across time and across populations. We begin in Section 6.2 by taking a close look at one of the longest available series of poverty rates, namely the official figures for the United States. This series is remarkable in that it has used a fixed poverty threshold for over half a century, despite major changes in the level and structure of the population's income and consumption over this period. Then in Section 6.3 we review more briefly the procedures used for assessing poverty trends in developing countries, in particular in China and India, the largest among them. Section 6.4 compares and contrasts the two different types of poverty thresholds used, namely the absolute and the relative thresholds. We observe that the distinction between absolute and relative conceptualisations of poverty is essentially determined by how the two levels – the minimum acceptable and the average levels of income in a society – are related. Finally, Section 6.5 presents a brief note concerning the effect of measurement errors on measures of longitudinal chronic or persistent poverty.

6.2 A look at the official estimates of poverty trends in the US

A close look at the official estimates of trends in poverty rates in the United States is helpful in formulating and illustrating some of the issues involved in the assessment of poverty trends. Let us begin by contrasting the official poverty measures used in European Union countries and the United States.

Usually, the official poverty threshold adopted in the EU is termed 'relative', while the one adopted in the United States is taken as an 'absolute' concept of poverty.

In Europe, poverty is officially defined in relative terms, as the percentage of individuals living in a household whose equivalent income is below the poverty threshold (equal to 60 per cent of the national median equivalent income). It is a dimension of national income inequality as distinct from the average standard of living. Countries with very different standards of living (and thus very different median equivalent income and hence different poverty thresholds) can have the same poverty rate. The justification of this approach is the conviction that the proper objective of social policy is to identify those at serious disadvantage in the *context of the particular society they live in*. These relative poverty rates are highly correlated with inequality measures, but are not the same thing. They focus at the lower end of the income distribution, and hence bring in an additional dimension:

> the literature on the measurement of inequality is both much broader and more comprehensive than that on poverty measurement, and some writers seem to feel that good measures of inequality may preclude the need for a separate, specific measure of poverty.... For many such writers, any consideration of poverty or of antipoverty policies automatically translates into a consideration of the distribution of income and wealth.... From the point of view of the policymaker, however, a concern about poverty does not necessarily imply any interest at all in broader issues of distribution. Many policymakers start instead with the idea that ... there is some minimum decent standard of living, and a just society must ensure that all its members have access to at least this level of economic well-being.
>
> (Ruggles, 1992)

As to an absolute poverty line, there are various bases for fixing an 'absolute' line: such as consumption on essentials, benefit levels for low incomes, or simply by starting with a relative poverty line but then keeping it 'anchored' in real terms to original level, i.e. to an arbitrarily fixed point in time.

The US established a set of official poverty thresholds nearly fifty years ago (1963), actually based on data over half a century old (1955). The original formulation of the poverty thresholds addressed the problem by starting with a set of official estimates of minimally adequate food budgets for families of various sizes and types. To obtain a poverty line, the author (Orshansky) simply multiplied these minimum food budgets by a factor of three, on the assumption that food typically represented about one-third of total expenditures of poorer families. Any family with income less than three times the cost of the minimum food budget was classified as poor. This one-third estimate in fact came from a 1955 food consumption survey, and was probably already outdated in 1963 when it was first used. According to Ruggles (1992), consumption data from 1960–61 indicate that food consumption was already closer to one-fourth of the typical budget by then.

This official poverty measure consists of a set of dollar amounts thresholds that vary by family size: a family of a given size with an income below the threshold for its size is considered poor; a family with income above the threshold is non-poor. Income, for the purpose of measuring poverty, consists of money income before taxes. It does not include non-cash forms of income such as food and medical subsidies. The poverty lines do not differ by state or region in the country. Recently, the government experimented with a new measure called the SPM – 'supplemental poverty measure'. This measures counts both disposable income and expenses differently from the official measure. It includes the buying power that state assistance adds to poor families' budgets. It also subtracts buying power for necessary costs other than food, like housing and transportation.

Two changes were made to the poverty definition in 1969. Thresholds for non-farm families were tied to annual changes in the *Consumer Price Index* (CPI) rather than changes in the cost of the economy food plan, and farm thresholds were raised from 70 to 85 per cent of the non-farm levels. In 1981, further minor changes were made, such as eliminating separate thresholds for farm and female-headed households, and making the largest family size category for reporting as 'nine persons or more'. Apart from these changes, the US government's approach to measuring poverty has remained static. The original measure has been updated for changes in prices since the 1960s, but *no adjustment has been made to take account of any other changes in needs or consumption patterns over this long period*. This conflicts with the fact that so much else has changed since then – especially how Americans spend their money. Today, 13 per cent of household income is spent on food, far less than the 33 per cent in the original standards. Meanwhile, nearly 35 per cent goes to housing, and 17 per cent goes to transportation, according to the *US Bureau of Labour Statistics*. But those costs, which now make up a majority of the budgets of households, are ignored in the adjustment of the poverty threshold over time.

In Table 6.1 are shown thresholds from Department of Health and Human Services. Alternative thresholds from the Census Bureau, used for official estimates, distinguish in addition by age in single-person households (15–64, 65+), and similarly by the householder's age in two-person households. Otherwise any differences between the two thresholds are minor. For lack of more detailed information, we have used a particularly simple equivalence scale: equivalent size equal to square-root of actual household size.

Incidentally, the threshold per adult equivalent, shown in the last column of Table 6.1, is fairly uniform for different family sizes.

The fundamental problem, as often noted (e.g. Schwarz, 2005: p. 194, note 13), is that

> the official poverty line today is essentially what it takes in today's dollars, adjusted for inflation, to purchase the same poverty-line level of living that was appropriate half a century ago, in 1955, for that year furnished the basic data for the formula for the very first poverty measure. Updated thereafter only for inflation, the poverty line lost all connection over time with current consumption patterns of the average family. Quite a few families then didn't have their own private telephone, or a car, or even a mixer in their kitchen.... The official poverty line has thus been allowed to fall substantially below a socially decent minimum, even though its intention was to measure such a minimum.

How have the poverty threshold and poverty rate changed over time? The poverty thresholds for families of all sizes have remained completely unchanged (Table 6.2). In Figure 6.1, trends in the poverty rate and the national median (equivalised) income are brought out by scaling the figures to equal 100 in 1983,

Table 6.1 US poverty thresholds 2012

Size of family unit (1)	*Poverty threshold (US$)* (2)	*One version of equivalised threshold* (3)=(2)/√(1)
1	11.170	11.170
2	15.130	10.699
3	19.090	11.022
4	23.050	11.525
5	27.010	12.079
6	30.970	12.643
7	34.930	13.202
8	38.890	13.750
Plus $3960 for each additional person, giving		
9	42.850	14.283
10	46.810	14.803
11	50.770	15.308
12	54.730	15.799
Higher thresholds for Hawaii and Alaska, e.g. for single person unit		
Hawaii	12.860	–
Alaska	13.970	–

Source: Department of Health and Human Services.

Table 6.2 Changes in poverty threshold (2)

Calendar year	*1961*	*1971*	*1981*	*1883*	*1991*	*2001*	*2011*	*mean*
Annual average CPI, all items (1)	30	41	91	100	136	177	225	–
Unrelated individuals, all ages	50	50	51	51	51	51	51	51
2-person households, all ages	65	65	65	65	65	65	65	65
3 people	80	80	80	80	80	80	80	80
4 people	102	102	102	102	102	102	102	102

Notes
1 1982–84 = 100.
2 In $100; base year 1983.

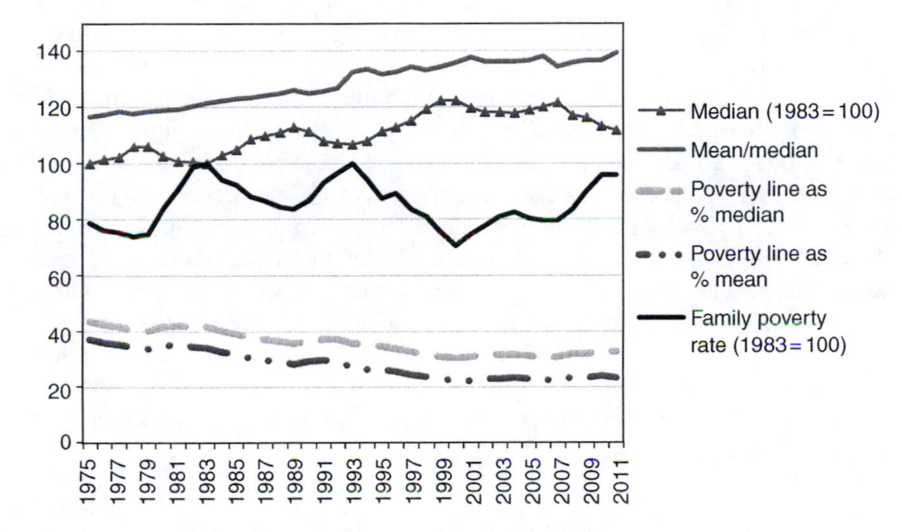

Figure 6.1 Trends in the US official poverty threshold and poverty rate.

the base year used in the series. (All figures are slightly approximate because of limitations in the data available to us.) The official poverty rate has hovered between 80–100 (with the 1983 figure taken as 100) – a remarkably stable picture. But this results from the average living standard rising slowly but persistently (as shown in the figure by the median value), but the poverty threshold, held constant in inflation-adjusted dollars, slipping constantly against the median, and especially against the mean level of income. The mean-to-median ratio has increased considerably over the period reflecting increasing skewness of the income distribution. As noted by Rosenberg (2012):

> None of this should be surprising. While the US economy as a whole has continued growing over the past 40 years – though at a slower pace than it did prior to 1973 – wages in general have largely remained stagnant.... In the US, from 1973 to 2010, the average income per tax unit was up just 12 per cent. Including capital gains, it was up 13 per cent. But for the bottom

> 90 per cent, on average, it was down 11 per cent, while for the 1 per cent it was up 153 per cent on average, and for the 1 per cent of the 1 per cent it was up 649 per cent.... In Sweden, the average income over this same time period was up 80 per cent – 93 per cent with capital gains. The bottom 90 per cent did roughly as well – up 78 per cent, while the top 1 per cent was up 124 per cent and the 1 per cent of the 1 per cent was up 564 per cent.... So, income gains were much higher overall, and though the wealthy did better than average, only the super-wealthy did dramatically better. For the most part, Sweden has managed to keep increased prosperity a broadly shared phenomenon.

To summarise, the income level in the US has been rising, albeit slowly, over a prolonged period of time. The official US poverty line has been kept at a fixed level in real terms despite rising income levels. This results in the poverty threshold constantly declining as a proportion of the median and mean income. For instance, by the mid-1990s the US poverty line had fallen to an equivalent of less than one-third of median household income, indicating not 'poverty' but 'extreme poverty'. In order to bring out the effect of this decline, Figure 6.2 shows what the poverty rate in Italy would be compared to the US poverty rate, when the poverty threshold in the two cases forms the same proportion of the median income. The US figures are historical (longitudinal), based on the actual poverty line and the median income each year. The Italian figures come from a single cross-section (for year 2007) and the computations have been made for different poverty lines. Each computation corresponds to a particular poverty threshold-to-median income ratio observed in the US series, and has been plotted in Figure 6.2 on the same vertical line.

6.3 Concerning poverty trends across developing countries

Measuring poverty consistently even in a single population and for a single time period is a challenging task. The task becomes more difficult for measurements extending over time, and undoubtedly extremely difficult when both variations across space and trends in time are to be assessed and compared. Some of the complications involved in this exercise in connection with the World Bank estimates of the levels and trends in absolute poverty in developing countries of the world are discussed by Stephan Klasen in Chapter 10 in this book. The central issue of interest in our present discussion is whether a poverty line separating the poor from the non-poor is invariant across space and time. The World Bank estimates take the international poverty line as invariant in both dimensions; the poverty line is adjusted only for differences in prices across space and time. It constitutes an *absolute* poverty line that tries to capture the subpopulation in extreme poverty where basic physical survival and health is at risk.

The World Bank starting point was always the national poverty lines of a large sample of developing countries in national currencies, converted to 'international $' based on International Price Comparison Project (ICP) results. On the basis of

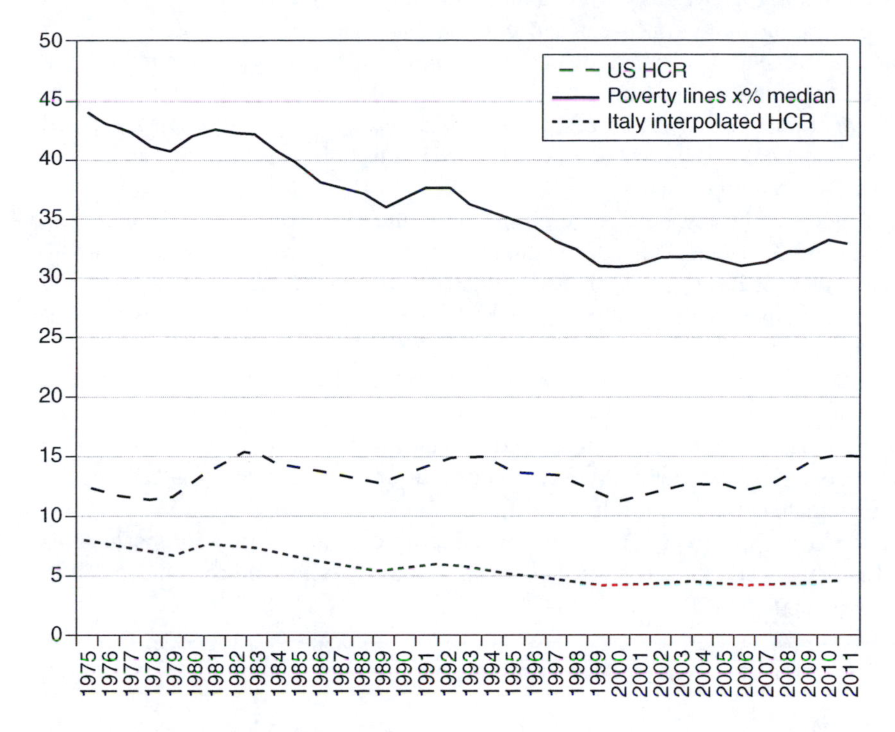

Figure 6.2 An indication of the impact on poverty trend of the changing poverty threshold-to-median income ratio in the US series.

the rather similar values of the line in 'international $' found in low income countries, a single poverty line is chosen and applied to all developing countries. The exercise has been repeated three times giving the following lines:

$1.02 from ICP round 1985 (published in 1990);
$1.08 from ICP round 1993 (published in 2000/01); and
$1.25 from ICP round 2005 (published in 2008).

The international poverty line is then turned into a poverty line in national currencies at the benchmark ICP year, and this poverty line is adjusted using national inflation rates to generate poverty lines in national currencies backwards and forward in time for all years since 1990 (or even since 1981). For any particular country, these reference poverty lines are at different levels, but have the same slope. Thus the predicted national poverty rates differ greatly from one standard to another, but the predicted poverty *trends* remain identical. The divergence in the predicted level arises from incompatibility between national inflation rates and ICP values. Developing countries tend to be 'disadvantaged' in terms of the latter, resulting in raising of the poverty line for them, and hence in higher predicted poverty rates.

Klasen in Chapter 10 provides a nice summary of some alternative approaches to generating more consistent and comparable time-series measuring extreme poverty at the global level. A widely used approach noted by the author 'consists of creating *national* poverty lines using a procedure that is internationally consistent so that the poverty measured in this consistent way could be aggregated across countries', the most common way being to link them in some form for a nutritional requirement.

Below we briefly note some points from the experience of the two largest countries, China and India. The national poverty lines have been based on some assessment of basic needs of the population but, unlike the example from the USA, have been periodically adjusted to reflect changes in the national socio-economic situation.

India and, especially, China have enjoyed rapid economic growth, with a median growth rate of 6 per cent and 10 per cent in the 1980–2010 period, respectively. Despite greatly increased inequality accompanying rapid growth in both countries, the marked reduction in reported poverty rates results from the use of unchanging poverty thresholds.

China has been very successful in reducing extreme deprivation, at least as measured in terms of proportion of the population living on less than (the PPS equivalent of) US$1.25 a day:

	Rural	*Urban*
Early 1980s	94	45
2005	26	2

The national poverty line used in China of course also indicates very sharp decline in the poverty rate:

1978	1985	1992	2000
30.7%	14.8%	of 8.8%	3.4%

The poverty threshold was estimated using the internationally accepted method based on income and consumption data from the National Rural Household Survey (RHS) in 1995. A National Rural Poverty Monitoring Survey covering 592 poor counties (out of a total of 2000 counties) listed in national poverty reduction programmes has been carried out since 1997. The third main source of poverty statistics is the regional statistics through administrative reports.

The National Bureau of Statistics has principally been using 'food share method' for constructing the poverty lines. It proceeds by first fixing the minimum food energy intake cut-off in calories, and then finding the basic food bundle and the relative consumption at which a person typically attains that food energy intake accordingly by using RHS data. Then the poverty line could be generated by dividing minimum food consumption by a 'reasonable' share

(60 per cent) of consumption going to food. CPI was used to update the poverty lines for subsequent years. As noted in China (2004),

> although the food poverty line could be obtained objectively using this method, the main disadvantage is that the non-food component cannot be measured precisely since it is hard to determine to what degree the share of consumption going to food is 'reasonable'.

An improved method used since 1995 has the following steps: select 2100 k-calories per person per day as the minimum nutrition need; determine the essential food expenditure per capita per day based on the actual food consumption bundle and food price of the low income group (the food poverty line); estimate the essential non-food expenditure through the regression model (the non-food poverty line); and finally derive the poverty line by adding the food and non-food poverty lines. In this connection, the poverty is measured by both income and consumption. The poor are defined as: (1) those with per capita net income below the poverty line and per capita consumption below 1.2 times poverty line; plus (2) people with per capita consumption below the poverty line and per capita net income below 1.2 times poverty line.

Despite these efforts, it remains true that China's official poverty lines have been derived based on a bundle of items dominated by food grains that have not been updated adequately to reflect changes in consumption patterns or adjusted to take into account inflationary trends in both food and non-food items. The result was one of the lowest rural poverty lines in the developing world (Niño-Zarazúa and Addison, 2012). In December 2011, the Chinese government announced it would lift the country's rural poverty line by over 80 per cent, to an equivalent of around US$1.80 per day, a threshold well above the US$1.25 used by the World Bank for international poverty comparisons.

As to India's poverty line, since the early 1990s official poverty estimates have been made on the basis of the methodology established in 1993 (Lakdawala Committee). These poverty lines are based on per capita consumption levels associated with a commodity bundle that yielded a specified level of caloric intake believed to be appropriate in 1973–74. Changes were recommended in 2009 (Tendulkar Committee) to base the poverty line in the consumption levels observed in the 2004–05 National Sample Survey after correcting for the rural–urban price differential. The new estimates increased the estimated poverty rates greatly, in fact bringing them close to those using the World Bank's international $1.25 per day (2005 based) poverty line:

	1993 methodology			*2009 methodology*			*World Bank*
	Rural	*Urban*	*Combined*	*Rural*	*Urban*	*Combined*	*US$1.25*
1993–94	37	32	36	50	32	45	49
2004–05	28	26	27	42	26	37	42

Source: India (2011); World Bank's PovCalnet database.

Certain limitations of the 2009 methodology have been noted (Niño-Zarazúa and Addison, 2012). The new methodology used an observed calorie intake that is quite low relative to the minimum dietary energy requirement recommended in the report of a joint Food and Agricultural Organization/United Nations University/World Health Organization Expert Consultation published in 2004. It did not consider changes in consumer preferences from the commodity bundle adopted in 1973–74. Also, differential rates of inflation for food and non-food items were not taken into account.

6.4 On the absolute and relative conceptualisations of the state of poverty

An extensively debated issue concerns the fact that measures of poverty and deprivation can be defined in 'relative' terms, or in terms of some 'absolute' standards, or even as a mixture of the two types (Citro and Michael, 1995). It is important to examine these apparently opposing concepts – of absolute and relative poverty – in depth and carefully. Are they clear, meaningful, useful for addressing relevant aspects of poverty and deprivation? Is one of them better than the other? These questions are important when we look at cross-sectional comparison of poverty; they are critical for deciphering poverty trends over time.

It is useful to look more closely into the concepts of absolute and relative poverty as employed in the construction and interpretation of poverty trends.

First of all, it has to be appreciated that the state of being in poverty can be defined and identified only within the specific context of the society in which people are living. The state of poverty has to be defined and recognised in relation to standards which cannot be some abstract or constant measures independent of the level of development of the society as a whole to which they relate.

> Servants, labourers and workmen of different kinds, make up the far greater part of every great political society. But what improves the circumstances of the greater part can never be regarded as an inconvenience to the whole. No society can surely be flourishing and happy, of which the far greater part of the members are poor and miserable. It is but equity, besides, that they who feed, clothe and lodge the whole body of the people, should have such a share of the produce of their own labour as to be themselves *tolerably well* fed, clothed and lodged.
>
> (Adam Smith, *The Wealth of Nations*, Book 1, Ch. 8, emphasis added)

What is 'tolerably well' is historically and socially defined. This notion is fundamentally rooted in classical economy.

> The value of labour-power is determined by the value of the *necessaries of life habitually required* by the average labourer. The quantity of these necessaries is known at any given epoch of a given society, and can

> therefore be treated as a constant magnitude. What changes, is the value of this quantity.
>
> (Karl Marx, *Capital*, Vol. I, Ch. XVII, emphasis added)

An acceptable level of living underlying the concept of poverty is therefore linked to the general (average) level of development and the standard of living of the society, albeit the relationship may be complex, even judgemental.

> As [the] literature demonstrates, poverty is ultimately a normative concept, not a statistical one … in the final analysis setting a poverty level requires a judgment about social norms. While analysis of statistical data can be very helpful in providing some basis for judgment, such a judgment cannot be made on statistical grounds alone. As Adam Smith put it more than two hundred years ago, poverty is a lack of those necessities that 'the custom of the country renders it indecent for creditable people, even of the lowest order, to be without.' Such necessities cannot be identified in some neutral, scientifically correct way – they do indeed depend on the 'custom of the country', and some notion of what that custom requires must enter into their selection.
>
> (Ruggles, 1992)

The relationship between the minimum acceptable and the average levels affects the estimation of cross-sectional levels of poverty, and even more critically the estimation of time-trends in poverty levels, as the examples above illustrate. We believe that *the distinction between an absolute and a relative conceptualisation of poverty is essentially determined by how these two levels are related.*

In an 'absolute' standard, the reference to some acceptable minimum level is explicit, while the average level enters only indirectly through its relationship to the acceptable minimum level. There is a 'historical tendency for absolute standards to be periodically revised,… [and] when the general standard of living rises, resources may need to be higher to achieve the same ends' (Foster 1998; also Atkinson, 1983).

It is a common error to confuse the concept of an absolute standard with a 'vulgar' or 'degenerate' form of it, namely a constant (unchanging) standard. In relation to the US official poverty figures, for instance, it cannot be doubted that the official poverty thresholds have long been quite outdated as indicators of real needs of the population. If adjusting for price changes is essentially the only adjustment made over time to the thresholds, no account is taken of changes in people's incomes and family structures or in the consumption goods and services available or needed. A set of minimum consumption needs established say a century ago and indexed merely to changing prices of that set alone would continue to disregard all the new conditions and what have *become* basic needs since that time. Furthermore, in so far as levels of real (inflation adjusted) income rise with economic development, the poverty thresholds fall farther and farther behind average standards of living.

In a 'relative' standard, by contrast, reference to some measure of the average level is explicit, while the acceptable minimum level enters indirectly through its relationship to the average level.

By taking different numerical values for the relationship between the relative poverty threshold and the average level, we are implicitly choosing different acceptable minimum standards. For instance, we may define several poverty lines at different 'thresholds', the 'poverty line threshold' signifying, for instance, the percentage of the median income defining the poverty line. Different values of the poverty rate are obtained using different poverty line thresholds. The substantive objective of introducing indicators based on different poverty line thresholds is to take more fully into account differences among populations in the shape at the lower end of the income distribution. Higher thresholds identify broad disadvantaged groups. Lower thresholds isolate the more severely poor and tend to be more sensitive in distinguishing among countries or other population groups being compared. As the threshold is raised, this sensitivity tends to fall: clearly in the extreme case when the median is taken as the poverty line, the poverty rate is always 50 per cent, by definition.

As noted by Foster (1998), there are several ways in which relative and absolute considerations enter into poverty measurement, and the author offers 'a taxonomy including the threshold and equivalence-scale choices in the identification step, and the treatment of population, scale, and individual deprivation in the aggregation step'. Below we will concentrate primarily on the basic choice of defining the income threshold for computing poverty rates (head-count ratios), and how this threshold is allowed to change over time, thus determining the poverty trends observed.

In the relative concept, poverty line changes proportionately to the change in level of income. Consequently, a constant level of poverty indicates that the poor are able to share proportionately in any improvements in the general living standard of the society. Increasing poverty means that the poor are missing out in this respect.[1]

In the 'constant' version of the absolute concept, the poverty line remains fixed in real terms (i.e. after adjustment only for price inflation). Consequently, a constant poverty rate indicates that the poor are being completely left out of any improvements in the general living standard of the society. An apparent reduction in poverty could still mean that they are losing out proportionately. When they are able to share proportionately in improvements in the general living standard of the society, the poverty rate keeps falling.

These factors make any poverty rates computed in this way, using the 'constant' version of the absolute concept, useless, at least over any extended period of time in so far as no account is taken of changing consumption needs and patterns in the society. The series may begin on a 'reasonable' basis (which in any case is often based on some notion of relative income), but that basis is lost over time.

Table 6.3 provides a numerical illustration of these points. The income distribution is taken from a round of the EU-SILC survey in Italy. This sample has been followed up over the next five years assuming the following combinations of nine variations over time:

1 The level of (median) income, and hence the poverty line taken at 60 per cent of the median (a) increases at a certain rate, or (b) remains the same, or (c) declines at a certain rate.
2 The income distribution (a) remains unchanged in shape, or (b) persons with equivalised incomes below the mean value lose in relation to those with incomes above the mean (or gain less if these are a general improvement in the levels of income), or (c) those with equivalised incomes below the mean value gain in relation to those with incomes above the mean (or lose less if these are a general drop in the levels of income).

Note that the poverty rate according to the relative standard is not affected by any change in the average income level. It depends only on changes in the income distribution.

Pattern 1(a) in combination with 2(b) seems to have been the most prevalent scenario in recent years.

6.4.1 No such thing as an 'absolute' poverty line

In fact, it is not possible to maintain an unchanging poverty threshold over a long time or under major economic changes. Surely, it makes no sense to insist that the standard be the same in the US and Bangladesh. In fact there is a

Table 6.3 Simulations of the effect of changes in income level and income distribution on poverty rates with relative and fixed poverty lines

Income distribution	*Income level*					
	Increasing		*Unchanging*		*Declining*	
	Poverty line					
	Relative	*Fixed*	*Relative*	*Fixed*	*Relative*	*Fixed*
Unchanging						
Year 1	0.19	0.19	*0.19*	*0.19*	0.19	0.19
Year 2	0.19	0.18	*0.19*	*0.19*	0.19	0.20
Year 6	0.19	0.06	*0.19*	*0.19*	0.19	0.48
Poor losing more/gaining less						
Year 1	0.19	0.19	0.19	0.19	0.19	0.19
Year 2	0.23	0.22	0.23	0.24	0.23	0.25
Year 6	0.38	0.27	0.38	0.40	0.38	0.51
Poor gaining more/losing less						
Year 1	0.19	0.19	0.19	0.19	0.19	0.19
Year 2	0.14	0.13	0.14	0.14	0.14	0.15
Year 6	0.06	0.02	0.04	0.04	0.04	0.38

historical tendency for absolute standards to be periodically revised – in order to retain some meaning of the figures. In this sense, there is no such thing as an absolute poverty line.

6.4.2 No such thing as purely 'relative' poverty either

The 'relative' concept of poverty as used is in fact not so relative. This can be seen, for instance, by considering poverty levels at the level of geographical regions of a country. It is important to distinguish between measures of disparities among regions from those of disparity within regions. A 'relative' poverty line defined in terms of the income distribution pooled for the whole country is not a purely relative line when we consider the resulting poverty measures at the level of individual regions. Relative poverty lines may in fact be defined at different levels of geographical aggregation. By the 'level of poverty line' we mean the population level to which the income distribution is pooled for the purpose of defining the relative poverty line – such as a poverty line defined in terms of the national median income, where the income distribution is considered at the level of the whole country, in relation to which a poverty line is defined and the number (and proportion) of poor computed. These numbers may then be disaggregated by region to obtain regional poverty rates – but still defined in terms of national poverty lines (Berthoud, 2004; Verma *et al*., 2005).

The use of regional poverty lines, i.e. a poverty line defined for each region based only on the income distribution within that region, can be very useful for constructing regional indicators. The numbers of poor persons identified with these lines can then be used to estimate regional poverty rates. They can also be aggregated upwards to give alternative national poverty rates – but in all cases they remain based on the regional poverty lines.

So defined, the poverty measures are not affected by disparities in mean levels of income among the regions. *The measures are more purely relative*, compared to regional poverty rates constructed using a national poverty line

In fact, different levels for the poverty line can be seen as implying a different mix of 'relative' and 'absolute' measures. For analysis at the country level, the use of national poverty lines provides a relative measure for each country, but the use of a country-level poverty line introduces quite a high degree of absoluteness into the measures for individual regions. By contrast, the use of the regional poverty line provides a more relative measure of poverty determined only by the income distribution within the region, independently of the degree of regional disparities in the country. Use of poverty lines defined at a higher level than individual regions, say groupings of regions or major partitions of the country, introduces an element of 'absoluteness' in the resulting regional figures, since the resulting poverty rate in an individual region now also depends on differences in income levels among regions in the same grouping of regions. The degree of absoluteness in the measure increases as the poverty line level is raised to country (and then in principle, even to groupings of countries), meaning that increasingly the resulting poverty rates reflect differences among regions in the level of mean income, in addition to the extent of disparity within each region.

We can mix any level of analysis or aggregation with any poverty line level. The former concerns the units for which the measures are computed; the latter refers to the population of which the income distribution is considered in defining the poverty line. The poverty line level chosen can make a major difference to the resulting poverty rates when it is higher than the level of analysis or aggregation. The extent depends on the degree of disparity between the units of analysis. However, we find that the poverty line level chosen often makes only a small difference to the resulting poverty rates when it is the same as or lower than the level of analysis or aggregation (Verma *et al.*, 2005).

Finally, we may take note of another important research finding. Relative indicators of inequality, even if confined to the purely monetary dimension, are found to be highly correlated with diverse aspects of the level of well-being of a society. Furthermore, in more unequal societies, it is not only those at the bottom but practically the whole population which suffers from lower levels of well-being. This is amply demonstrated, for example, by Wilkinson and Pickett (2010) in the book *The Spirit Level*, both for European and other developed countries, and among regions (states) of the US.

6.5 A note concerning longitudinal measurement

In this concluding section we briefly comment on some measurement issues in the assessment of longitudinal poverty. Longitudinal profiles of individuals can be constructed from two types of data: from retrospective information on the individual's present and past levels of income; or from repeated measurement of the income situation using a panel design in which the same households and/or persons remain in the sample over a certain time period.

Both types of measurements are of course subject to errors. Retrospective information is particularly prone to response bias – systematic errors which affect the measurement in a survey. Information cumulated from repeated measurements in a panel is prone particularly to response variability – random errors to which repeated measurements are subject.

Measurement problems tend to be more serious in retrospective data. Therefore the construction of longitudinal measures of poverty is based mostly on panel data.

But then, response variability can be a disturbing problem. The main danger is that, because of response variability affecting the repeated measurements in a panel at random, correlation over time may be attenuated. We may over-estimate mobility across the poverty line, from the state of poverty to non-poverty and from non-poverty to poverty and, by the same token, under-estimate persistent or chronic poverty.

Consider, as an example, movements in and out of poverty reported in a panel survey. In a study several years ago, Jarvis and Jenkins (1995) estimated that in the UK a high proportion of the population moved across the poverty line from one year to the next. While around 10 per cent stayed in the state of poverty and around three-quarters in the state of non-poverty throughout the two years, almost one-in-seven reported moving out of poverty into non-poverty, or into poverty from non-poverty.

> These results indicate that there is a significant degree of low income turnover in Britain during the early 1990s. Almost one quarter (23.6%) of our sample had incomes below half 1991 average income [the chosen poverty threshold in this study] at either one or both of the times at which they were interviewed in 1991 and 1992. Thus longitudinal data reveal that the number of people experiencing low income at least once during the two years is greater than the proportion with low income which is revealed by cross-sectional snapshots of either the 1991 distribution or the 1992 distribution (about 17% each year).

In a follow-up study covering four years of longitudinal observations, Jarvis and Jenkins (1996) reported the following picture of persistence or lack of persistent of poverty:

Poor during all 4 years	4.3%
Poor during 3 out of 4 years	5.6%
Poor during 2 out of 4 years	9.1%
Poor during 1 out of 4 years	13.3%
Non-poor during all 4 years	68.7%

If we define those in the state of poverty for at least three of the four years observed as persistently or chronically poor, then 10 per cent of the population fall in this category. Only two-thirds report never having been in the state of poverty during the four observation years.

An important issue is the extent to which the levels of mobility of the population in the income distribution may have a tendency to be over-estimated as a result of random errors in the reporting of incomes. The following are some results from a simple simulation. Data from two successive years of the EU-SILC survey in Italy were used to estimate the proportion in persistent poverty during the period. Then individual incomes in both the years were subject to random perturbations of a certain magnitude, and the proportions in persistent poverty were re-estimated using these data. A sample of the results is in Table 6.4.

It can be seen that as a result of random perturbation in the reported income, there is small reduction in the proportion in persistent poverty, and a more substantial reduction in the proportion never poor. The proportions in transient poverty – in poverty during one but not both the years – are much increased.

We may also note that the cross-sectional poverty rates at both the years are also increased as an effect of the perturbations. This has resulted from the particular type of perturbations applied in this simulation. The perturbations in individuals' incomes were taken from a normal distribution with zero mean and

Table 6.4 Simulation of the effect of response variance on estimates of poverty persistence

	Year 1 non-poor / *Year 2 non-poor*	*Year 1 non-poor* / *Year 2 poor*	*Year 1 poor* / *Year 2 non-poor*	*Year 1 poor* / *Year 2 poor*	*Anytime poverty*	*Cross-sectional (current) poverty year 1*	*Cross-sectional (current) poverty year 2*
Original (unperturbed) data	76.1	5.4	5.8	12.8	**23.9**	18.6	18.2
After random perturbations							
Moderate perturbation (parameter $k=0.25$)	69.5	9.6	9.9	11.1	**30.5**	21.0	20.6
Strong perturbation (parameter $k=0.50$)	56.8	15.9	15.7	11.6	**43.2**	27.3	27.5
% Change from original (without perturbation)							
Moderate perturbation (parameter $k=0.25$)	–6.6	**4.2**	**4.1**	**–1.7**	**6.6**	2.4	2.5
Strong perturbation (parameter $k=0.50$)	–19.3	10.5	9.9	–1.2	**19.3**	8.7	9.3

standard deviation equal to parameter k given in the table (with a cut-off to avoid negative values resulting from the perturbation). The perturbation applied to an income was in proportion to the size of that income, more precisely the form being the following:

$$Y_i' = Y_i(1 + k\delta_i)$$

where δ_i is taken from the above-mentioned normal distribution. This form may be a reasonable modelling of response errors in a survey. However, larger incomes above the poverty line have a greater chance of receiving a large enough negative perturbation to move to a position below the poverty line, than the chance smaller incomes below the poverty line have of receiving a large enough positive perturbation to move to a position above the poverty line. This results in increased level of cross-sectional poverty.

Note

1 We may note that, more precisely, the above applies when the mean is taken as the measure of the average level. With the median taken as the measure of the average level, disproportionate increase in income above the median is not reflected in trends of the poverty rate. This point has become particularly important in recent years with explosive increases in incomes at the top end of the distribution.

References

Atkinson, A. (1983). *The Economics of Inequality* (2nd edn). Oxford: Clarendon.

Atkinson, A. (1987). On the measurement of poverty. *Econometrica*, 55(4), pp. 749–64.

Berthoud, R. (2004). *Patterns of Poverty across Europe*. Bristol: Policy Press.

China (2004). Poverty statistics in China. A report from Rural Survey Organization of National Bureau of Statistics, China.

Citro, C., Michael, R. (1995). *Measuring Poverty: A New Approach*. Washington DC: National Academy Press.

Foster, J. (1998). Absolute versus relative poverty. *American Economic Review, Papers and Proceedings*, 88(2), pp. 335–41.

India (2011). Government of India Planning Commission, press note on poverty estimates.

Jarvis, S., Jenkins, S. (1995). Do the poor stay poor? New evidence about income dynamics from the British Household Panel Survey. *Occasional Paper* 95-2, University of Essex: ESRC Research Centre on Micro-Social Change.

Jarvis, S., Jenkins, S. (1996). Changing places: income mobility and poverty dynamics in Britain. *Working Paper* 96-19, University of Essex: ESRC Research Centre on Micro-Social Change.

Niño-Zarazúa, M., Addison, T. (2012). Redefining poverty in China and India: economic development. WIDER Angle, United Nations University.

Rosenberg, P. (2012). Wages largely remain stagnant. Al Jazeera, 16 December.

Ruggles P. (1992). Measuring poverty. *Focus*, 14(1), University of Wisconsin-Madison Institute for Research on Poverty.

Schwarz, J. (2005). *Freedom Reclaimed: Rediscovering the American Vision*. Baltimore, MD: Johns Hopkins University.

Verma, V., Betti, G., Lemmi, A., Mulas, A., Natilli, M., Neri, L., Salvati, N. (2005). *Regional Indicators to Reflect Social Exclusion and Poverty*. Brussels: European Commission.

Wilkinson, R., Pickett, K. (2010), *The Spirit Level: Why Equality is Better for Everyone*. London: Penguin Books.

7 Intertemporal material deprivation[1]

Walter Bossert, Lidia Ceriani, Satya R. Chakravarty and Conchita D'Ambrosio

7.1 Introduction

Individual well-being is multidimensional and various aspects of the quality of life of an individual need to be jointly considered in its measurement. The axiomatic literature on the subject has proposed many indices of multidimensional poverty and deprivation and explored the properties that are at the basis of these measures; see, for example, Chakravarty *et al.* (1998), Tsui (2002), Bourguignon and Chakravarty (2003), Diez *et al.* (2008), Bossert *et al.* (2009), and Alkire and Foster (2011).

The intertemporal aspect of deprivation has received relatively little attention so far. Most of the studies in the literature have been atemporal. At the same time, many contributions on unidimensional poverty have shown that chronic poverty and persistent periods of poverty are worse, in a number of ways, for individuals than sporadic episodes. For surveys of this literature, see, among others, Rodgers and Rodgers (1993) and Jenkins (2000). These considerations gave impetus to some recent theoretical contributions on measuring income poverty over time, such as Calvo and Dercon (2009), Foster (2009), Hojman and Kast (2009), Hoy and Zheng (2011), Dutta *et al.* (2011) and Bossert *et al.* (2012). The *Journal of Economic Inequality* has recently published a special issue on measuring poverty over time. We refer the reader to its introduction (Christiaensen and Shorrocks, 2012) for an exhaustive summary of the literature. See also Hoy *et al.* (2012), Gradin *et al.* (2012) and Mendola and Busetta (2012).

The purpose of this chapter is to bring these two strands of the poverty and deprivation literature together by employing the EU-SILC panel data set, which includes information on different aspects of well-being over time. We analyze the role of intertemporal considerations in material deprivation and compare EU countries based on this additional information. The only other paper similar in spirit that we are aware of is Nicholas and Ray (2012). These authors propose generalizations of the contributions of Foster (2009) and Bossert *et al.* (2012) and apply the indices to the study of multidimensional deprivation in Australia during the period from 2001 to 2008.

The distinction between multidimensional poverty and material deprivation we adopt in this chapter is that endorsed by the EU. In particular, a multidimensional poverty measure takes into consideration all dimensions of well-being that may be of relevance (including non-material attributes such as health status and political participation), whereas an index of material deprivation restricts attention to functioning failures regarding material living conditions. According to EU policy, indices of material deprivation are to be combined with income-based poverty measures and indicators of low employment.

In the multidimensional framework, each person is assigned a vector of several attributes that represent different dimensions of well-being. For measuring multidimensional poverty, it then becomes necessary to check whether a person has "minimally acceptable levels" of these attributes; see Sen (1992, p. 139). These minimally acceptable quantities of the attributes represent their threshold values or cut-offs that are necessary for an adequate standard of living. Therefore, a person is treated as deprived or poor in a dimension if the requisite observed level falls below this cut-off. In this case we say that the individual is experiencing a functioning failure. Material deprivation at the individual level is an increasing function of these failures.

The identification of the poor in a multivariate framework can be performed with different methods. One possible way of regarding a person as poor is if the individual experiences a functioning failure in every dimension, which identifies the poor as those who are poor in all dimensions. This is known as the *intersection* method of identification of the poor. But if a person is poor in one dimension and non-poor in another, then trading off between the two dimensions may not be possible. Lack of access to essential durables, say, cannot be compensated by housing. In view of this, a person may be treated as poor if she is poor in at least one dimension. This is the *union* method of identifying the poor; see Tsui (2002) and Bourguignon and Chakravarty (2003). In between these two extremes lies the *intermediate* identification method which regards a person as poor if she is deprived in at least $m \in \{1,\ldots, M\}$ dimensions, where M is the number of dimensions on which human well-being depends. Our approach to identification follows the union method: a person is considered poor if she is poor in at least one dimension.

We follow what Atkinson (2003) refers to as the *counting* approach. The counting measure of individual poverty consists of the number of dimensions in which a person is poor, that is, the number of the individual functioning failures. Since some of the dimensions may be more important than others, an alternative counting measure can be obtained by assigning different weights to different dimensions and then adding these weights for the dimensions in which functioning failure is observed. In this chapter we follow both suggestions and produce results for two different weighing schemes: equal weights and Eurobarometer weights, where the latter's reflect EU citizens' views on the importance of the dimension of well-being under consideration. For a discussion of weighing schemes in EU indicators, see Guio *et al*. (2009). A survey on the use of weights in multidimensional indices of well-being can be found in Decancq and Lugo (2013).

The intertemporal aspect is included in three alternative specifications following the proposals of Foster (2009), Bossert *et al*. (2009), and Hojman and Kast (2009) for income poverty.

The measures proposed by Foster (2009) are generalizations of the Foster–Greer–Thorbecke (1984) class and allow for time to matter. The individual Foster index is the arithmetic mean over time of per-period Foster–Greer–Thorbecke indices. In a similar spirit, the corresponding individual intertemporal index of material deprivation is the average material deprivation experienced by the individual over time.

Bossert *et al.* (2012) take into account persistence in the state of poverty. Their measure pays attention to the length of individual poverty spells by assigning a higher level of poverty to situations where, *ceteris paribus*, poverty is experienced in consecutive rather than separated periods. The individual index is calculated as the weighted average of the individual per-period poverty values where, for each period, the weight is given by the length of the spell to which this period belongs. Similarly, the corresponding individual intertemporal index of material deprivation is calculated as the weighted average of the individual indices of material deprivation where, for each period, the weight is given by the length of the spell to which this period belongs.

Hojman and Kast's (2009) index of poverty dynamics trades off poverty levels and changes (gains and losses) over time and is consistent with loss aversion. The loss aversion property captures the idea that, given income streams with the same levels of deprivation but in a different sequence, an individual is better off with an increasing sequence of outcomes than a decreasing one. The individual measure characterized by Hojman and Kast (2009) is an increasing function of absolute levels of poverty at each period and also of changes in poverty. We follow a similar proposal for material deprivation according to which the individual intertemporal index of material deprivation is the sum of two components: the first is the average material deprivation experienced by the individual over time, the same index applied in the approach inspired by Foster (2009), the second component is the average of the weighted changes in material deprivation experienced over time, where the weights can be consistent with loss aversion. While the first two approaches can be found in Nicholas and Ray (2012), Hojman and Kast's (2009) approach has, so far, not been generalized in this manner.

The remainder of the chapter proceeds as follows. Section 7.2 contains a description of the intertemporal indices of material deprivation. The application of these measures to illustrate the evolution of material deprivation in the European Union using the EU-SILC dataset can be found in Section 7.3. Section 7.4 provides some brief concluding remarks.

7.2 Measuring material deprivation

Suppose there are $N \in \{\mathbb{N}\} \setminus \{1\}$ individuals in a society, $M \in \{\mathbb{N}\} \setminus \{1\}$ characteristics (or dimensions of material deprivation) and $T \in \{\mathbb{N}\} \setminus \{1\}$ time periods. For each individual $n \in \{1, \ldots, N\}$, for each time period $t \in \{1, \ldots, T\}$ and for each characteristic $m \in \{1, \ldots, M\}$, we observe a binary variable $x_m^{nt} \in \{0, 1\}$. A value of one indicates that individual n is poor with respect to dimension m in period t, a value of zero identifies a characteristic with respect to which the individual is not poor in that period. For all $n \in \{1, \ldots, N\}$ and for all $t \in \{1, \ldots, T\}$, we let $x^{nt} = (x_1^{nt}, \ldots, x_M^{nt}) \in \{0, 1\}^M$. For all $n \in \{1, \ldots, N\}$, we define $x^n = (x^{n1}, \ldots, x^{nT}) \in (\{0, 1\}^M)^T$. Furthermore, we let $x = (x^1, \ldots, x^N) \in ((\{0, 1\}^M)^T)^N$.

For each individual $n \in \{1, \ldots, N\}$ and each time period $t \in \{1, \ldots, T\}$, individual's n's material deprivation in t is given by

$$\sum_{m=1}^{M} x_m^{nt} \alpha_m$$

where $\alpha_m \in \{R\}_{++}$ is a parameter assigned to dimension $m \in \{1, \ldots, M\}$. In the applied part of the chapter, we examine two different weighing schemes – one with identical weights for all dimensions, one with weights that are derived from the Eurobarometer survey. See Section 7.3 for details.

A measure of intertemporal material deprivation for individual $n \in \{1, \ldots, N\}$ is a function D^n: $(\{0, 1\}^M)^T \rightarrow R_+$ which assigns a non-negative individual intertemporal material deprivation value to each x^n in its domain. A measure of aggregate intertemporal material deprivation is a function D: $((\{0, 1\}^M)^T)^N \rightarrow \{R\}_+$ that assigns a non-negative intertemporal material deprivation value to each x in its domain.

The first approach to be analyzed here is inspired by Foster (2009). For each individual n, intertemporal material deprivation F^n is the average material deprivation experienced throughout the T periods. That is, for all $x^n \in (\{0, 1\}^M)^T$,

$$F^n(x^n) = \frac{1}{T} \sum_{t=1}^{T} \sum_{m=1}^{M} x_m^{nt} \alpha_m.$$

Aggregate intertemporal material deprivation F is the arithmetic mean of the individual intertemporal material deprivation values. Thus, we obtain, for all $x \in ((\{0, 1\}^M)^T)^N$,

$$F(x) = \frac{1}{N} \sum_{n=1}^{N} F^n(x^n) = \frac{1}{N} \frac{1}{T} \sum_{i=1}^{N} \sum_{t=1}^{T} \sum_{m=1}^{M} x_m^{nt} \alpha_m.$$

In order to discuss our adaptation of Bossert *et al.*'s (2012) approach to the intertemporal setting, we require some additional definitions. Let $n \in \{1, \ldots, N\}$ and $x^n \in (\{0, 1\}^M)^T$. We say that n is deprived in period $t \in \{1, \ldots, T\}$ in x^n if and only if there exists $m \in \{1, \ldots, M\}$ such that $x_m^{nt} = 1$. That is, in order to be deprived in period t in x_n, individual n must be deprived with respect to at least one dimension in this period. This corresponds to the union method of identifying the deprived. Thus, individual n is not deprived in period t in x^n if and only if $x_m^{nt} = 0$ for all $m \in \{1, \ldots, M\}$.

To capture the notion of persistence in a state of material deprivation, we introduce functions P^{nt}: $(\{0, 1\}^M)^T \rightarrow \{1, \ldots, T\}$ for each $n \in \{1, \ldots, N\}$ and for each $t \in \{1, \ldots, T\}$. If n is deprived in period t in x^n, we let $P^{nt}(x^n)$ be the maximal number of consecutive periods including t in which n is deprived. Analogously, if n is not deprived in period t in x^n, $P^{nt}(x^n)$ is the maximal number of consecutive periods including t in which n is not deprived. To illustrate this definition, suppose $T=7$ and x^n is such that n is deprived in periods one, four, five, and seven. The length of the first spell of material deprivation is one and, thus, $P^{n1}(x^n)=1$. This is followed by a spell out of deprivation of length two (in periods two and three), which implies $P^{n2}(x^n)=P^{n3}(x^n)=2$. The next two periods are periods with deprivation and we obtain $P^{n4}(x^n)=P^{n5}(x^n)=2$. Period six is a single period without deprivation and, thus, $P^{n6}(x^n)=1$. Finally, there is a one-period spell of material deprivation and we have $P^{n7}(x^n)=1$.

Following Bossert *et al.* (2012), intertemporal material deprivation BCD^n for individual $n \in \{1,\ldots, N\}$ is a weighted mean of the individual material deprivation values where, for each period, the weight is given by the length of the spell to which this period belongs. Thus, according to this approach, individual intertemporal material deprivation BCD^n is given by

$$BCD^n(x^n) = \frac{1}{T}\sum_{t=1}^{T} p^{nt}(x^n)\sum_{m=1}^{M} x_m^{nt}\alpha_m$$

for all $x^n \in (\{0, 1\}^M)^T$. Again, aggregate intertemporal material deprivation BCD is the arithmetic mean of the individual intertemporal material deprivation values. Thus $x \in ((\{0, 1\}^M)^T)^N$.

$$BCD(x) = \frac{1}{N}\sum_{n=1}^{N} BCD^n(x^n) = \frac{1}{N}\frac{1}{T}\sum_{i=1}^{N}\sum_{t=1}^{T} p^{nt}(x^n)\sum_{m=1}^{M} x_m^{nt}\alpha_m .$$

Hojman and Kast (2009) propose to include variability as a determinant of individual intertemporal material deprivation. Their individual measure HK^n has two components: the level of individual intertemporal material deprivation and the changes of individual material deprivation over time. The level is measured by means of F^n and the changes are given by the weighted sum of upward and downward movements of individual material deprivation over time. In the terminology of Hojman and Kast (2009), there is poverty creation whenever deprivation increases and poverty destruction whenever deprivation decreases.

To illustrate, consider a situation with $T=3$ and $x^n, y^n \in ((\{0, 1\}^M)^T$ such that n is deprived in periods one and three in x^n, and in periods two and three in y^n. According to the Hojman and Kast (2009) approach, n is intertemporally more deprived in y^n than in x^n. The levels of individual intertemporal material deprivation are the same in x^n and in y^n. However, in x^n, there is poverty destruction (in the move from period one to period two) and poverty creation (in the move from period two to period three), whereas in y^n, there is only poverty creation (in the move from period one to period two).

In general, for a fixed level of individual material deprivation, each movement that decreases material deprivation decreases the overall index and each movement that increases material deprivation increases the index. To provide a formal definition, we introduce two sets of functions g^{nt}: $\{0, 1\}^M \rightarrow \{0, 1\}$ and ℓ^{nt}: $\{0, 1\}^M \rightarrow \{0, 1\}$ for $n \in \{1,\ldots, N\}$ and $t \in \{1,\ldots, T-1\}$ that are intended to capture gains (decreases in individual material deprivation) and losses (increases in material deprivation). They are defined by letting, for all $x^{nt} \in \{0, 1\}^M$,

$$\ell^{nt}(x^{nt}) = \begin{cases} 1 & \text{if } \sum_{m=1}^{M} x_m^{nt}\alpha_m < \sum_{m=1}^{M} x_m^{n(t+1)}\alpha_m \\ 0 & \text{otherwise.} \end{cases}$$

and

$$g^{nt}(x^{nt}) = \begin{cases} 1 & \text{if } \sum_{m=1}^{M} x_m^{nt}\alpha_m > \sum_{m=1}^{M} x_m^{n(t+1)}\alpha_m \\ 0 & \text{otherwise} \end{cases}$$

For each individual n, intertemporal material deprivation HK^n is given by

$$HK^n(x^n) = \frac{1}{T}\sum_{t=1}^{T}\sum_{m=1}^{M} x_m^{nt}\alpha_m + \frac{1}{T}\sum_{t=1}^{T}(\gamma\ell^{nt}(x^{nt}) - \delta_t g^{nt}(x^{nt}))$$

for all $x^{nt} \in (\{0, 1\}^M)^T$, where γ_t, $\delta_t \in \mathrm{R}_{++}$ are parameters such that $\gamma_t \geq \delta_t$ for all $t \in \{1,\ldots, T-1\}$. When $\gamma_t = \delta_t$, gains and losses are perfect substitutes: any increase in deprivation can be compensated by any decrease of the same amount. When $\gamma_t \geq \delta_t$, losses weigh more than gains. This second possibility is what we assume for the application presented in the following section.

Finally, aggregate intertemporal material deprivation HK is the arithmetic mean of the individual intertemporal material deprivation values. Thus,

$$HK(x) = \frac{1}{N} HK^n(x^n)$$
$$= \frac{1}{N}\frac{1}{T}\sum_{n=1}^{N}\sum_{t=1}^{T}\sum_{m=1}^{M} x_m^{nt}\alpha_m + \frac{1}{N}\frac{1}{T}\sum_{n=1}^{N}\sum_{t=1}^{T}(\gamma\ell^{nt}(x^{nt}) - \delta_t g^{nt}(x^{nt}))$$

for all $x \in ((\{0, 1\}^M)^T)^N$.

7.3 Data and results

In this section, we apply the indices defined above to measure material deprivation over time in the EU. The dataset we use is EU-SILC, which is employed by European Union member states and the Commission to monitor national and EU progress towards key objectives for the social inclusion process and the Europe 2020 growth strategy. Our analysis covers the years from 2006 to 2009 and, since we are interested in intertemporal material deprivation, we focus only on the longitudinal component of the dataset. The variables that may be used in the measurement of material deprivation are available mainly at the household level. We follow a conservative approach in the sense that we treat the households reporting a missing value like those reporting not to experience the functioning failure. As a result, we may be underestimating material deprivation since we are attributing a functioning failure exclusively to households who explicitly claim to have the failure. The unit of our analysis is the individual, that is, the household failure is attributed to each household member and we analyze the distribution of functioning failures among individuals.

The variables at the basis of the measures of material deprivation are listed in Table 7.1.

They are grouped according to three domains of quality of life: financial difficulties, housing conditions and durables, for a total of twelve indicators. These are the same variables chosen by Fusco *et al.* (2010). For other EU studies on material deprivation on different dimensions of well-being see, among others, Guio (2009), Guio *et al.* (2009).

We use two weighing schemes: identical weights for all dimensions and weights that are constructed from the views of EU citizens as surveyed in 2007 in the special Eurobarometer 279 on poverty and social exclusion (see TNS Opinion & Social, 2007). This weighing method has first been proposed by Guio *et al.* (2009). For each variable, with this weighing scheme, we use as weight the percentage of the EU27 citizens answering "absolutely necessary, no one should have to do without" to the requisite question as expressed by these instructions:

Table 7.1 Material deprivation variables

Financial difficulties
1 Has been in arrears at any time in the last 12 months on:
• mortgage or rent payments (hs010)
• utility bills (hs020)
• hire purchase instalments or other loan payments (hs030)
2 Cannot afford paying for one week annual holiday away from home (hs040)
3 Cannot afford a meal with meat, chicken, fish (or vegetarian equivalent) every other day (hs050)
4 Lacks the capacity to face unexpected required expenses (hs060)
Durables
5 Cannot afford a telephone (including mobile phone) (hs070)
6 Cannot afford a colour tv (hs080)
7 Cannot afford a computer (hs090)
8 Cannot afford a washing machine (hs100)
9 Cannot afford to have a car (hs110)
Housing Conditions
10 Lacks the ability to keep the home adequately warm (hh050)

Source: EU-SILC dataset.

Note
For a selected number of countries in years 2007, 2008, and 2009, variable hs010, hs020, and hs030 has been replaced by new variables labeled hs011, hs021, hs031 respectively. The two set of variables measure the same dimensions. While hs010, hs020, and hs030 are binary variables (1-yes, 2-no), variable hs011, hs021, and hs031 take three values (1-yes, once; 2-yes, twice or more times; 3-no). We recode hs011, hs021 and hs031 as binary and use them in place of hs010, hs020 and hs030.

> In the following questions, we would like to understand better what, in your view, is necessary for people to have what can be considered as an acceptable or decent standard of living in (OUR COUNTRY). For a person to have a decent standard of living in (OUR COUNTRY), please tell me how necessary do you think it is ... (if one wants to).

The possible answers also included "necessary," "desirable but not necessary," and "not at all necessary."

The results of the intertemporal indices are reported in Tables 7.2, 7.3, and 7.4. In each table we include the value of the index and the rankings of the countries (where 1 indicates the country with minimum deprivation) using both weighing schemes. In Figures 7.1 and 7.2 we plot, for each weighing scheme, the rankings of the intertemporal material deprivation indices. As a benchmark, we also compute the indices of material deprivation for each year. These are contained in Table 7.5 (results with equal weights) and in Table 7.6 (results with Eurobarometer weights).

A clear message is conveyed from looking at the figures: the rankings obtained by applying the HK indices are very different from the other two. F and BCD agree more with the Eurobarometer weighing scheme with differences being observed only for the extremes.

When time is not taken into consideration, in all the years under analysis and for both weighing schemes, the Netherlands is the least deprived country

Table 7.2 Intertemporal material deprivation: F

Country	*F*	*rank_F*	*F_Eu*	*rank_F_Eu*
AT	0.068	7	0.042	6
BE	0.065	6	0.043	7
BG	0.287	20	0.247	20
CY	0.119	13	0.105	14
CZ	0.101	11	0.056	9
EE	0.102	12	0.076	12
ES	0.075	8	0.051	8
FI	0.059	4	0.037	5
HU	0.188	19	0.138	18
IT	0.086	9	0.056	10
LT	0.157	16	0.115	15
LU	0.031	2	0.019	2
LV	0.187	18	0.147	19
NL	0.037	3	0.019	1
PL	0.181	17	0.131	17
PT	0.128	14	0.099	13
SE	0.030	1	0.019	3
SI	0.090	10	0.064	11
SK	0.151	15	0.115	16
UK	0.059	5	0.036	4

Source: our calculations based on EU-SILC 2009 longitudinal dataset.

Table 7.3 Intertemporal material deprivation: BCD

Country	*BCD*	*rank_BCD*	*BCD_Eu*	*rank_BCD_Eu*
AT	0.143	7	0.089	6
BE	0.140	6	0.092	7
BG	0.648	20	0.568	20
CY	0.261	13	0.231	14
CZ	0.225	12	0.124	9
EE	0.220	11	0.164	12
ES	0.156	8	0.109	8
FI	0.126	5	0.080	5
HU	0.445	19	0.331	19
IT	0.185	9	0.124	10
LT	0.340	16	0.254	15
LU	0.064	2	0.040	2
LV	0.394	17	0.327	18
NL	0.077	3	0.040	3
PL	0.413	18	0.302	17
PT	0.285	14	0.228	13
SE	0.058	1	0.038	1
SI	0.194	10	0.140	11
SK	0.337	15	0.258	16
UK	0.123	4	0.074	4

Source: our calculations based on EU-SILC 2009 longitudinal dataset.

Table 7.4 Intertemporal material deprivation: HK (γ_t=1, δ_t=0.5)

Country	*HK*	*rank_HK*	*HK_Eu*	*rank_HK_Eu*
AT	0.137	7	0.118	8
BE	0.110	5	0.087	5
BG	0.376	20	0.350	20
CY	0.185	11	0.180	16
CZ	0.154	8	0.108	7
EE	0.188	12	0.165	12
ES	0.166	10	0.147	10
FI	0.102	4	0.081	4
HU	0.339	19	0.304	19
IT	0.163	9	0.136	9
LT	0.205	14	0.173	13
LU	0.069	2	0.056	2
LV	0.279	18	0.245	18
NL	0.072	3	0.057	3
PL	0.210	16	0.163	11
PT	0.209	15	0.180	15
SE	0.064	1	0.053	1
SI	0.196	13	0.177	14
SK	0.217	17	0.183	17
UK	0.116	6	0.096	6

Source: our calculations based on EU-SILC 2009 longitudinal dataset.

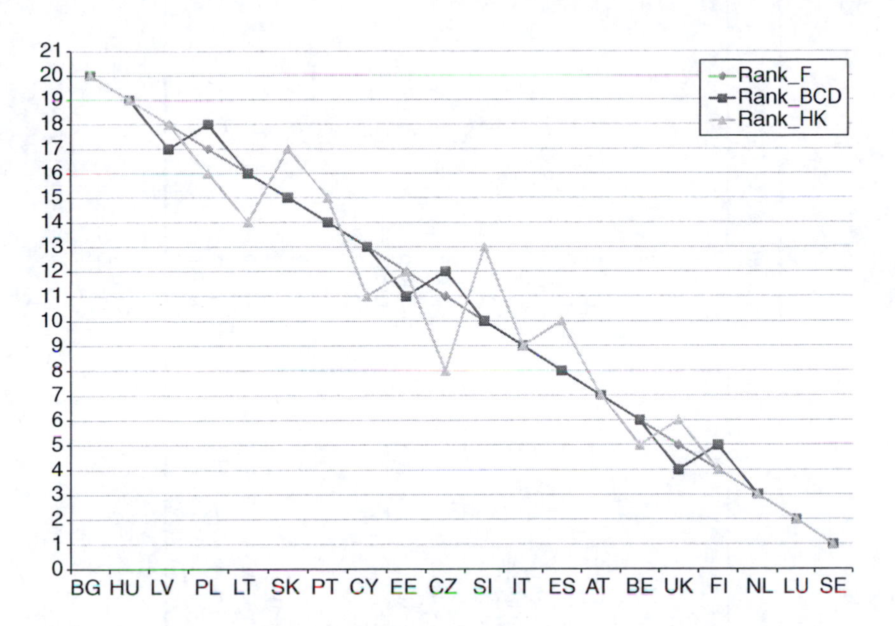

Figure 7.1 Rank comparison between F, BCD, and HK, symmetric weights.

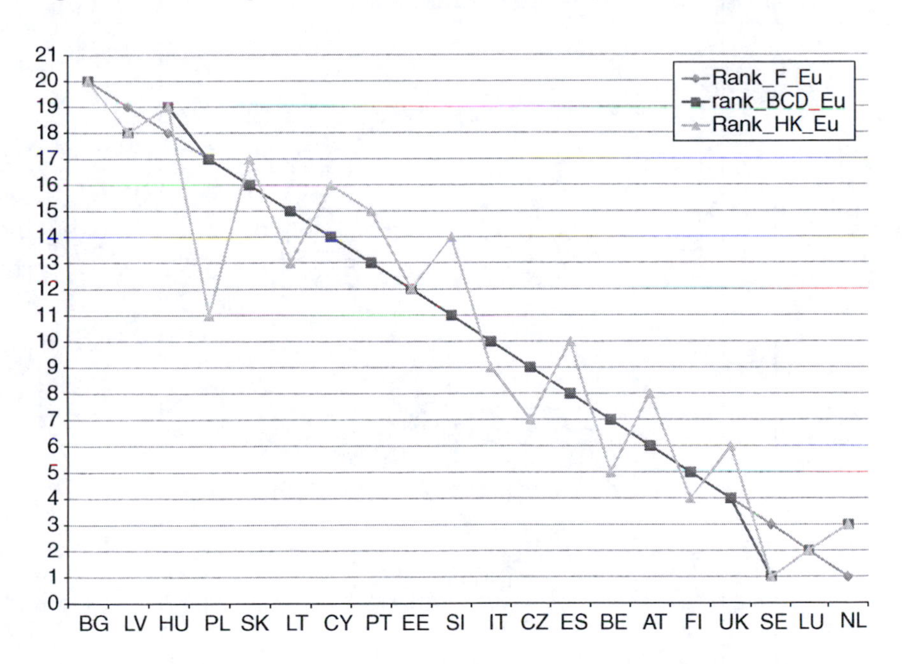

Figure 7.2 Rank comparison between F, BCD, and HK, Eurobarometer weights.

Table 7.5 Yearly material deprivation: equal weights

Country	*I_2006*	*rank_2006*	*I_2007*	*rank_2007*	*I_2008*	*rank_2008*	*I_2009*	*rank_2009*
AT	0.07	6	0.075	7	0.086	8	0.071	7
BE	0.085	7	0.073	6	0.071	6	0.066	6
BG	0.407	20	0.369	20	0.303	20	0.343	20
CY	0.169	14	0.179	16	0.132	13	0.142	14
CZ	0.123	11	0.103	9	0.1	10	0.102	9
EE	0.127	12	0.113	11	0.096	9	0.119	11
ES	0.098	8	0.093	8	0.084	7	0.098	8
FI	0.058	4	0.052	4	0.052	4	0.052	4
HU	0.208	16	0.211	17	0.204	18	0.228	19
IT	0.103	9	0.107	10	0.106	11	0.105	10
LT	0.21	17	0.168	15	0.159	16	0.157	16
LU	0.046	3	0.049	3	0.049	3	0.046	3
LV	0.267	19	0.232	19	0.21	19	0.225	18
NL	0.039	1	0.036	1	0.034	1	0.03	1
PL	0.243	18	0.212	18	0.186	17	0.177	17
PT	0.152	13	0.153	13	0.149	14	0.142	13
SE	0.042	2	0.038	2	0.038	2	0.034	2
SI	0.113	10	0.116	12	0.119	12	0.119	12
SK	0.19	15	0.162	14	0.15	15	0.155	15
UK	0.067	5	0.059	5	0.062	5	0.058	5

Source: our calculations based on EU-SILC 2009 longitudinal dataset.

Table 7.6 Yearly material deprivation: Eurobarometer weights

Country	*I_Eu2006*	*rank_2006*	*I_Eu2007*	*rank_2007*	*I_Eu2008*	*rank_2008*	*I_Eu2009*	*rank_2009*
AT	0.042	5	0.046	6	0.054	7	0.045	7
BE	0.057	7	0.048	7	0.045	6	0.041	6
BG	0.342	20	0.305	20	0.256	20	0.295	20
CY	0.148	15	0.156	18	0.114	16	0.125	16
CZ	0.071	10	0.058	8	0.055	8	0.056	8
EE	0.092	12	0.081	11	0.071	10	0.09	12
ES	0.067	8	0.064	9	0.058	9	0.07	9
FI	0.037	4	0.034	4	0.035	4	0.035	4
HU	0.151	16	0.154	17	0.15	18	0.171	18
IT	0.071	9	0.074	10	0.073	11	0.074	10
LT	0.151	17	0.119	14	0.112	15	0.117	14
LU	0.03	3	0.032	3	0.031	3	0.029	3
LV	0.198	19	0.171	19	0.156	19	0.171	19
NL	0.021	1	0.02	1	0.018	1	0.016	1
PL	0.177	18	0.153	16	0.132	17	0.126	17
PT	0.114	13	0.115	13	0.112	14	0.107	13
SE	0.027	2	0.023	2	0.024	2	0.022	2
SI	0.081	11	0.085	12	0.086	12	0.088	11
SK	0.141	14	0.119	15	0.109	13	0.117	15
UK	0.043	6	0.038	5	0.039	5	0.037	5

Source: our calculations based on EU-SILC 2009 longitudinal dataset.

followed by Sweden, Luxembourg, Finland, and the UK. When time is taken into account, the picture that emerges is very different. The weighing scheme has now an impact on the rankings of the countries. When the dimensions are weighed equally, in all three approaches Sweden is the least deprived country, followed by Luxembourg and the Netherlands. When we use Eurobarometer weights, the Netherlands gains back the best position only according to F while BCD and HK confirm the rankings of equal weighing. This fact indicates that there is more persistence of poverty in the Netherlands compared to Sweden and Luxembourg and that, on average, the material deprivation profiles of individuals are not decreasing over time. Finland and the UK follow these three countries. The position of the UK worsens for both weighing schemes according to the HK index suggesting that the improvements in material deprivation observed over time are not enough to compensate the losses.

At the opposite side of the rankings, the worst position in the yearly material deprivation index is occupied by Bulgaria, for both weighing schemes. Latvia follows in the ranking in all the years, but in 2009, for equal weights when this position is occupied by Hungary. Hungary is a country whose position is clearly worsening over time. Poland is the third-worst country in 2006 and 2007 for equal weights, and the fourth worst in the following two years. When we use Eurobarometer weights, in 2007 Poland gains two positions against Hungary and Cyprus. When time is taken into account, the picture that emerges is, as before, very different. Latvia and Hungary follow Bulgaria. Poland is the country whose ranking improves the most for Eurobarometer weights, when material deprivation over time is measured according to the HK index, indicating that the individual paths have sufficient improvements to overcome the trends and the losses. A similar picture emerges for mid-ranked countries such as Lithuania and the Czech Republic for both weighing schemes. Slovenia, on the contrary, sees its position worsening when movements in material deprivation are taken into account.

7.4 Concluding remarks

In this chapter we analyze the role of intertemporal considerations in material deprivation and compare EU countries based on this additional information. If we follow the path of material deprivation experienced by each individual over time we obtain a different picture from the yearly results. We analyze three alternative indices inspired by some recent proposals on the measurement of poverty over time. The generalization of the proposals by Foster (2009) and Bossert *et al.* (2012) tend to produce a similar ranking of countries. The approach based on Hojman and Kast (2009) conveys a different picture and tends to advantage countries whose individuals experience an improvement in their material deprivation scores. Since the measurement of material deprivation is used by the EU member states and the European Commission to monitor national and EU progress in the fight against poverty and social exclusion, these results suggest that time cannot be neglected. Countries should not only be compared based on their yearly results but additional information is gained by following individuals over time and producing an aggregate measure once time is taken into account.

Note

1 Financial support from the Fonds de recherche Société et culture of Québec is gratefully acknowledged.

References

Alkire, S. and J. Foster, Counting and multidimensional poverty measurement. *Journal of Public Economics*, 95:476–487, 2011.

Atkinson, A.B., Multidimensional deprivation: contrasting social welfare and counting approaches. *Journal of Economic Inequality*, 1:51–65, 2003.

Bossert, W., S.R. Chakravarty, and C. D'Ambrosio, Multidimensional poverty and material deprivation. Working Paper 12-2009, CIREQ, 2009.

Bossert, W., S.R. Chakravarty, and C. D'Ambrosio, Poverty and time. *Journal of Economic Inequality*, 10:145–162, 2012.

Bourguignon, F. and S.R. Chakravarty, The measurement of multidimensional poverty. *Journal of Economic Inequality*, 1:25–49, 2003.

Calvo, C. and S. Dercon, Chronic poverty and all that: the measurement of poverty over time. In: T. Addison, D. Hulme, and R. Kanbur, editors, *Poverty dynamics: interdisciplinary perspectives*, 29–58. Oxford University Press, Oxford, 2009.

Chakravarty, S.R., D. Mukherjee, and R. Ranade, On the family of subgroup and factor decomposable measures of multidimensional poverty. *Research on Economic Inequality*, 8:175–194, 1998.

Christiaensen, L. and T. Shorrocks, Measuring poverty over time. *Journal of Economic Inequality*, 10:137–143, 2012.

Decancq, K. and M. Lugo, Weights in multidimensional indices of well-being: an overview. *Econometric Reviews*, 32:7–34, 2013.

Diez, H., C. Lasso de la Vega, and A. Urrutia, Multidimensional unit- and sub-group consistent inequality and poverty measures: some characterization results. *Research on Economic Inequality*, 16:189–211, 2008.

Dutta, I., L. Roope, and H. Zank, On intertemporal poverty: affluence-dependent measures. EDP 1112, University of Manchester, 2011.

Foster, J., A class of chronic poverty measures. In: T. Addison, D. Hulme, and R. Kanbur, editors, *Poverty dynamics: interdisciplinary perspectives*, 59–76. Oxford University Press, Oxford, 2009.

Foster, J., J. Greer, and E. Thorbecke, A class of decomposable poverty indices. *Econometrica*, 52:761–766, 1984.

Fusco, A., A.-C. Guio, and E. Marlier, Income poverty and material deprivation in European countries: population and social conditions, Methodologies and working papers, Eurostat, 2010.

Gradin, C., C. del Rio, and O. Cantó, Measuring poverty accounting for time. *Review of Income and Wealth*, 58:330–354, 2012.

Guio, A.-C., What can be learned from deprivation indicators in Europe. Population and social conditions, Methodologies and working papers, Eurostat, 2009.

Guio, A.-C., A. Fusco, and E. Marlier, An EU approach to material deprivation using EU-SILC and Eurobarometer data. Working Paper 2009-19, IRISS, 2009.

Hojman, D. and F. Kast, On the measurement of poverty dynamics. Faculty Research Working Paper Series RWP09-35, Harvard Kennedy School, 2009.

Hoy, M., B.S. Thompson, and B. Zheng, Empirical issues in lifetime poverty measurement. *Journal of Economic Inequality*, 10:163–189, 2012.

Hoy, M. and B. Zheng, Measuring lifetime poverty. *Journal of Economic Theory*, 146:2544–2562, 2011.

Jenkins, S.P., Modelling household income dynamics. *Journal of Population Economics*, 13:529–567, 2000.

Mendola, D. and A. Busetta, The importance of consecutive spells of poverty: a path-dependent index of longitudinal poverty. *Review of Income and Wealth*, 58:355–374, 2012.

Nicholas, A. and R. Ray, Duration and persistence in multidimensional deprivation: methodology and an Australian application. *Economic Record*, 88:106–126, 2012.

Rodgers, J. and J. Rodgers, Chronic poverty in the United States. *Journal of Human Resources*, 28:25–54, 1993.

Sen, A., *Inequality re-examined*. Harvard University Press, Cambridge, MA, 1992.

TNS Opinion & Social, Poverty and exclusion. Report on the Special Eurobarometer No. 279/Wave 67.1, 2007. Available at: http://ec.europa.eu/public opinion/archives/ebs/ebs 279.pdf.

Tsui, K.-Y., Multidimensional poverty indices. *Social Choice and Welfare*, 19:69–93, 2002.

8 Measuring chronic poverty

James E. Foster and Maria Emma Santos

> One of the least remarked-on problems of living with two dollars a day is that you do not literally get that amount each day. The two dollars a day is just an average over time. You make more on some days, less on others, and often get no income at all.
>
> Collins *et al.*, 2009, p. 2

8.1 Introduction

Time is an important additional dimension for understanding poverty and informing policy design.[1] A common way of incorporating time into the analysis of poverty is by separating the poor into two groups: the chronically poor and the transiently poor.[2] Hulme and Shepherd (2003) have described the chronically poor as follows: "intuitively, we are talking about people who remain poor for much of their life course, and who may 'pass on' their poverty to subsequent generations" (p. 405). However, the specific criterion used to identify the chronically poor (and hence the transiently poor) is a subject of continuing debate. Two identification approaches can be distinguished: a *counting approach* and a *permanent income approach*.[3]

In the counting approach, also called the *spells approach*, the chronically poor are identified based on the number of periods or the proportion of time they are observed to be in poverty. This approach goes back to Levy (1977), among others, and is also used by Duncan and Rodgers (1991).[4] More recently, Foster (2009) proposed a new family of chronic poverty measures within this general approach.[5] A related variant is that of Bane and Ellwood (1986) who estimate the exit probabilities associated with continuous poverty spells of different lengths.

The permanent income approach, also called the *components approach*, compares the resources a person has over time to the poverty line. Lillard and Willis (1978) and Duncan and Rodgers (1991) estimate permanent income as a person's intercept in a fixed-effects earnings model, while the transitory component is given by the error term. Persistent poverty is measured as the proportion of individuals with permanent income below the poverty line. A different method proposed in Ravallion (1988) and later used by Jalan and Ravallion (1998) defines people as chronically poor when their mean resources through time are below the poverty line, and measures chronic poverty using a traditional static measure applied to the distribution of means.[6] This method effectively maps the problem of multi-period poverty assessment into the traditional static framework using an income standard.[7]

These two approaches for identifying the chronically poor make very different assumptions regarding substitutability across periods. The counting approach assumes that resources observed in a time period are consumed in that time period, and are not transferred across periods; the permanent income approach freely averages up resources, effectively assuming perfect substitutability over time.[8] In view of this, Rodgers and Rodgers (1993) expand upon the permanent income approach by explicitly accounting for the individual's potential saving and borrowing behavior. Their proposed measure of an individual's permanent income is "the maximum sustainable annual consumption level that the agent could achieve with his or her actual income stream over the same T years, if the agent could save and borrow at prevailing interest rates" (p. 31). When positive interest rates are explicitly considered in the present value calculation, the permanent income level is below the mean income.

Yet modeling permanent income using interest rates may not reflect the full complexity of the transaction costs the poor face in transferring income and other resources over time. In their account of over 250 financial diaries of poor households across India, Bangladesh and South Africa, Collins *et al.* (2009, p. 3) find that poor households use a host of different methods to save and borrow, namely:

> storing savings at home, with others, and with banking institutions, joining savings clubs, savings-and-lean clubs, and insurance clubs, and borrowing from neighbors, relatives, employers, moneylenders, or financial institutions. At any one time the average poor household has a fistful of financial institutions relationships on the go.

Not only are the effective interest rate spreads faced by the poor far greater than the spreads in the formal market, but the other transaction costs of shifting resources can also be much higher. These vary from the extra time they must spend in a long queue, to the cost of the bus ride to reach the local bank, to an implicit obligation to work some days at a low wage in return for financial services (Collins *et al.*, 2009, p. 135).

We propose a new methodology for chronic poverty measurement that follows the permanent income approach, but explicitly allows an imperfect degree of substitutability across periods. As a result, volatility or inequality in the distribution of a person's resources over time is reflected in a lower measured level of permanent income. The methodology is based on a well-known class of income standards – namely, Atkinson's (1970) parametric family of "equally distributed equivalent income" functions, also known as the general means of order β – which exhibits lower levels of substitutability as its parameter

β falls from 1 (the usual mean) to $-\infty$ (the limiting case of perfect complements). The general mean is used to convert each person's resource stream over time into a permanent income standard, and then a corresponding member of the Clark *et al.* (1981) decomposable family of poverty measures is applied to the distribution of permanent income standards. The resulting class of chronic poverty measures is shown to have many attractive properties.

The chapter is organized as follows. Section 8.2 presents the notation used in the chapter, while Section 8.3 reviews previous chronic poverty measures. Section 8.4 presents the new class of chronic poverty measures and Section 8.5 describes the properties satisfied by this class. The transient component of poverty is the focus of Section 8.6, while Section 8.7 provides an empirical application using panel data from urban Argentina. Section 8.8 concludes.

8.2 Notation

Let $M^{n,T}$ denote the set of all $n \times T$ matrices with positive entries, and interpret a typical element $Y \in M^{n,T}$ as containing a panel of income observations for n different individuals over T periods.[9] Where N denotes the positive integers, the set $M = U_{n,T \in N} M^{n,T}$ contains all possible panels of data for any finite number of individuals, while $M^n = U_{T \in N} M^{n,T}$ contains all possible arrays across n individuals and $M^{\cdot T} = U_{n \in N} M^{n,T}$ is the set of T-period arrays. The population size and horizon associated with a given distribution Y are denoted by $n(Y)$ and $T(Y)$, respectively, or by n and T when fixed. For every $i = 1, 2, \ldots, n$ and $t = 1, 2, \ldots, T$, the typical entry y_{it} of Y is individual i's income in period t, where we assume that $y_{it} \in R_{++} := (0, \infty)$.[10] The row vector $y_i = (y_{i1}, y_{i2}, \ldots, y_{iT})$ contains individual i's incomes across time; the column vector $y_{\cdot t}, = (y_{1t}, y_{2t}, \ldots, y_{nt})'$ gives the income distribution across individuals in period t. The sum of entries in any given vector or matrix v will be denoted by $|v|$, while $\mu(v)$ will be used to represent the mean of v (or $|v|$ divided by the number of entries in v). It is assumed that all incomes have been adjusted by price differences over time and by the demographic characteristics of the individual, so that the same poverty line $z \in R_{++}$ can be used for all individuals and periods.[11]

The measurement of chronic poverty can be broken down into an identification step and an aggregation step analogous to Sen's (1976) presentation in the single period case. The first step results in an *identification function* $\rho(y_i; z)$, which determines whether individual i with income stream y_i is chronically poor given the poverty line z. The identification function indicates that individual i is in chronic poverty when $\rho(y_i; z) = 1$, while $\rho(y_i; z) = 0$ signals that the individual is not chronically poor. In contrast to the one period case, which entails a straightforward comparison of the income to the poverty line, the identification step here must consider the entire income stream to determine the chronic poverty status of the individual. Even so, the solution is immediate in two cases. When individual i is "never poor" so that $y_{it} \geq z$ for all t, every reasonable identification method would conclude $\rho(y_i; z) = 0$. Likewise, $\rho(y_i; z) = 1$ naturally arises when individual i is "always poor" so that $y_{it} < z$ for all t. Identification methods can significantly differ when individual i is "sometimes poor" so that $y_{it}<z$ for some t and $y_{it}\geq z$ at some other, and the issue then revolves around how poor and non-poor spells are to be compared.

The aggregation step builds upon the identification step to construct a *measure of chronic poverty* $P(Y; z)$ that combines the data of the chronically poor to obtain an overall level of chronic poverty. The basic *headcount ratio* $H(Y; z)=Q(Y; z)/n(Y)$ is found by counting the number $Q(Y; z)$ of chronically poor individuals and dividing by the total population size $n(Y)$. This is a useful partial index of chronic poverty, but like its single period version, it is rather insensitive to certain basic changes in the distribution that should arguably change the measured level of chronic poverty. Alternative chronic poverty measures have been proposed, with all methods up to now making use of standard one-period poverty measures. Solutions to both steps determine the *chronic poverty methodology* $\mathcal{M}=(\rho, P)$, where $\rho: M^1 \times R_{++} \rightarrow \{0, 1\}$ identifies the chronically poor individuals and $P: M \times R_{++} \rightarrow R_+$ aggregates the data into an overall level of chronic poverty in $R_+ := [0, \infty)$. The next section presents two existing chronic poverty methodologies.

8.3 Previous chronic poverty measures

There are two main approaches to measuring chronic poverty, distinguished primarily by their methods of identifying people who are chronically poor. One approach, exemplified by Foster (2009), is based on the number of periods that an individual is poor, and implicitly assumes that the observed income is not subsequently transferred across periods. A second method proposed by Ravallion (1988) and used by Jalan and Ravallion (1998) compares a person's mean income across time to the poverty line, which presumes that the resource variable can be transferred freely across periods.[12]

Foster (2009) begins by counting the periods of poverty experienced by individual i or, equivalently, the number of dates t for which $y_{it}<z$, and then expresses this poverty duration as a fraction d_i of the T periods. The identification function $\rho_\tau(y_i; z)$ is based on a fixed cutoff $\tau \in (0, 1]$, with an individual being chronically poor if the individual is poor at least τ share of the time. In symbols, $\rho_\tau(y_i; z)=1$ if $d_i \geq \tau$, and $\rho_\tau(y_i; z)=0$ if $d_i<\tau$.

As for the aggregation method, it is noted that the headcount ratio $H(Y; z)$ is insensitive to duration in that H does not change if the fraction d_i of time a chronically poor individual spends in poverty rises. This problem can be addressed by adjusting H and other single period measures to account for duration. The static measures used are from the Foster, Greer, and Thorbecke (1984) – or *FGT* – class, which is defined as follows. Let $w \in M^{n,1}$ be a distribution of income over a single period. For any $\alpha \geq 0$, the ith entry of the vector $g^\alpha(z)=(g_1^\alpha(z),\ldots,g_n^\alpha(z))$ is given by $g_i^\alpha(z)=0$ if $w_i \geq z$, and $g_i^\alpha(z) = ((z-w_i)/z)^\alpha$ if $w_i < z$. In words, $g_i^\alpha(z)$ is the α power of the normalized income shortfall if individual i's income falls below the poverty line, and 0 if not. The *FGT* class of measures is then $F_\alpha(w; z)=\mu(g^\alpha(z))$, or the mean of the vector $g^\alpha(z)$, with F_0 being the standard headcount ratio, F_1 being the per capita poverty gap, and F_2 being the squared gap *FGT* index.

The associated chronic poverty indices are defined in a similar fashion, but taking into account the duration cutoff τ defined above. For any matrix $Y \in M$, define the normalized gap matrix $G^\alpha(z) = [g_{it}^\alpha(z)]$ by $g_{it}^\alpha(z) = 0$ if $y_{it} \geq z$, and $g_{it}^\alpha(z) = ((z - y_{it})/z)^\alpha$ if $y_{it} < z$, and note that $G^\alpha(z)$ gives the α power of the normalized gaps across all individuals and periods, irrespective of whether an individual has been identified as being chronically poor. Identification is incorporated into the censored matrix $G^\alpha(z, \tau)$, whose typical entry is $g_{it}^\alpha(z,\tau) = g_{it}^\alpha(z)\rho_\tau(y_i; z)$; in other words, the entries are unchanged for the chronically poor, while the entries for the remaining persons are censored to zero.

The *duration adjusted FGT indices* are then defined as $K_\alpha(Y; z, \tau) = \mu(G^\alpha(z, \tau))$, or the mean of the censored matrix $G^\alpha(z, \tau)$. When $\alpha = 0$, the measure becomes the *duration adjusted headcount ratio* $K_0 = HD$, which is the product of the headcount ratio $H(Y; z) = Q/n$ and the average duration of poverty among the chronically poor, given by $D(Y; z, \tau) = |G^0(z, \tau)|/(QT)$. For $\alpha = 1$, the measure becomes the *duration adjusted poverty gap* $K_1 = HDA$, where $A(Y; z, \tau) = |G^1(z, \tau)|/|G^0(z, \tau)|$ is the average gap across the poverty spells of the chronically poor. Finally, $\alpha = 2$ yields the *duration adjusted squared poverty gap* $K_2 = HDS$, where $S(Y; z, \tau) = |G^2(z, \tau)|/|G^0(z, \tau)|$ is the average squared gap derived from the poverty spells of the chronically poor. The resulting methodology (ρ_τ, K_α) for evaluating chronic poverty satisfies a range of useful properties including *population decomposability*.[13]

In contrast, the chronic poverty methodology of Jalan and Ravallion (1998) uses an identification approach that ignores per-period poverty status and focuses on a specific income standard: the mean across periods. Their identification function $\rho_\mu(y_i, z)$ is defined by $\rho_\mu(y_i, z) = 1$ if $\mu(y_i) < z$, and $\rho_\mu(y_i, z) = 0$ if $\mu(y_i) \geq z$; in other words, an individual is chronically poor when the mean income is below the poverty line. For the aggregation step, they apply the single period *FGT* measure F_2 to the distribution $\bar{y} = (\bar{y}_1, \ldots, \bar{y}_n)$ of mean incomes $\bar{y}_i = \mu(y_i)$ drawn from Y, and hence their chronic poverty measure is simply $J(Y; z) = F_2(\bar{y}; z)$.

The methodology (ρ_μ, J) of Jalan and Ravallion (1998) can be readily linked to the Foster (2009) methodology. Consider the "smoothed" matrix $\bar{Y}$ defined by $y_{it} = \bar{y}_i$ for all i and t. In words, the individual's income in a given period is replaced by the mean income across all periods, as might be expected if an individual could freely transfer income through time. For matrices like $\bar{Y}$, chronic poverty identification becomes as simple as the single period case, since every individual is seen as either "never poor" or "always poor" while the contested category "sometimes poor" is absent. Every reasonable identification method would in fact agree on the set of chronically poor individuals in this situation, including the functions ρ_τ used by Foster (2009). Moreover, since the mean squared normalized gaps is the same for the vector $\bar{y}$ and the matrix $\bar{Y}$, we see that the measure J is just the duration adjusted *FGT* measure K_2 applied to the smoothed matrix $\bar{Y}$, that is, $J(Y; z) = K_2(\bar{Y}; z)$. Hence the chronic poverty methodology (ρ_μ, J) of Jalan and Ravallion (1998) can be viewed as the Foster (2009) methodology (ρ_τ, K_2) applied to the smoothed matrix $\bar{Y}$ rather than Y itself.

The counting approach of Foster (2009) interprets the observed data in a given period as the amount that is actually consumed in that period, with no subsequent resource movement across periods. Of course, this is more applicable to consumption data than for income data (which is presumably more fungible across periods), but it does provide one point of view from which the data may be evaluated. In contrast, by using the mean over time as the income standard, Jalan and Ravallion (1998) implicitly assume that resources can be costlessly transferred across periods. Such an assumption would seem to be more applicable when using income as the resource variable.

We argue that, in many circumstances, an intermediate level of substitutability between these two extremes may be more relevant, whether the resource variable is consumption or income. For consumption, some smoothing can be presumed to take place across time via storage or durable goods. For income, Collins *et al.* (2009, p. 3) note that households living on less than a dollar a day per person manage their money by saving when they can and borrowing when they need to, but due to their use of informal financial institutions, face high and variable interest rates as well as other transaction costs.[14] These cases are consistent with imperfect substitutability, whereby less than the full amount of what is drawn from one period will be available in another, as if the income was being carried in a "leaky bucket" (Atkinson, 1973; Okun, 1975). Variability lowers the effective pool of resources available to the family in accordance with the extent of the imperfection. In the next section, we develop a new class of chronic poverty measures whose identification step is consistent with an intermediate level of substitutability across time periods.

8.4 A new class of chronic poverty measures

The methodology we propose is analogous to Jalan and Ravallion's chronic poverty measure in that it makes use of a permanent income standard and a single period poverty measure. The income standard used here is a general mean (of order β), while the static poverty measure used is Atkinson's (1987) decomposable version of the Clark *et al.* (1981) – or CHU – class of poverty measures.[15] We introduce each of these components in turn.

Given an individual's income distribution y_i, the general mean income over time is defined as:

$$\mu_\beta(y_i) = \begin{cases} \left(\sum_{t=1}^{T} y_{it}^{\beta} \big/ T\right)^{1/\beta} & \beta \neq 0 \\ \prod_{t=1}^{T} y_{it}^{1/T} & \beta = 0 \end{cases} \qquad (8.1)$$

Each general mean can be interpreted as a permanent income standard, which summarizes y_i in a single income level.[16] When $\beta=1$, the general mean reduces to the *arithmetic mean*. For $\beta>1$, more weight is placed on higher incomes and the general mean is higher than the arithmetic mean, approaching the maximum income as β tends to ∞. For $\beta<1$ more weight is placed on lower incomes, and the general mean is lower than the arithmetic mean, approaching the minimum income as β tends to $-\infty$. The case of $\beta=0$ is known as the *geometric mean* and $\beta=-1$ is the *harmonic mean.*

As noted by Foster and Szekely (2008), this class of income standards satisfies a number of desirable properties.[17] For $\beta<1$ the standard will decrease when a mean-preserving income transfer from a period of lower income to a period of higher income increases dispersion; for $\beta=1$ it will be unaffected; and for $\beta>1$ it will increase. The range $\beta\leq 1$ yields the family of *equally distributed equivalent incomes* introduced by Atkinson (1970). The quantity $(1-\beta)$ represents the income standard's aversion to inequality over time or, equivalently, the cost or "leakage" when income is transferred across periods; $1/(1-\beta)$ is the standard's elasticity of substitution. We focus on the case $\beta<1$ for which income is imperfectly substitutable over time, so that a more unequal income stream will produce a lower income standard. We also include the case $\beta=1$ for which income is perfectly substitutable and inequality is ignored.

The general mean $\mu_\beta(y_i)$ with $\beta\leq 1$ can be used to construct an identification function $\rho_\beta(y_i, z)$ as follows: $\rho_\beta(y_i, z)=1$ if $\mu_\beta(y_i)<z$, and $\rho_\beta(y_i, z)= 0$ if $\mu_\beta(y_i)\geq z$. In words, an individual is identified as chronically poor when that person's permanent income level is below the poverty line. Note that when $\beta=1$ the identification function ρ_β becomes ρ_μ, which corresponds to the case of Jalan and Ravallion (1998). The general means, however, allow for imperfect substitutability, and the possibility that a person having an arithmetic mean above the poverty line will be identified as chronically poor due to variations in income and the costs of transferring it across time.[18]

We now turn to the class of static poverty measures that will be used. Let $w\in M^{n,1}$ be any distribution of income over a single period and let w^* denote the associated censored distribution given by $w_i^* = w_i$ for $w_i<z$, and $w_i^* = z$ otherwise. The CHU class of indices is defined by:

$$C_\beta(w;z)=\begin{cases}\dfrac{1}{\beta}\left[1-\left(\mu_\beta(w^*/z)\right)^\beta\right] & \beta\leq 1;\beta\neq 0\\ -\ln\left(\mu_0(w^*/z)\right) & \beta=0\end{cases} \tag{8.2}$$

Just as the FGT measures are based on income gaps among the poor, the CHU measures can be seen as being based on utility gaps, or the difference between the utility level of the poverty line income and the utility of the poor person's actual income. The utility function is given by $U_\beta(w_i)=w_i^\beta/\beta$ when $\beta\leq 1$ and $\beta\neq 0$, or by $U_\beta(w_i)=\ln w_i$ when $\beta=0$. The indices can be interpreted as measuring the average utility loss due to poverty where parameter $(1-\beta)$ indicates the underlying aversion to inequality among the poor.[19]

The new class of chronic poverty measures can now be defined. From the original income matrix Y, a vector of permanent income standards $\bar{y}_\beta=(\bar{y}_\beta^1,\ldots,\bar{y}_\beta^n)$

is constructed, where $\bar{y}^i_\beta = \mu_\beta(y_i)$ is the ith person's general mean across time. The proposed family P_β of chronic poverty measures is defined by:

$$P_\beta(Y;z) = C_\beta(\bar{y}_\beta, z) \quad \beta \leq 1 \tag{8.3}$$

In words, P_β is the decomposable CHU measure C_β applied to the vector $\bar{y}_\beta$ of general means with the same value of β. Our overall methodology for measuring chronic poverty is then given by (ρ_β, P_β). Requiring the same β for both components ensures that the chronic poverty measure has the same degree of aversion to inequality among the poor as to inequality over time.

It also leads to a simplified, matrix-based definition. Let Y^* be the censored matrix whose typical element is given by $y^*_{it} = y_{it}$ if $\mu_\beta(y_i) < z$, and $y^*_{it} = z$ otherwise; in other words, the incomes of the chronically poor, whether above or below the poverty line, are left unchanged, while the incomes of those not chronically poor are replaced by the poverty line. The new class P_β can equivalently be defined as:

$$P_\beta(Y;z) = \begin{cases} \frac{1}{\beta}\left[1 - \left(\mu_\beta(Y^*/z)\right)^\beta\right] & \beta \leq 1; \beta \neq 0 \\ -\ln\left(\mu_0(Y^*/z)\right) & \beta = 0 \end{cases} \tag{8.4}$$

The resulting formula resembles the original CHU formula, but uses a censored matrix instead of a censored vector.

Each of the measures also has the following intuitive interpretation in terms of utility gaps

$$P_\beta(Y;z) = A(z)\frac{1}{n}\sum_{i=1}^{n}\left[U_\beta(z) - \frac{1}{T}\sum_{t=1}^{T} U_\beta(y^*_{it})\right] \tag{8.5}$$

where $A(z)>0$ is a normalization factor.[20] If person i is not chronically poor, then $y^*_{it} = z$ and the utility gap – or the expression in brackets – is 0. If i is chronically poor, then $y^*_{it} = y_{it}$ and since the average utility over time is below the utility at the poverty line, the utility gap is positive. The use of a utility function with diminishing marginal utility ensures that variance across time lowers the average utility. Indeed, $(1-\beta)$ is the elasticity of the marginal utility with respect to the resource, so that a higher $(1-\beta)$ means that marginal utility falls more quickly as income rises and there is greater sensitivity of the average utility to variations over time.

8.5 Properties

Which properties are satisfied by this family of chronic poverty measures given the identification functions? We now present several properties that are natural extensions of requirements for static poverty measures. Our first set of properties requires chronic poverty to be unchanged under certain basic transformations.

We say that X is obtained from Y by a *permutation of incomes across people (across time)* if $X=\Pi Y$ (resp. $X=Y\Pi$), where Π is a $n\times n$ (resp. $T\times T$) permutation matrix. Matrix X is matrix Y with rows (respectively, columns) interchanged according to the particular permutation matrix. A permutation of incomes across people implies that entire distributions of income over time are switched among persons, while a permutation of incomes across time implies that, for every individual in the population, incomes are switched across periods.

P1 – Population Symmetry: if X is obtained from Y by a permutation of incomes across people then $P(X; z)=P(Y; z)$.

P2 – Time Symmetry: if X is obtained from Y by a permutation of incomes across time, then $P(X; z)=P(Y; z)$.

Population symmetry corresponds to the standard anonymity property for static poverty measures; time symmetry requires that the order in which one receives income should not affect the measure.

At first glance, time symmetry may appear to be quite restrictive. In particular, one could argue that incomes from later periods should be discounted to obtain a present value measure (Calvo and Dercon, 2009; Hoy and Zheng, 2011; Aaberge and Mogstad, 2007) or, alternatively, from an *ex-post* point of view, that incomes in earlier periods should be discounted (Calvo and Dercon, 2009). Rather than entering into this discussion, the time symmetry axiom takes the middle ground and treats each period's income with equal importance. Alternatively, one might assume that Y is composed of incomes already discounted for time, in which case time symmetry may be less controversial.

A second critique of time symmetry is that it ignores the specific sequencing of incomes and hence of poverty spells. For example, one might argue that consecutive spells should be disproportionally weighted in measuring poverty over time.[21] Note, however, that even the richest panel data sets do not contain information on the *lifetime* income stream. Information is necessarily truncated. Any penalization of consecutiveness is at risk of being misleading. Time symmetry reflects this concern and ensures that the poverty measure is not unduly reliant on the observed sequencing of incomes.

We say that X is obtained from Y by a *population (time) replication* if $X=[Y';\ldots; Y']'$ is constructed from multiple copies of Y stacked on top of one another, where X is in $M^{kn,T}$ and Y is in $M^{n,T}$ for integers $n\geq 1$ and $k\geq 2$ (respectively, if $X=[Y;\ldots; Y]$ is constructed from multiple copies of Y placed adjacent to one another where X is in $M^{n,kT}$ and Y is in $M^{n,T}$ for integers $T\geq 1$ and $k\geq 2$).

P3 – Population Replication Invariance: if X is obtained from Y by a population replication, then $P(X; z)=P(Y; z)$.

P4 – Time Replication Invariance: if X is obtained from Y by a time replication, then $P(X; z)=P(Y; z)$.

The two replication invariance axioms ensure comparability across different population sizes and number of periods.

We say that $(Y'; z')$ is obtained from $(Y; z)$ by a *proportional change* if $(Y'; z')=(\alpha Y; \alpha z)$ for some $\alpha>0$.

P5 – Scale Invariance: if $(Y'; z')$ is obtained from $(Y; z)$ by a proportional change, then $P(Y'; z')=P(Y; z)$.

Scale invariance ensures that chronic poverty is unchanged when incomes and poverty lines are altered, so long as incomes expressed in poverty line units remain unchanged.

We say that X is obtained from Y by a *simple increment among the non-chronically poor* if $x_{kj}>y_{kj}$ for a given (k, j) with $\rho(y_k; z)=0$, while $x_{it}=y_{it}$ for every other $(i, t)\neq(k, j)$.

P6 – Focus: if X is obtained from Y by an increment among the non-chronically poor, then $P(X; z)=P(Y; z)$.

The focus axiom ensures that the measure is not affected by income changes among the non-chronically poor. Note that changes in non-poor incomes of the chronically poor can have an impact on the level of chronic poverty. This is consistent with an identification step that allows substitutability of income over time.

We now turn to a set of properties that require the chronic poverty measure to move in a specific direction under certain transformations.

We say that X is obtained from Y by an *increment among the chronically poor* if $x_{kj}>y_{kj}$ for a given (k, j) with $\rho(y_k; z)=1$, while $x_{it}=y_{it}$ for every other $(i, t)\neq(k, j)$.

P7 – Monotonicity: if X is obtained from Y by an increment among the chronically poor, then $P(X; z)<P(Y; z)$.

Monotonicity ensures that chronic poverty falls when an income of a chronically poor person rises. Note that this is true for *any* income of a chronically poor person – either below or above the poverty line – which is consistent with the assumed substitutability of income over time.

We now introduce a Kolm (1977) transformation that smooths each period's income distribution in the same way using a bistochastic matrix.[22] We say that X is obtained from Y by a *smoothing of incomes among the chronically poor* if $X=BY$ for an $n\times n$ bistochastic matrix B satisfying $b_{ii}=1$ for every non-chronically poor household i where X is not a permutation of Y. The condition on B ensures that only the chronically poor are affected.

P8 – Weak Transfer: if X is obtained from Y by a smoothing of incomes among the chronically poor, then $P(X ; z)\leq P(Y ; z)$.

Under weak transfer, when incomes are smoothed so that inequality among the chronically poor unambiguously falls, chronic poverty should not increase.

The next axiom ensures that if all incomes are z or higher then there will be no chronic poverty.

P9 – Normalization: if Y is completely equal with $y_{it}=z$ for all i and t, then $P(Y; z)=0$.

Next is a strong form of continuity, which rules out abrupt changes in chronic poverty as incomes gradually increase or decrease, even when the number of chronically poor persons is altered.

P10 – Continuity: for every fixed $z\in R_{++}$, the chronic poverty measure $P(Y; z)$ is continuous as a function of Y on M.

This is a convenient property for measures, especially given the errors inherent in data and the essential arbitrariness of a poverty line. Other chronic poverty measures that reflect the number of persons in chronic poverty, or the extent of their duration, typically will not satisfy this property, although alternative restrictive versions of continuity may well be satisfied.[23]

Finally, there is a well-accepted property for traditional poverty measures that requires overall poverty to be expressible as a population-share weighted sum of subgroup poverty levels. The following is the analogous property for chronic poverty measures.

P11 – Population Decomposability: for all X and Y satisfying $T(X)=T(Y)$, we have

$$P(X,Y;z)=\frac{n(X)}{n(X,Y)}P(X;z)+\frac{n(Y)}{n(X,Y)}P(Y;z) \tag{8.6}$$

Measures satisfying this property allow a subgroup's contribution to overall chronic poverty to be evaluated.

It can be verified that the proposed family of measures satisfies all the above properties (P1 to P11). One may also consider a "time decomposability" axiom, but since identification depends on every period's income, it is not possible to obtain a full decomposition in terms of the original income matrix. However, after identification, when the censored matrix Y^* has been constructed, overall chronic poverty can be broken down period by period as follows:

$$P_\beta(Y;z)=\begin{cases}(1/T)\sum_{t=1}^{T}(1/\beta)\left(1-[\mu_\beta(y^*_{\cdot t}/z)]^\beta\right) & \beta\leq 1,\beta\neq 0\\ (1/T)\sum_{t=1}^{T}-\ln\left(\mu_0(y^*_{\cdot t}/z)\right) & \beta=0\end{cases} \tag{8.7}$$

where $y^*_{\cdot t}$ is the t-th column of the censored matrix Y^*. This "time breakdown" over time can be useful in an *ex-post* evaluation of the relative contribution of each time period to overall chronic poverty.[24] Note that such a breakdown is possible because the β values of μ_β and C_β coincide; if they did not, the measure would lose this additional feature. Similar issues ensure that Jalan and Ravallion's chronic poverty measure cannot be broken down in an analogous way. However, the variant $F_1(\bar{y};z)$ of their approach, which matches the poverty gap to their arithmetic mean, is just our $C_1(\bar{y}_1;z)$ and hence has a breakdown over time. Foster's (2007) measures K_α also can be broken down over time after identification.

8.6 Transient poverty

A measure of chronic poverty P requires panel data linked through time; the associated static poverty measure F does not. How far off would we be if we simply evaluated static poverty period-by-period and averaged up? The answer is provided by a transient poverty measure $P^T = P^A - P$ where P^A is average static poverty, or

$$\sum_{t=1}^{T} F(y_{\bullet t};z)/T,$$

and P is chronic poverty. We now present the transient poverty measure for our class of chronic poverty measures and contrast it to P^T for previous measures.

The transient poverty measure for Foster's (2007) chronic poverty measure is

$$K_\alpha^T(Y;z,\tau) = \sum_{t=1}^{T} F_\alpha(y_{\bullet t};z)/T - K_\alpha(Y;z,\tau)$$

where the static poverty measure F_α is an FGT index. Using the notation of Section 8.3, it follows that $K_\alpha^T(Y;z,\tau) = \mu(G^\alpha(z) - G^\alpha(z,\tau))$, where the positive entries of the matrix $G^\alpha(z) - G^\alpha(z,\tau)$ arise only from the "transiently poor" or people with spells of poverty who do not have enough spells to be chronically poor. Transient poverty is composed of the poverty episodes of the transiently poor, while chronic poverty is all that is generated by the chronically poor. The two notions are mutually exclusive in the approach of Foster (2009).

The transient poverty measure for Jalan and Ravallian (1998) is

$$J^T(Y;z) = \sum_{t=1}^{T} F_2(y_{\bullet t};z)/T - J(Y;z)$$

where the FGT measure F_2 is the static poverty measure. Note that the average poverty measure can be expressed as $\mu(G^2(z))$, the mean of the squared gap matrix, while the chronic poverty measure can be viewed as $\mu(\bar{G}^2(z))$, the mean of the matrix of squared gaps obtained from the smoothed matrix $\bar{Y}$ (so that the entries of $\bar{G}^2(z)$ are $\bar{g}_{it}^2(z) = ((z - \bar{y}_i)/z)^2$ if $\bar{y}_i < z$ and $\bar{g}_{it}^2(z) = 0$ otherwise). It follows that $J^T(Y;z) = \mu(G^2(z) - \bar{G}^2(z))$ where the entries in the matrix $G^2(z) - \bar{G}^2(z)$ may be positive, negative, or zero. $J^T(Y;z) = \mu(G^2(z) - \bar{G}^2(z))$

A chronically poor person will have a mean income below the poverty line and hence all entries in the person's row in $\bar{G}^2(z)$ are an identical positive number. A person having all zeroes in $\bar{G}^2(z)$ and at least one positive entry in $G^2(z)$ has a spell of poverty but is not chronically poor, and hence can be called transiently poor. A person with all zero entries in both $\bar{G}^2(z)$ and $G^2(z)$ is neither. All chronic poverty originates from the chronically poor, but in this approach not all transient poverty is due to the transiently poor; instead there are two additional components associated with the chronically poor. First, it is possible for the average gap to exceed the gap of the average, due to the censoring of incomes above the poverty line. Second, even if the average gap was the same as the gap of the average, the average squared gap can exceed the squared gap of the average, due to variation in incomes over time.

Finally, for our chronic poverty measure, the transient poverty measure is

$$P_\beta^T(Y;z) = \sum_{t=1}^{T} C_\beta(y_{\bullet t};z)/T - P_\beta(Y;z)$$

where the CHU measure C_β is used to measure static poverty. Note that the average poverty measure can be expressed as $\mu(V^\beta(z))$, the mean of the normalized utility gap matrix $V^\beta(z)$ whose entries for the case of $\beta\leq 1$, $\beta\neq 0$ are $v_{it}^\beta = (1-(y_{it}/z)^\beta)/\beta$ for $y_{it}<z$ and $v_{it}^\beta = 0$ for $y_{it}\geq z$, while for the case of $\beta=0$ are $v_{it}^\beta = -\ln(y_{it}/z)$ for $y_{it}<z$ and $v_{it}^\beta = 0$ for $y_{it}\geq z$. In contrast, the chronic poverty measure is the mean of the matrix of utility gaps obtained from the censored matrix Y^*, or $\mu(V^{*\beta}(z))$ where the entries of $V^{*\beta}(z)$ are given by $v_{it}^{*\beta}(z) = (1-(y_{it}^*/z)^\beta)/\beta$ for $\beta\leq 1$, $\beta\neq 0$, and $v_{it}^{*\beta}(z) = -\ln(y_{it}^*/z)$ for $\beta=0$. It follows that $P^T(Y;z)=\mu(V^\beta(z)-V^{*\beta}(z))$ where (as we show below) the entries in the matrix $V^\beta(z)-V^{*\beta}(z)$ may be positive or zero.

Let i be a chronically poor person so that $\mu_\beta(y_i)<z$ and the entry $v_{it}^{*\beta}(z)$ from $V_{it}^{*\beta}(z)$ is always based on y_{it}. Now if $y_{it}<z$, we have $v_{it}^\beta(z)-v_{it}^{*\beta}(z)=0$ since in this case both gaps are based on y_{it}. On the other hand, if $y_{it}\geq z$, we have $v_{it}^\beta(z)=0$ while $v_{it}^{*\beta}(z) < 0$ which implies that $v_{it}^\beta(z)-v_{it}^{*\beta}(z)=-v_{it}^{*\beta}(z)$ is positive. The static poverty term censors income to the poverty line, while the chronic poverty term views this income as lowering chronic poverty, and so the difference between the two is positive. Now suppose that i is a transiently poor person, so that $\mu_\beta(y_i)\geq z$ with $y_{it'}<z$ for some t'. Clearly, for all t we have $v_{it}^{*\beta}(z)=0$ and $v_{it}^\beta(z)\geq 0$ and hence $v_{it}^\beta(z)-v_{it}^{*\beta}(z)\geq 0$; for $t=t'$ we have $v_{it}^{*\beta}(z)=0$ and $v_{it}^\beta(z)>0$ and so $v_{it}^\beta(z)-v_{it}^{*\beta}(z)>0$. The transiently poor have at least one period of positive static poverty while their chronic poverty is always zero, and hence the difference is positive. Finally, a person who is neither chronically nor transiently poor has zero entries for both terms.

As with the Jalan and Ravallion approach, not all transient poverty here arises from the transiently poor. However, in the present case only one component of transient poverty is associated with the chronically poor. It arises from the censoring that occurs in the average poverty term and reflects the absence of information linking observations across time. There is no separate term reflecting the variations in income; instead, variations are incorporated into the permanent income standard and are evaluated using a matched poverty measure.[25]

Our measure of transient poverty can be viewed as the extent to which average CHU poverty from cross-sections would fall if our chronic poverty measure were applied to the associated panel, thereby linking observations over time. The reduction has two sources: (i) a person who is viewed as poor in a given period may have sufficient excess resources in other periods to avoid being chronically poor; (ii) a person who is viewed as poor in a given period may have enough excess resources in other periods to moderate the level of chronic poverty. These are the sources of transient poverty for our measure.

8.7 Empirical illustration for Argentina

We now illustrate the family of chronic poverty measures proposed in this chapter using data from urban Argentina. The data correspond to the *Encuesta Permanente de Hogares* (Permanent Individual Survey, or EPH from now on), a survey that was conducted twice a year in the main urban areas of Argentina by the National Institute of Statistics and Census (INDEC) in the months of May and October until May 2003. The survey represents 61 percent of the country's population and 70 percent of the urban population according to the 1991 Census. The survey used a rotating panel where 25 percent of the sample was replaced in each wave. Thus, it is possible to observe households along four waves of the survey, following them for about a year-and-a-half. For this illustration we work with the panel formed by the observations of October 2001, May 2002, October 2002, and May 2003. Clearly, four observations over a year-and-a-half is a short period of time to evaluate chronic poverty, but it allows us to illustrate the new methodology. The panel is formed with observations from 27 urban agglomerations. Total sample size is 2387 households and 8704 people. However, only households with complete and valid information on income were considered, so the sample was reduced to 1731 households and 6362 people.

Estimates of chronic poverty are obtained using the household's equivalent income as the resource variable, normalized by the poverty line for the corresponding time period and region. We use the poverty lines and equivalent adult scale provided by INDEC (2003). To obtain population-based estimates, households are weighted by their size.[26]

Figure 8.1 presents the normalized general mean incomes $\mu_\beta(y_i)/z$ for an individual who is *never* poor, three individuals who are *sometimes* poor in a different number of periods, and one who is *always* poor. The figure illustrates how the general mean income increases with the value of β. For the individual who is never poor the general mean income lies above the poverty line for all values of β, whereas for the individual who is always poor, the general mean income lies

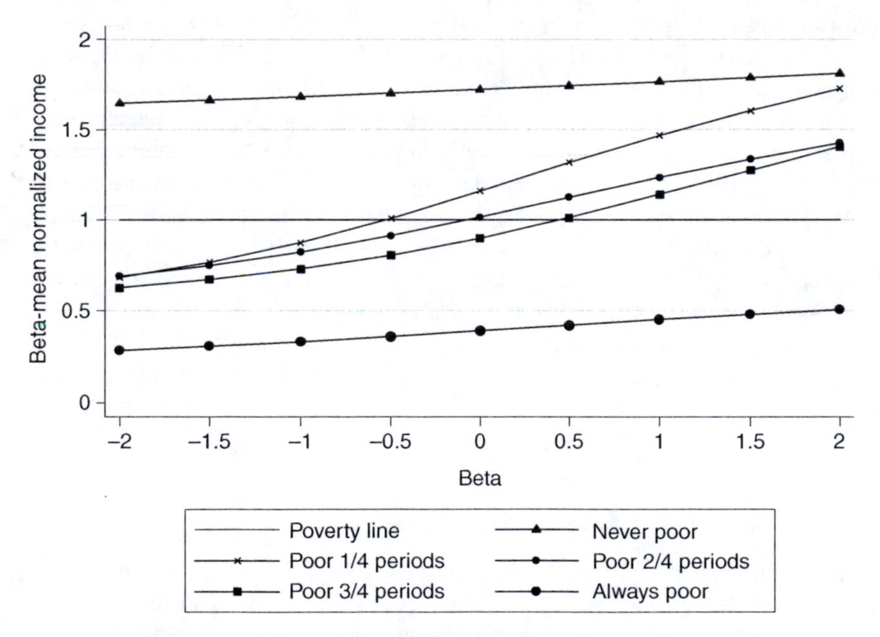

Figure 8.1 Normalized general mean incomes for five persons as β varies.

below the poverty line for all β. For each of the three persons who are sometimes poor, the general mean income crosses the poverty line at a different β value. This β cutoff identifies the lowest parameter value at which the person would escape chronic poverty. If it is above 1, the person will be identified as chronically poor by all our measures. If it is below 1, then the person avoids chronic poverty over the range $[\beta, 1]$ – a range that expands as the person's resources rise or become more evenly spread over time.[27]

The second column in Table 8.1 presents the percentage of chronically poor people for different values of β. By definition, such percentage increases as β decreases, in this case from 53 to 58.5 percent. The fourth column in the table contains the estimates of the chronic poverty measure. Again, the lower the income substitutability across periods, the higher the chronic poverty estimates. The table also presents the estimates of average poverty in the four periods (third column), and transient poverty (sixth column), obtained as the difference between average and chronic poverty. For all β values, chronic poverty accouts for most of average poverty in the panel (between 89 and 98 percent). This percentage naturally increases as β decreases, since part of the transiently poor start to be considered chronically poor.

It is interesting to compare the chronic poverty estimates presented in Table 8.1 with per-period CHU poverty estimates, which are presented in Table 8.2.[28] It is worth recalling that the year 2002 was of deep economic recession, after the collapse of the economy in December 2001. This is reflected in the substantially higher per-period poverty estimates of May and October 2002 and May 2003. Interestingly, the chronic poverty estimates, which consider the income in the four points in time and allow varying degrees of income substitutability across periods, lie in between the lower estimates of the CHU index in October 2001, and the higher estimates of the subsequent observations.

One of the advantages of the proposed measure is that it can be broken down by population subgroup and (post identification) by time period. Tables 8.3 and 8.4 present examples of both types of analyses. Table 8.3 presents the chronic poverty estimates in each of the (main urban areas of the) six geographic regions of the country, their contribution to overall chronic poverty alongside their population share. Notice that the northern regions are the two bigger contributors to overall chronic poverty. Although the northeast contributes only 15 percent of the urban population, it makes up 22 to 25 percent of overall urban chronic poverty, depending on the value of β. At the other extreme, while the Patagonia region represents 13.4 percent of the urban population, it contributes 6 percent or less of overall chronic poverty.

Table 8.4 shows that the first period of the panel contributed only 12.8 to 14.5 percent of overall chronic poverty (varying with the value of β), whereas, the other three periods – associated with the crisis and subsequent recession – contributed 26 to 32 percent each.

Table 8.1 Average, chronic and transient poverty for different values of β Argentina October 2001 to May 2003

Beta	*% of chronically poor people*	*Average poverty* $P_\beta^A(Y; z)$	*Chronic poverty*		*Transient poverty*	
			Measure $P_\beta^C(Y; z)$	*Percent contribution*	*Measure* $P_\beta^T(Y; z)$	*Percent contribution*
(1)	*(2)*	*(3)*	*(4)*	*(5)*	*(6)*	*(7)*
1	52.9	0.27	0.24	*88.9*	0.03	*11.1*
0.5	54.1	0.34	0.31	*91.2*	0.03	*8.8*
0	55.6	0.45	0.43	*95.5*	0.02	*4.5*
–0.5	57.1	0.66	0.63	*95.6*	0.03	*4.4*
–1	58.5	1.09	1.07	*98.2*	0.02	*1.8*

Table 8.2 One-period CHU poverty estimates Argentina, October 2001 to May 2003

Beta	*October 2001*	*May 2002*	*October 2002*	*May 2003*
1	0.18	0.29	0.31	0.29
0.5	0.22	0.38	0.39	0.36
0	0.29	0.52	0.52	0.48
–0.5	0.40	0.79	0.76	0.70
–1	0.59	1.40	1.22	1.16

Table 8.3 Decomposition of chronic poverty by region

Region	*Population share**	*Beta*				
		1	*0.5*	*0*	*–0.5*	*–1*
Northeast	15%	0.36	0.47	0.66	1.00	1.77
Percentage contribution		*22.3*	*22.6*	*22.9*	*23.6*	*24.7*
Northwest	30%	0.29	0.38	0.51	0.75	1.26
Percentage contribution		*36.1*	*35.8*	*33.4*	*35*	*28*
GBA	8.2%	0.17	0.23	0.30	0.43	0.65
Percentage contribution		*6*	*6*	*5.8*	*5.6*	*5*
Pampeana	21%	0.22	0.29	0.40	0.62	1.06
Percentage contribution		*19.6*	*20*	*20.3*	*20.9*	*21.3*
Cuyo	12.4%	0.21	0.27	0.36	0.51	0.84
Percentage contribution		*11.4*	*11*	*10.4*	*10*	*9.7*
Patagonia	13.4%	0.09	0.12	0.16	0.23	0.35
Percentage contribution		*6*	*5.1*	*5.1*	*4.9*	*4.36*

Note

* The population shares correspond to those of the sample of the conformed panel, and thus may differ from the population shares when all urban and rural areas are considered.

Table 8.4 Decomposition of chronic poverty by period

Period	*Beta*				
	1	*0.5*	*0*	*–0.5*	*–1*
October 2001	0.14	0.18	0.24	0.35	0.55
Percentage contribution	*14.5*	*14.3*	*14.1*	*13.7*	*12.8*
May 2002	0.27	0.36	0.50	0.77	1.39
Percentage contribution	*28.5*	*28.8*	*29.4*	*30.5*	*32.4*
October 2002	0.28	0.37	0.50	0.74	1.20
Percentage contribution	*29.4*	*29.4*	*29.3*	*29*	*28.0*
May 2003	0.27	0.34	0.46	0.68	1.15
Percentage contribution	*27.6*	*27.5*	*27.2*	*26.8*	*26.8*

8.8 Concluding remarks

This chapter has introduced a new class of chronic poverty measures having two components: a permanent income standard for identification purposes, and a static poverty measure for aggregation. We summarize the resource stream over time using a general mean μ_β with parameter $\beta \leq 1$, and a person is defined to be chronically poor if the permanent income standard falls below the poverty line. The structure accommodates perfect substitutability ($\beta = 1$) and imperfect substitutability ($\beta < 1$) of resources across time, and the parameter may be adjusted to fit the conditions facing the poor and the particular resource variable (e.g., consumption or income) at hand. As our static poverty measure we use the decomposable CHU index C_β having the same parameter value β as the general mean. Our class P_β of chronic poverty measures is obtained by applying the CHU poverty measure to the distribution of general means.

The resulting methodology for identifying and measuring chronic poverty has several convenient features. It satisfies a set of properties that are natural extensions of the static case, including population decomposability. It has a policy relevant breakdown by time period, after identification of the chronically poor, by which the contribution of a given time period to overall chronic poverty can be ascertained. This helpful property arises because the same parameter value is used in the general mean and the CHU poverty measure. The chronic poverty measure also has a concise definition as the mean of a matrix, which should facilitate its calculation and the application of statistical tests.

The measure has a natural welfare interpretation in terms of utility shortfalls, or the difference between the utility of the poverty line and the average utility from the resource stream over time. The utility loss is greater for higher elasticity of the marginal utility with respect to the resource $(1-\beta)$, hence for lower values of β. An alternative interpretation of $(1-\beta)$ is that it represents the cost or "leakage" of transferring income over time. When this value is *assumed*, empirical implementations of P_β should consider a range of values to test the robustness of the estimations, as was done in the illustration above.

An interesting exercise might be to calibrate β in line with the explicit and implicit costs of transferring income through time in the particular context under analysis. If such calibration were possible, β could then perhaps become a variable affected by policies: a reduction in the cost of intertemporal transfers for poor people would be reflected in a higher β value and this in turn could result in a reduction of chronic poverty. The policy relevance of β is in line with recommendations based on studies from the ground: "if poor households enjoyed assured access to a handful of better financial tools, their chances of improving their lives would surely be much higher" (Collins *et al.*, 2009, p. 4).

The focus of this chapter has been chronic poverty, with transient poverty being a residual component between the average poverty level from cross-sections and the level of chronic poverty. Some of the transient poverty originates with the chronically poor. This differs from the spells approach in which chronic (transient) poverty is associated exclusively with the chronically (transiently) poor. It is also distinct from Jalan and Ravallion's approach in which transient poverty may additionally arise from the variations in resource levels below the poverty line. It would be interesting to study further the different forms of transient poverty and, indeed, to identify properties that a reasonable measure of transient poverty should display.

Acknowledgements

The authors are grateful for comments from Chico Ferreira and participants at: the 2006 LACEA Meetings at ITAM, Mexico; and the Oxford Poverty and Human Development Initiative (OPHI) Seminar Series, the Inaugural Conference of the Courant Research Centre on Poverty Equity and Growth in Developing Countries, University of Gottingen, the III Meeting of the Society for the Study of Economic Inequality, Universidad Torcuato Di Tella, Argentina, and the Annual Meeting of the Brazilian Econometric Society, all in 2009. Completion of the project was facilitated by a Visiting Fellowship from Magdalen College to Foster in Trinity Term and by visits to OPHI in the summer of 2012. Foster thanks the Institute for International Economic Policy at the Elliott School of International Affairs and Santos thanks ANPCyT-PICT 1888 for research support.

Notes

1 The present chapter focuses purely on income (or consumption) and lets multiple observations create a kind of multidimensionality over time. Many recent papers have considered dimensions beyond income (or consumption) in identifying and measuring poverty. See, for example, Atkinson (2003), Bourguignon and Chakravarty (2003), Alkire and Foster (2007, 2011), and Alkire and Santos (2010). In the future, both aspects of poverty – its dynamics and its multidimensionality – may be combined into a single analysis.

2 A related study of poverty dynamics does not focus on the chronically poor but rather constructs measures of lifetime or intertemporal poverty. See, for example, Calvo and Dercon (2009), Hoy and Zheng (2011), Bossert *et al.* (2012), Porter and Quinn (2008), Dutta *et al.* (2011), Gradin *et al.* (2012), and Mendola and Busetta (2012). Duclos *et al.* (2010) propose measures of chronic and transient poverty but these categories apply to aggregates and not to individuals; there is no identification criterion for the chronically poor.

3 Yaqub (2000) uses the terms "spells approach" and "components approach".

4 For further references on early uses of this approach see the excellent discussion in Rodgers and Rodgers (1993).

5 Alkire and Foster (2011) propose a related methodology for multidimensional poverty measurement.

6 See also Rodgers and Rodgers (1993) whose chronic poverty measure reduces to this case when the interest rate is zero.

7 An "income standard" is a function that reduces a distribution or vector of some resource variable to a representative level of that variable; see Foster and Szekely (2008) for a general definition including properties. When the aggregation is over time we will call it a "permanent income standard."

8 Note, though, that both implicitly assume perfect substitutability *within* each time period.

9 The term "income" is meant to represent a generic resource variable, which may in fact be income, consumption, or expenditure.
10 In what follows we make the practical assumption that income (or consumption) levels are positive, which is needed for some of the measures we discuss.
11 Equivalently, one could reflect differences in local prices or demographics in the poverty line and normalize; see Foster (1998). Our example incorporates demographics into the income variable and price change into the poverty line.
12 See also Jalan and Ravallion (2000).
13 For a list of these properties and their definitions, see Foster (2007).
14 For a detailed discussion on the transaction costs of informal risk sharing see Morduch (1999).
15 Also included are the measures of Watts (1969) and Chakravarty (1983).
16 Note that $\mu_\beta(y_i)$ is weakly increasing in β and also that

$$\lim_{\beta \to 0} \mu_\beta(y_i) = \mu_0(y_i).$$

17 In fact, it is the only class of income standards satisfying symmetry, replication invariance, linear homogeneity, normalization, continuity, *and* subgroup consistency (Foster and Szekely, 2008, p. 1149). Foster and Shneyerov (1999, 2000) highlight the special role of the general means in inequality measurement.
18 The general mean over time is also used in the identification step of other studies. Cruces and Wodon (2007) combine it with the FGT measure F_2 to create a measure of "risk-adjusted poverty." Aaberge and Mogstad (2007) use an "equally-allocated equivalent income" to identify the chronically poor, but do not have an aggregation step. See also Porter and Quinn (2008).
19 The utility-loss interpretation is discussed extensively in Foster and Jin (1998), who also provide a characterization of the CHU measures. Note that C_β converges to C_0, the Watts (1969) poverty index, as β tends to 0.
20 This follows immediately from equation (8.5) and the definition of U_β, where $A(z)=1$ for the case of $\beta=0$ and $A(z)=1/\beta U_\beta(z)$ for the case $\beta \leq 1$ and $\beta \neq 0$.
21 See for example Bossert *et al.* (2012), Hoy and Zheng (2011), Guenther and Maier (2008), Dutta *et al.* (2011), Gradin *et al.* (2012), and Mendola and Busetta (2012).
22 A bistochastic matrix is a square matrix whose entries are nonnegative and each column and row sums to 1; it is a convex combination of permutation matrices. Multiplying by such a matrix gives each person a convex combination of the income streams in society (Bourguignon and Chakravarty, 2003, p. 30–31).
23 See Foster and Shorrocks (1991) and Foster (2006) for discussions of restricted and unrestricted continuity in the context of traditional poverty measures.
24 The censored vector y_i^* can include entries that exceed z and diminish the contribution of period t to overall chronic poverty. In the unlikely case where enough entries exceed z, the contribution of period t could be negative, indicating that the period was one of general affluence that reduced chronic poverty via the assumed substitution across time.
25 In both Jalan and Ravallion's approach and ours, when there is an increase in a non-poor income of a chronically poor individual, average poverty is unaffected, chronic poverty falls, and transient poverty rises.
26 For families in which the household size changed over the four observations of the panel, all household members were considered to calculate the equivalent income in each period, but only household members that were present in the four waves of the panel were considered to compute the poverty estimates.
27 In this way, the value of the parameter β at which each individual starts to be considered chronically poor can be nicely linked to the fuzzy sets approach to poverty measurement (Cerioli and Zani, 1990; Cheli and Lemmi, 1995); in fact $1/(2-\beta)$

may be interpreted as a measure of the degree of membership to the group of the chronically poor. Alternatively, such value may be seen as a measure of vulnerability to poverty.

28 Note that because of the decomposability property of the CHU indices, for each value of β, the (row) average of the per-period CHU estimates presented in Table 8.2 equals the average poverty reported in the third column of Table 8.1.

References

Aaberge, R. and Mogstad, M. (2007) "On the Definition and Measurement of Chronic Poverty," IZA Discussion Paper Series No. 2659.

Alkire, S. and Foster, J. E. (2007) "Counting and Multidimensional Poverty Measurement," OPHI Working Paper No. 7.

Alkire, S. and Foster, J. E. (2011) "Counting and Multidimensional Poverty Measurement," *Journal of Public Economics* 95: 476–487.

Alkire, S. and Santos, M. E. (2010) "Acute Multidimensional Poverty: A New Index for Developing Countries," OPHI Working Paper No. 38.

Atkinson, A. B. (1970) "On the Measurement of Inequality," *Journal of Economic Theory* 2: 244–263.

Atkinson, A. B. (1973) *Wealth, Income and Inequality*, Penguin, London.

Atkinson, A. B. (1987) "On the Measurement of Poverty," *Econometrica* 55: 749–764.

Atkinson, A. B. (2003) "Multidimensional Deprivation: Contrasting Social Welfare and Counting Approaches," *Journal of Economic Inequality* 1: 51–65.

Bane, M. J. and Ellwood, D. (1986) "Slipping Into and Out of Poverty: The Dynamics of Spells," *Journal of Human Resources* 21: 1–23.

Bossert, W., Chakravarty, S. R., and D'Ambrosio, C. (2012) "Poverty and Time," *Journal of Economic Inequality* 10: 145–162.

Bourguignon, F. and Chakravarty, S. (2003) "The Measurement of Multidimensional Poverty," *Journal of Economic Inequality* 1: 25–19.

Calvo, C. and Dercon, S. (2009) "Chronic Poverty and All That: The Measurement of Poverty Over Time," in Addison, T., D. Hulme, and R. Kanbur (eds.), *Poverty Dynamics: Interdisciplinary Perspectives*: 29–58. Oxford University Press, Oxford.

Cerioli, A. and Zani, S. (1990) "A Fuzzy Approach to the Measurement of Poverty," in Dagum, C. and M. Zenga (eds.), *Income and Wealth Distribution, Inequality and Poverty*: 272–84. Springer Verlag, Berlin.

Chakravarty, S. (1983) "A New Index of Poverty," *Mathematical Social Sciences* 6: 307–13.

Cheli, B. and A. Lemmi (1995) "A Totally Fuzzy and Relative Approach to the Multidimensional Analysis of Poverty," *Economic Notes* 24: 115–133.

Clark, S., Hemming, R., and Ulph, D. (1981) "On indices for the measurement of poverty," *Economic Journal* 91, 515–26.

Collins, D., Morduch, J., Rutherford, S., and Ruthven, O. (2009) *Portfolios of the Poor: How the World's Poor Live on $2 a Day*. Princeton University Press, Princeton NJ and Oxford.

Cruces, G. and Wodon, Q. T. (2007) "Risk-Adjusted Poverty in Argentina: Measurement and Determinants," *Journal of Development Studies* 43: 1189–1214.

Duclos, J., Araar, A., and Giles, J. (2010) "Chronic and Transient Poverty: Measurement and Estimation, with Evidence from China," *Journal of Development Economics* 91: 266–277.

Duncan, G. J. and Rodgers, W. (1991) "Has Children's Poverty Become More Persistent?" *American Sociological Review* 56(August): 538–550.

Dutta, I. L., Roope, L., and Zank, H. (2011) "On Intertemporal Poverty: Affluence-Dependent Measures," School of Economics Discussion Paper Series 1112, Economics, University of Manchester.

Foster, J. E. (1998) "Absolute vs. Relative Poverty," *American Economic Review* 88: 335–341.

Foster, J. E. (2006) "Poverty Indices," in de Janvry, A. and R. Kanbur (eds.), *Poverty, Inequality and Development: Essays in Honor to Erik Thorbecke*: 41–66. New York: Springer Science.

Foster, J. E. (2007) "A Class of Chronic Poverty Measures," Working Paper No. 07-W01, Department of Economics, Vanderbilt University.

Foster, J. E. (2009) "A Class of Chronic Poverty Measures," in Addison, T., D. Hulme, and R. Kanbur (eds.), *Poverty Dynamics: Interdisciplinary Perspectives*: 59–76. Oxford University Press, Oxford.

Foster, J. E. and Jin, Y. (1998) "Poverty Orderings for the Dalton Utility-Gap Measures," in Jenkins, S. P., A. Kapteyn, and B. M. S. van Praag (eds.), *The Distribution of Welfare and Households Production*: 268–285. Cambridge University Press, Cambridge.

Foster, J. E. and Shneyerov, A. (1999) "A General Class of Additively Decomposable Inequality Measures," *Economic Theory* 14: 89–111.

Foster, J. E. and Shneyerov, A. (2000) "Path Independent Inequality Measures," *Journal of Economic Theory* 91: 199–202.

Foster, J. E. and Shorrocks, A. (1991) "Subgroup Consistent Poverty Indices," *Econometrica* 59: 687–709.

Foster, J. E. and Szekely, M. (2008) "Is Economic Growth Good for the Poor? Tracking Low Incomes Using General Means," *International Economic Review* 49: 1143–1172.

Foster, J. E., J. Greer and Thorbecke, E. (1984) "A Class of Decomposable Poverty Indices," *Econometrica* 52: 761–766.

Gradin, C., Del Rio, C., and Cantó, O. (2012) "Measuring Poverty Accounting for Time," *Review of Income and Wealth* 58: 330–354.

Guenther, I. and Maier, J. (2008) "Poverty, Vulnerability and Loss Aversion," General Conference of the International Association for Research in Income and Wealth, Slovenia, August 24–30.

Hoy, M. and Zheng, B. (2011) "Measuring Lifetime Poverty," *Journal of Economic Theory* 146: 2544–2562.

Hulme, D. and Shepherd, A. (2003) "Conceptualizing Chronic Poverty," *World Development* 31: 403–423.

Instituto Nacional de Estadísticas y Censos (INDEC) (2003) "Incidencia de la Pobreza y la Indigencia en los aglomerados urbanos," Información de Prensa, Mayo.

Jalan, J. and Ravallion, M. (1998) "Transient Poverty in Postreform Rural China," *Journal of Comparative Economics* 26: 338–357.

Jalan, J. and Ravallion, M. (2000) "Is Transient Poverty Different? Evidence for Rural China," *Journal of Development Studies* 36: 82–89.

Kolm, S.-C. (1977) "Multidimensional Egalitarisms," *Quarterly Journal of Economics* 91: 1–13.

Levy, F. (1977) "How Big is the American Underclass?" Working Paper 0090-1, Urban Institute, Washington DC.

Lillard, L. and Willis, R. (1978) "Dynamic Aspects of Earning Mobility," *Econometrica* 46: 985–1012.

Mendola, D. and Busetta, A. (2012) "The Importance of Consecutive Spells of Poverty: A Path Dependent Index of Longitudinal Poverty," *Review of Income and Wealth* 58: 355–374.

Morduch, J. (1999) "Between the State and the Market: Can Informal Insurance Patch the Safety Net?" *World Bank Research Observer* 14: 187–207.

Okun, A. M. (1975) *Equality and Efficiency: The Big Tradeoff*, Brookings Institution, Washington DC.

Porter, C. and Quinn, N. N. (2008) "Intertemporal Poverty Measurement: Tradeoffs and Policy Options," CSAE Working Paper 2008-21.

Ravallion, M. (1988) "Expected Poverty Under Risk-Induced Welfare Variability," *Economic Journal* 98: 1171–1182.

Rodgers, J. R. and Rodgers, J. L. (1993) "Chronic Poverty in the United States," *Journal of Human Resources* 28: 25–54.

Sen, A. K. (1976) "Poverty: An Ordinal Approach to Measurement," *Econometrica* 44: 219–231.

Watts, H. W. (1969) "An Economic Definition of Poverty," in Moynihan, D. P. (ed.), *On Understanding Poverty*: 316–329. Basic Books, New York.

Yaqub, S. (2000) *Poverty Dynamics in Developing Countries*. Institute of Development Studies, Development Bibliography, University of Sussex.

9 Measuring intertemporal poverty

Policy options for the poverty analyst

Catherine Porter and Natalie Naïri Quinn

9.1 Introduction

This chapter presents an analytical review of the recent technical literature on intertemporal poverty measurement (also known as chronic poverty measurement or lifetime poverty measurement) in which an individual measure of well-being (or the lack of) is aggregated both over time and across people. The chapter has a practical emphasis. Its main aims are to make the intertemporal poverty measures which have been introduced in the technical literature accessible to applied practitioners conducting poverty analysis and to encourage their appropriate application. Different measures have different properties which reflect alternative normative principles or judgements. We present intuitive motivations for and explanations of these properties and give a comprehensive review of the properties satisfied by different measures. We argue that there is no 'best' measure; different measures may be more suited to particular types of data or concepts of poverty.

A fairly clear consensus has emerged on desirable properties of 'snapshot' or cross-sectional, unidimensional poverty measures, led by work such as that of Sen (1976). Sen distinguished between the crucial steps of *identification* of the poor, and of *aggregation* of information about individuals' poverty across a society. Whilst a number of measures have been proposed in this cross-sectional literature, Foster, Greer and Thorbecke (1984) made what has probably been the most utilised contribution, with their class of decomposable poverty measures, known in the literature as the FGT measures or p-alpha measures. The measures are based on individuals' shortfalls of income or consumption x_i from an poverty line z, which may be set according to policy requirements, based for example on the cost of basic needs, or a relative proportion of median income in society. This shortfall may be weighted according to the policy need again, using a parameter α. The headcount measure, poverty gap and squared poverty gap are all special cases of the p-alpha family of measures.

In the past decade, economists have paid attention to the concepts of both multidimensionality of poverty, for example poverty in the dimension of income, but also health and access to education, and duration of poverty over time, which is the concern of this chapter. When the dimension of time is introduced to poverty measurement, a new set of issues emerges, some of which are deeply philosophically challenging. It is not straightforward to extend Sen's concepts of *identification* or *aggregation* to the intertemporal context. Time-relevant facets of poverty include duration or chronicity, systematic changes or shifts, variability or uncertainty and risk or vulnerability. Some of these have been explored in the literature while others are yet to be attempted. Several interesting measures have recently been suggested in the literature and whilst they have commonalities, there are also substantive differences in both identification and aggregation approaches used. This leads to a number of choices that an applied poverty analyst must make when undertaking empirical analysis of intertemporal poverty. The chapter outlines some advice in terms of choosing an appropriate measure in different contexts or for diverse policy purposes; indeed, it may be that more than one measure is needed to capture the salient issues in any one country.

The chapter proceeds as follows. In Section 9.2 we outline notation and discuss practical issues relating to the data that may be used for intertemporal poverty measurement. In Section 9.3 we conduct a systematic analysis of potentially desirable properties of intertemporal poverty measures. We start with a recap of the consensus on the properties of cross-sectional poverty measures and then discuss the alternative ways in which they may be extended to the intertemporal context. The literature has not reached a consensus on properties for intertemporal poverty measures and we refer to the findings of Porter and Quinn (2012), emphasising that some desirable properties are not compatible with one another, so the applied practitioner should make an informed choice of a measure which has properties appropriate for the context of application and the particular facets of poverty she wishes to evaluate. Section 9.4 surveys a variety of measures suggested in the recent literature, identifying which properties each has, to facilitate this informed choice of measure. Section 9.5 applies the various measures suggested to panel data from rural Ethiopia and discusses the contrasting evaluation of poverty that the different measures provide. Section 9.6 concludes.

9.2 Data and measurement framework

All of the intertemporal and chronic poverty measures suggested in the literature aggregate data representing the wellbeing of a group of individuals in a number of discrete time periods.[1] This is appropriate for practical application as it reflects the structure of the relevant data sources, individual (or household) panel or longitudinal surveys.

In practice, we will never have a direct measure of wellbeing. As the data are to be aggregated both over individuals and over time it is important that the indicator of wellbeing is comparable in these dimensions. In practice this may be achieved by adjusting for price changes over time and for relevant characteristics of the individual. If household survey data is used then a real, per-adult-equivalent value of income or consumption may be the best indicator. The practical and conceptual issues are discussed (in the context of cross-sectional poverty measurement but equally relevant here) in Ravallion (1994). It is important to note that a poverty analyst may wish to choose a measure with different intertemporal substitutability properties depending on the wellbeing indicator chosen; this issue is discussed further in the next section. Throughout this chapter we shall assume that the indicator is accurately measured and that there are no missing data. Appropriate methods to deal with measurement error and missing data are a subject of ongoing research; Calvo and Fernandez (2012) explore the issues that arise in the multidimensional context.

We may label the individuals $i=1, 2,\ldots, n$ and the time periods $t=1, 2,\ldots, T$. If the measure is based on a single indicator of wellbeing, for example income or consumption, the value of the indicator for person i in time period t may be represented by x_{it}. If the poverty line is z then $g_{it}=z-x_{it}$ is the *poverty gap*. The values of the indicator for all of the individuals in the group in all time periods may be represented by a matrix:

$$X = \begin{pmatrix} x_{11} & x_{12} & \cdots & x_{1T} \\ x_{21} & x_{22} & \cdots & x_{2T} \\ \vdots & \vdots & \ddots & \vdots \\ x_{n1} & x_{n2} & \cdots & x_{nT} \end{pmatrix} \tag{9.1}$$

which we shall describe as a *wellbeing profile* or just *profile*. The number of individuals is $n(X)$ and the number of time periods is $T(X)$. An *intertemporal poverty measure* is a real-valued function $\mathbb{P}(X)$ where X may be any profile of this form.

When conceptualising poverty over time, it is useful to think about the *sequence* of wellbeings experienced by a particular individual i, that is, the ith row of the data matrix $\mathbf{x}_i=(x_{i1}, x_{i2},\ldots, x_{iT})$. We shall refer to this as i's *trajectory* of wellbeings. A trajectory in which the individual's indicator of wellbeing takes the same value in every time period, $\mathbf{x}_i=(x_{i1}, x_{i1},\ldots, x_{i1})$ is referred to as a *constant-wellbeing trajectory*. A profile in which every individual's trajectory is a constant-wellbeing trajectory is a *profile of constant-wellbeing trajectories* (in general in such a profile different individuals may experience different wellbeing levels).

A profile with $T=1$ corresponds to cross-sectional measurement of poverty,

$$\mathbf{x} = \begin{pmatrix} x_{11} \\ x_{21} \\ \vdots \\ x_{n1} \end{pmatrix} = \begin{pmatrix} x_1 \\ x_2 \\ \vdots \\ x_n \end{pmatrix}, \tag{9.2}$$

and may be described as a *distribution*. A cross-sectional or static poverty measure is then a real-valued function $P(\mathbf{x})$ where $\mathbf{x}$ is any distribution of this form.

Any intertemporal poverty measure $\mathbb{P}$ represents an ordering[2] of alternative profiles.[3] A poverty analyst might be interested in comparing the degree of poverty in profiles representing different populations, or the same population under alternative policy choices. For such comparisons to be made, it is the *ordering* of profiles represented by the measure, rather than the numerical value of the measure, which is important. This ordering is preserved under any strictly increasing transformation of the measure, so measures which are strictly increasing transformations of one another may be regarded as equivalent.

There is a caveat to this principle; if the poverty analyst wishes to compare the amount of poverty in a profile as evaluated by *different* measures, it is necessary for those measures to have an equivalent normalisation. This may be achieved by specifying particular profiles which the poverty analyst wishes each of the different measures to evaluate as equivalent. In Section 9.3.2 we formalise this, for measures which evaluate profiles of constant-wellbeing trajectories identically. A further caveat is that, as noted by Gradin *et al.* (2011), in the intertemporal context it may only make sense to compare profiles of the same duration or number of time periods. Most papers in the literature assume a fixed number of time periods and the poverty analyst should be very cautious if applying any intertemporal measure to compare profiles of different duration.

9.3 Normative principles and properties of the measures

All poverty measures reflect normative principles or judgements about the concept of poverty and the distribution of the underlying indicators in the population being studied. These normative principles are embodied in the mathematical properties of the measure, so an applied practitioner should choose a measure whose properties reflect the normative judgements that she wishes to make. In this section we present and explain alternative properties for intertemporal poverty measures. The aim is to enable the applied practitioner to make a fully informed choice between measures with different properties.

In order to establish clearly the relevant concepts we start with a recap of the established consensus for properties of cross-sectional poverty measures; for identification the names of these properties are prefixed with S-. We then argue that some of these properties extend directly into the intertemporal case, with similar motivations. We observe that these more straightforward properties ensure that an intertemporal poverty measure must be constructed by aggregating the data over time, establishing an ordering of trajectories, before aggregating over individuals. This provides a structure to the intertemporal poverty measure and enables us to explore the less straightforward properties, distinguishing between properties relating to the ordering of trajectories and properties of the aggregation over the population. We highlight the conflicts that exist between certain desirable properties and emphasise that the applied practitioner should carefully consider properties that are appropriate for the context in which she is measuring poverty.

9.3.1 Recap: properties of cross-sectional poverty measures

Sen's (1976) seminal paper distinguishes between *identification* of the poor in a population and *aggregation*, constructing an index of poverty using the available information about those identified as poor. Sen followed the conventional approach to identification, selecting a 'poverty line' z and identifying as poor those people whose income is no more than z. Although he did not formalise it in this way, a simple formalisation invoked by later authors (see, for example, Foster and Shorrocks, 1991), of a property which Sen's measure satisfies and which is fundamental to the distinction between 'poor' and 'non-poor', is the following.

(1) S-Focus – *The poverty measure evaluates as equivalent any pair of distributions which differ only by an increase in wellbeing for a person* i *whose wellbeing* x_i *is greater than* z.

Another property which Sen's measure satisfies and which was specified formally by later authors is anonymity (also referred to as symmetry).

(2) S-Anonymity – *The poverty measure evaluates as equivalent distributions which differ only by a permutation of wellbeings among individuals.*

This property means that the only information which is taken into account is that captured by the wellbeing indicator; it is therefore important for the poverty analyst to choose an appropriate indicator which captures all information that she considers relevant. It also means that all individuals are treated equivalently or symmetrically; any poverty measure which satisfies this property will be a symmetric function of the wellbeings of the individuals in the population.

Sen introduced two properties that have been fundamental to the subsequent literature on poverty measures.

(3) S-Strict Monotonicity – *Given other things, a reduction in wellbeing for a person below the poverty line must increase the poverty measure.*

This property ensures that the measure of poverty is actually sensitive to the levels of wellbeing among the poor. Most measures suggested in the literature satisfy this property, although the frequently applied headcount measure which is not sensitive to depth of poverty satisfies just a weak version, S-Weak Monotonicity in which 'increase' is replaced by 'not decrease'.

(4) S-Strict Transfer – *Given other things, a pure transfer of wellbeing from a person below the poverty line to anyone with higher wellbeing must increase the poverty measure.*

This property ensures that the poverty measure is sensitive to inequality among the poor; given the average level of wellbeing among the poor, a less-equal distribution of wellbeings is evaluated as containing greater poverty. Many but not all of the measures suggested in the literature satisfy this property. The head-count measure and the average-poverty-gap measure both satisfy S-WEAK TRANSFER in which 'increase' is replaced by 'not decrease'.

Sen actually introduced stronger properties and characterised the unique measure that satisfies them. But as pointed out by Foster *et al.* (1984) his measure, and all others which depend on ranks, give a perverse evaluation of poverty when comparing certain distributions in which poverty has increased for a subset of the population but remains unchanged for the rest of the population. Foster *et al.* (1984) introduced a *subgroup monotonicity* property which (together with *subgroup consistency* introduced by Foster and Shorrocks (1991)) is closely related to the following, slightly stronger property. Its motivation exactly the same as that of *subgroup monotonicity* or *consistency*.

(5) S-SUBSET CONSISTENCY – *If the measure of poverty increases for a subset of the population (whose size may change) while the distribution of wellbeings remains unchanged for the rest of the population (whose size is fixed) then overall poverty must increase.*

A further property is needed to reflect the poverty analyst's approach to comparisons across different population sizes. The conventional *replication invariance* property under which poverty remains the same under a replication of the distribution (that is, adding a replica of each individual) is essentially equivalent to the following, which may be applied more straightforwardly to distributions of any size.

(6) S-POPULATION SIZE NEUTRALITY – *The poverty measure evaluates as equivalent any two distributions in which every individual has the same level of wellbeing* x, *regardless of population size.*

This property ensures that the measure has a 'per-capita' interpretation, which is conventional in poverty measurement. If poverty analyst desires that the measure takes account of population size, the analysis of Blackorby *et al.* (2005) should be consulted.

One property which is satisfied by many, though not all, poverty measures is continuity.

(7) S-CONTINUITY – *An infinitesimal change in any individual's wellbeing leads to no more than an infinitesimal change in the value of the poverty measure.*

Provided that the number of individuals is finite, S-CONTINUITY means that the poverty measure must be a continuous function of the individuals' wellbeings.

The p-alpha or FGT measures introduced by Foster *et al.* (1984) have the form

$$P_\alpha(\mathbf{x}) = \frac{1}{n(\mathbf{x})} \sum_i \left(\frac{z - x_i}{z} \right)^\alpha \mathbb{I}(x_i \leq z) \tag{9.3}$$

for $\alpha \geq 0$. These measures satisfy S-FOCUS, S-ANONYMITY, S-WEAK MONOTONICITY, S-WEAK TRANSFER, S-SUBSET CONSISTENCY and S-POPULATION SIZE NEUTRALITY. In addition they satisfy S-STRICT MONOTONICITY and S-CONTINUITY for $\alpha > 0$ and S-STRICT TRANSFER for $\alpha > 1$. The FGT measures are not the only measures to satisfy these properties; the measure suggested by Chakravarty (1983) also does, while Foster and Shorrocks (1991) fully characterise the class of measures satisfying all of the above properties except S-WEAK or STRICT TRANSFER. However, the FGT measures have become the most well-known and widely applied measures; P_2 in particular has become a standard for cross-sectional poverty measurement while P_0 or the headcount remains widely used in the policy literature.

In general (an extension of Foster and Shorrocks' 1991 results) any cross-sectional measure of the form

$$P(\mathbf{x}) = \frac{1}{n(\mathbf{x})} \sum_i g(x_i), \tag{9.4}$$

satisfies the properties S-ANONYMITY, S-SUBSET CONSISTENCY and S-POPULATION SIZE NEUTRALITY. It satisfies S-FOCUS if g is zero above z, S-WEAK MONOTONICITY if g is non-increasing below z and S-STRICT MONOTONICITY if g is strictly decreasing below z. It satisfies S-WEAK TRANSFER if g is non-concave below z and S-STRICT TRANSFER if g is strictly convex below z. It satisfies S-CONTINUITY provided g is a continuous function.

The properties and measures above reflect the consensus which has been reached in the literature on cross-sectional poverty measurement, in which, in our notation, $T=1$. Extending to the intertemporal context in which $T>1$ is not entirely straightforward. Some of the properties have direct analogues which may be similarly motivated. Others are more complex, and we must distinguish between properties which relate to the intertemporal aggregation and properties which relate to the aggregation over the population.

Intertemporal poverty measures are equivalent to cross-sectional measures when the number of time periods $T=1$ and when evaluating profiles of constant-wellbeing trajectories. As P_2 has become a standard for cross-sectional poverty measurement we shall normalise all of the measures discussed in this chapter to P_2 for constant-wellbeing trajectories, in order to facilitate comparisons of different measures. Any other cross-sectional measure with attractive properties could have been used.

9.3.2 Elementary properties for intertemporal measures

We start with the more straightforward properties which extend directly to an intertemporal framework. Anonymity simply requires permutation of trajectories rather than instantaneous wellbeings over individuals.

(8) ANONYMITY – *The poverty measure evaluates as equivalent profiles which differ only by a permutation of trajectories of wellbeings among individuals.*

As above, this means that individuals are treated equivalently and that all relevant information is contained in the wellbeing indicator. Similarly, subset consistency may be expressed in terms of trajectories rather than instantaneous wellbeings.

(9) SUBSET CONSISTENCY – *If the measure of poverty increases in a subset of the population (whose size may change) while the profile of wellbeing trajectories remains unchanged in the rest of the population (whose size is fixed) then overall poverty must increase.*

This is a basic consistency property which is necessary to avoid perverse evaluation, applying to poverty measurement in an intertemporal context just as much as in a cross-sectional context.

Porter and Quinn (2012, Theorem 1) show that the properties ANONYMITY and SUBSET CONSISTENCY are satisfied by precisely those intertemporal poverty measures that have the following form:

$$\mathbb{P}(X) = G_{n(X)}\left(\sum_{i=1}^{n(X)} f(p(x_{i1}, x_{i2}, \ldots, x_{iT}))\right) \tag{9.5}$$

where p is any function, f, and each G_n is a strictly increasing transformation.

Population size neutrality also extends naturally to the intertemporal context.

(10) POPULATION SIZE NEUTRALITY – *The poverty measure evaluates as equivalent any two profiles in which every individual has the same trajectory of wellbeings* $\mathbf{x}_1$, *regardless of population size.*

It is a straightforward extension of Porter and Quinn's result to show that adding POPULATION SIZE NEUTRALITY means that the poverty measure must have the form

$$\mathbb{P}(X) = \frac{1}{n(X)} \sum_{i=1}^{n(X)} f(p(x_{i1}, x_{i2}, \ldots, x_{iT})) \tag{9.6}$$

where p is any function and f is a strictly increasing transformation. It is clear from the mathematical form of this general measure that any such intertemporal poverty measure embodies an unambiguous ordering of the space of possible wellbeing trajectories, which we shall call the *trajectory ordering*. This ordering is represented by the function $p(x_{i1}, x_{i2}, \ldots, x_{iT})$ which we shall call a *trajectory measure*.

S-FOCUS, S-MONOTONICITY and S-TRANSFER do not extend quite so straightforwardly to the intertemporal case; for these properties the time dimension is pertinent and the concepts may be applied over time as well as across individuals. Of course, the motivations discussed above may or may not be relevant in the time dimension. A simple starting point is to restrict attention to profiles of constant-wellbeing trajectories so that we have a direct analogy with the cross-sectional case and need no further motivation for properties which reflect the aggregation over individuals. We may introduce constant-wellbeing equivalents of the properties which are entirely analogous to the cross-sectional properties:

(11) CW-FOCUS – *The poverty measure evaluates as equivalent profiles of constant-wellbeing trajectories which differ only by an increase in wellbeing for a person whose (constant) level of wellbeing* x_i *is greater than* z.

(12) CW-STRICT MONOTONICITY – *Given other things, a reduction in (constant) wellbeing for a person below the poverty line must increase the poverty measure.*

As in the cross-sectional case, CW-WEAK MONOTONICITY is similar, replacing 'increase' by 'not decrease'.

(13) CW-STRICT TRANSFER – *Given other things, a pure transfer of (constant) wellbeing from a person below the poverty line to anyone with higher wellbeing must increase the poverty measure.*

As in the cross-sectional case, CW-WEAK TRANSFER is similar, replacing 'increase' by 'not decrease'.

Provided each trajectory has a constant-wellbeing equivalent[4] the functional forms (9.6) and (9.4) may be combined to obtain a general form for an intertemporal poverty measure,

$$\mathbb{P}(X) = \frac{1}{n(X)} \sum_{i=1}^{n(X)} g(c_p(p(x_{i1}, x_{i2}, ..., x_{iT})) = P(\mathbf{c}_p(p(x_{i1}, x_{i2}, ..., x_{iT}))), \quad (9.7)$$

where P is any cross-sectional poverty measure, p represents a trajectory ordering and c_p gives the corresponding constant-wellbeing equivalent of a trajectory. c_p is defined such that for any constant-wellbeing trajectory $(x, x, \ldots, x)$, $c_p(p(x, x, \ldots, x) = x$ for $x \leq z$ and $c_p(p(x, x, \ldots, x) = z$ for $x > z$. All of the intertemporal poverty measures discussed in Section 9.4 may be expressed in this form.

We noted above that ANONYMITY, SUBSET CONSISTENCY and POPULATION SIZE NEUTRALITY follow immediately from the functional form of such a measure. The CW properties follow from the transformation function g in the same way as the analogous properties of cross-sectional measures discussed in Section 9.3.1. In particular if $P = P_2$ (9.3) then CW-FOCUS, CW-STRICT MONOTONICITY and CW-STRICT TRANSFER are satisfied. Applying $P = P_2$ as we shall do in the empirical application below gives the following form,

$$\mathbb{P}(X) = \frac{1}{n(X)} \sum_{i=1}^{n(X)} (1 - c_p(p(x_{i1}, x_{i2}, ..., x_{iT})) / z)^2. \quad (9.8)$$

Equation (9.7) provides a general functional form (and (9.8) a more specific form) for an intertemporal poverty measure which satisfies uncontroversial properties.

9.3.3 Properties of the trajectory ordering

It remains to discuss properties of the trajectory ordering, represented by the trajectory measure p. We distinguish below between *identification* of poor trajectories of wellbeing and *aggregation* of information about an individual's wellbeing over time. There is a clear analogy with identification and aggregation in cross-sectional poverty measurement but it is not quite direct, as we identify poor trajectories rather than poor time periods.[5] As the trajectory measure evaluates a single individual's trajectory of wellbeings we may drop the individual label i, so $\mathbf{x}$ represents a trajectory $\mathbf{x}=(x_1, x_2,\ldots, x_T)$. (The distinction between trajectory and distribution should be clear from the context.)

Identification of poor and non-poor trajectories is not straightforward for trajectories in which the wellbeing level is above the poverty line in some periods but below it in others. The issues that arise and the alternative approaches that have been applied are discussed in Section 9.3.4.

The concept of *chronicity*, that is, extended duration or persistence of poverty, has been emphasised in the policy debate; see, for example, the work of the Chronic Poverty Research Centre.[6] In fact, the early contributions to the literature on intertemporal poverty measures (Jalan and Ravallion, 2000; Calvo and Dercon, 2009; Foster, 2009) referred to their measures as 'chronic poverty' measures, even though not all of them were sensitive to duration or persistence per se.

It seems intuitive that longer total length, or greater proportion, of time spent in poverty should be reflected in a higher value of the trajectory measure. We may postulate that intrinsically, longer is simply worse when you are in a difficult situation. Further, it may be that spending longer in poverty reduces the chances of being able to climb out of poverty; assets are depleted (including body mass), morale is lower (Narayan-Parker *et al*., 2000), and it may be that the probability of exiting poverty is reduced, the longer is the 'spell' in poverty (Bane and Ellwood, 1986; Stevens 1999). This parallels the literature on unemployment, where state dependence has been observed due to signalling effects and loss of morale.[7] Where greater duration of poverty arises from lower wellbeing in some periods, the intertemporal poverty measure should evaluate the trajectory as more poor; the relevant properties are formalised in Section 9.3.5.

However, *depth* of deprivation experienced by an individual is also important: sustaining consumption just below the poverty line for quite a long time may arguably not be as difficult as experiencing a severe (albeit more short-lived) drop to extremely low levels of consumption. There is a tradeoff here; when greater duration of poverty arises from transfer of wellbeing between periods, the way in which this is evaluated by the trajectory measure may vary and the applied practitioner should choose a measure with properties which reflect her choice of the more important factor. Most analysts would of course agree that a person who is in deeper poverty for a longer time should be considered as worse off; the question is how to compare two trajectories which may have the same amount of total poverty over time, but spread across periods differently. This tradeoff and alternative properties are discussed further in Section 9.3.6.

Finally, in Section 9.3.7 we discuss the response of the trajectory ordering to sequencing of the wellbeing levels in a trajectory. It has been suggested that contiguous periods in poverty should weigh more heavily than periods of the same depth which are separated by periods above the poverty line, an interpretation of chronicity which reflects persistence of poverty rather than total length or proportion of time spent in poverty.

9.3.4 Identification of poor trajectories

In the cross-sectional, unidimensional context, since Sen (1976) it has been taken as axiomatic that poverty measures must not be sensitive to the wellbeings of those who are not poor. To identify the poor, a poverty line z is chosen according to the context (for procedures, see Ravallion, 1994) and any measure that satisfies the property S-Focus is not sensitive to changes in the wellbeing indicator for any individual whose level starts, and remains, above z.

In the intertemporal context, when multiple time periods are under consideration, it is not straightforward to extend this concept. We have assumed that the indicator of wellbeing is comparable over time so it is natural to suppose that the poverty line z does not vary over time. CW-Focus extends naturally to trajectories of varying wellbeing such that an individual whose wellbeing lies above z in every time-period must be identified as non-poor and an individual whose wellbeing lies below z in every period must be identified as poor. But what if an individual's wellbeing lies below z in some time periods, and above z in others? Different authors have taken different approaches to this identification question.

The *weak focus* property, which has been discussed in the literature and which is satisfied by all intertemporal poverty measures suggested, simply reflects the statement above that an individual whose wellbeing is above z in every period is non-poor; however, it does not specify who is poor. *Weak focus* has been contrasted with *strong focus*; that, however, does not apply to identification of the poor but rather sensitivity of the poverty measure to certain changes. This is more closely related to aggregation and we shall return to it below. A extension of CW-Focus which is more closely analogous to the concept of focus in the cross-sectional context is the following:

(14) General Focus – *The trajectory measure evaluates as equivalent trajectories which differ only by an increase in wellbeing* x_t *in any period* t *for a person whose trajectory of wellbeings* $\mathbf{x}$ *is identified as non-poor.*

The idea here is that once we can identify a non-poor trajectory (which may or may not have some poverty episodes in it), then the poverty measure should not be sensitive to changes in wellbeing for that person (so long as their trajectory remains non-poor as identified by the trajectory measure across the whole time period under consideration). GENERAL FOCUS is satisfied by all of the intertemporal poverty measures suggested in the literature. However, as indicated above, they differ in their approach to identification of trajectories as poor or non-poor. We shall describe here three approaches that have emerged in the literature. The first is to treat any person who has ever experienced poverty as being intertemporally poor.

(15) WEAK IDENTIFICATION – *An individual's trajectory of wellbeings* $\mathbf{x}$ *is identified as non-poor if and only if the wellbeings* x_t *are greater than* z *in every period.*

This is the approach taken by most authors who have adopted a 'spells' approach, for example Bossert *et al.* (2012) and Gradin *et al.* (2011). Note that this means every person who has ever experienced one period in poverty is included in the group of poor, and so does not distinguish in the identification between those poor once, and those who experience poverty in multiple periods. An approach that distinguishes between these two groups is to specify a duration cutoff, so that only those individuals who experience more than a certain proportion of time below the poverty line are considered intertemporally poor.

(16) DURATION-CUTOFF IDENTIFICATION – *An individual's trajectory of wellbeings* $\mathbf{x}$ *is identified as non-poor if and only if the wellbeings* x_t *are greater than* z *in a proportion of periods greater than or equal to some parameter* τ.

This is the approach introduced and applied by Foster (2009); the trajectories identified as poor are a subset of those identified as poor under WEAK IDENTIFICATION. As discussed below a measure with this property will typically have discontinuities at the boundary between poor and non-poor trajectories, such that an infinitesimal change in wellbeing corresponds to a finite change in the value of the poverty measure.

The third approach is to first aggregate the intertemporal trajectory of wellbeing into a representative level of wellbeing, and then focus on those whose aggregated wellbeing lies below the poverty line.

(17) ISOQUANT IDENTIFICATION – *The boundary between poor and non-poor trajectories is an isoquant of a function of* $\mathbf{x}$ *which is strictly decreasing in each period's wellbeing* x_t.

This is the approach taken by authors who have adopted a 'components' approach, for example Jalan and Ravallion (2000), Foster and Santos (Chapter 8, this volume) and Porter and Quinn (2012). Combining with CW-FOCUS, the boundary must pass through the trajectory which comprises wellbeings at the poverty line z in every period. As with DURATION-CUTOFF IDENTIFICATION the trajectories identified as poor are a subset of those identified as poor under WEAK IDENTIFICATION, but the 'shape' of the boundary may be quite different.

9.3.5 Aggregation: sensitivity to depth of poverty

We now explore how CW-STRICT MONOTONICITY may be extended to reflect sensitivity to depth of poverty for variable-wellbeing trajectories, focusing on response to changes in one period. We start with a weak monotonicity property to reflect the idea that increases in wellbeing in any time period should never increase the measure of poverty. This applies equally to poor and non-poor trajectories (although it is already implied by GENERAL FOCUS for non-poor trajectories).

(18) WEAK MONOTONICITY – *The trajectory measure does not decrease when, all other things equal, the individual's wellbeing* x_t *decreases in any one time period* t.[8]

This may be illustrated by comparing two trajectories **x** and **y** which comprise identical wellbeings in all time periods except one, as illustrated in Figure 9.1. If the only difference in their time profile is that **y** has lower wellbeing than **x** in a single time period, then **y** should be considered as (weakly) more poor than **x**.

As in the cross-sectional context, the poverty analyst will want the measure to be sensitive to depth of poverty. The measures proposed by different authors extend CW-STRICT MONOTONICITY to variable-wellbeing trajectories in different ways. One possibility is for the trajectory measure to be sensitive to changes in wellbeing in any period, for any poor trajectory.

(19) STRICT MONOTONICITY – *The trajectory measure increases whenever for an individual with a poor trajectory* **x**, *all other things equal, wellbeing* x_t *(in any one time period t) decreases.*

The level of wellbeing in the period which changes may or may not be below the poverty line *z*. Some authors have argued that changes in wellbeing in periods in which the wellbeing is above and remains above *z* should not have any effect on the poverty measure, as that this would imply that periods above the poverty line can compensate for periods below, which they reject. This restriction is captured by the following property, which is not compatible with STRICT MONOTONICITY.

(20) STRONG FOCUS – *The trajectory measure does not change whenever for an individual with a poor trajectory* **x**, *all other things equal, wellbeing* x_t *(in any time period* t *in which it is greater than* z*) decreases but remains above* z.

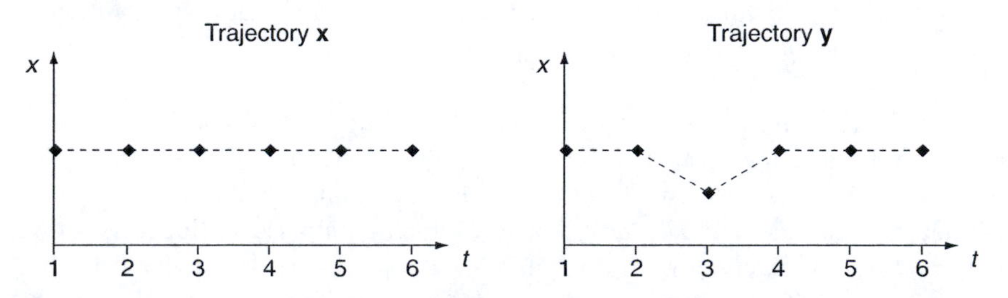

Figure 9.1 Weak monotonicity.

This does not preclude sensitivity to depth of poverty in periods in which the wellbeing is below the poverty line.

(21) RESTRICTED STRICT MONOTONICITY – *The trajectory measure increases whenever for an individual with a poor trajectory* **x**, *all other things equal, well-being* x_t *(in any one time period* t *in which it is less than* z*) decreases.*

Note that there is a certain amount of ambiguity regarding monotonicity properties in the current literature. For example, Foster (2009) and Mendola *et al.* (2011) specify a *time monotonicity* property, whereby increasing the number of periods 'in poverty' (in our framework, with $x_t<z$) increases the trajectory measure. This property is in fact entailed by RESTRICTED STRICT MONOTONICITY (and therefore by STRICT MONOTONICITY which itself entails RESTRICTED STRICT MONOTONICITY) and so does not need to be specified as a separate property. We should emphasise that RESTRICTED STRICT MONOTONICITY is sufficient to ensure that the poverty measure is sensitive to increases in the duration of poverty, or proportion of time spent in poverty, when that increase is due to loss of wellbeing in one or more periods. The alternative possibility in which an increase in duration due to *transfer* of wellbeing between periods is discussed below.

Having discussed properties which ensure sensitivity of the measure to changes in wellbeing, we conclude this section with a property that restricts the degree of sensitivity.

(22) CONTINUITY – *An infinitesimal change in wellbeing* x_t *in any period* t *leads to no more than an infinitesimal change in the value of the trajectory measure.*

A motivation for continuity is that marginal changes in wellbeing should have a marginal effect on the evaluation of poverty. If the trajectory ordering is not continuous then we may always find trajectories which are ordered in a perverse way. For empirical applications this is also extremely important: a discontinuous measure would be excessively sensitive to measurement error, at any point of discontinuity. Despite this, several measures suggested in the recent literature exhibit discontinuities, including Foster (2009), Bossert *et al.* (2012) and Gradin *et al.* (2011).

9.3.6 Aggregation: changes in multiple periods

A fundamental aspect of the trajectory measure is the way in which it orders trajectories with different levels of wellbeing in more than one time period, the extension of CW-TRANSFER. These properties may be expressed in terms of transfers of wellbeing from one period to another. It is valuable to consider whether a transfer of a particular amount of the indicator of wellbeing from one period to another[9] should increase or decrease the trajectory measure. Alternatively we may consider the marginal rate of compensation (MRC), that is, the increase in wellbeing in one period that is necessary to compensate for a decrease in wellbeing in another period.[10] It may also be valuable to consider the elasticity of compensation associated with the MRC.

Consider the trajectories **a** and **b** illustrated in Figure 9.2. Suppose that $a_L < b < z < a_H$ and that $(a_L + a_H)/2 = b$ so that **a** and **b** have the same average level of wellbeing.

If the trajectory measure reflects perfect intertemporal compensation then it will evaluate these two trajectories as equivalent. However, it may be argued that for trajectory **a**, the elevated wellbeings in the even periods are not considered sufficient to compensate for the depth of poverty in the odd periods. This is reflected in the following property.

(23) INTERTEMPORAL TRANSFER – *For any poor trajectory, a transfer of wellbeing from a more poor to a less poor period must increase the trajectory measure.*

For any trajectory measure which satisfies INTERTEMPORAL TRANSFER the MRC between a more poor and a less poor period must be less than one. The applied practitioner may wish to choose the actual degree of intertemporal compensation between periods to reflect the characteristics of the indicator of wellbeing. For example, with income data, we may be quite happy to allow a high degree of intertemporal substitution between two time periods for people who have some degree of access to borrowing and lending, even if in the informal sector. With consumption data we might expect that any smoothing has already taken place, and consider that if a person consumes below the poverty line in any given period, it is because she was unable to reallocate income from another period in order to avoid this. Accordingly the degree of intertemporal compensation reflected in the measure should be lower.

As mentioned briefly in Section 9.3.3, it may alternatively be argued that the trajectory measure should be sensitive to duration of poverty, even when greater duration of poverty arises from transfer of wellbeing between periods.

(24) DURATION SENSITIVITY – *For any poor trajectory, a transfer of wellbeing between periods such that the number of periods below z increases must increase the trajectory measure.*

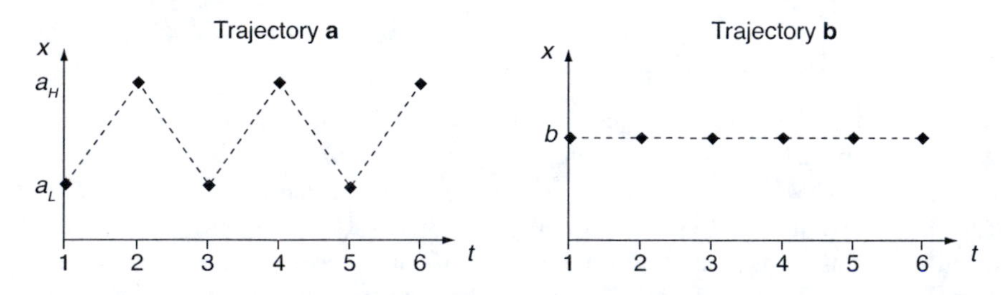

Figure 9.2 Intertemporal compensation.

This reflects the idea that that we should care more about people who spend a greater amount of time under the poverty line, whether or not that poverty is deep (being under the poverty line is arguably already bad enough); it does not, however, incorporate any judgement about the sequencing of periods under and above the poverty line. Porter and Quinn (2012) show that INTERTEMPORAL TRANSFER and DURATION SENSITIVITY are not compatible; a measure may satisfy one or the other but not both of these properties and the choice between them should be made explicitly.

Provided that the marginal rate of compensation between a more poor and less poor period is non-zero, a loss of wellbeing in one period may be compensated by greater wellbeing in another. Is this reasonable? To what extent should the measure permit this compensation? If deeper poverty is intrinsically very bad then we may wish to limit the extent to which it can be compensated for by higher wellbeings in other periods. We have suggested (Porter and Quinn, 2008, 2012) that when wellbeings in two periods are increased in proportion, the MRC between them should not decrease as this would perversely indicate that it is easier to compensate for a period of deeper poverty than a period of less deep poverty.

(25) NON-DECREASING COMPENSATION – *Given a poor trajectory, the marginal rate of compensation between the wellbeings in any two periods should not decrease, as the period wellbeings increase in proportion.*

We have shown (Porter and Quinn, 2012) that this property is not consistent with STRONG FOCUS; it is satisfied by the measures suggested by Jalan and Ravallion (2000), Foster and Santos (Chapter 8, this volume) and Porter and Quinn (2008). We have also strengthened this property (Porter and Quinn, 2008) to ensure that the MRC is lower for periods of extreme poverty compared to mild poverty.

(26) INCREASING COMPENSATION – *Given a poor trajectory, the marginal rate of compensation between the wellbeings in any two periods should increase, as the period wellbeings increase in proportion.*

We showed (Porter and Quinn, 2008) that a trajectory measure which satisfies INCREASING COMPENSATION must not be homothetic; the space of constant MRS trajectories must diverge as wellbeings increase. We suggested a family of measures which satisfy this property, one of which is illustrated in the next section.

9.3.7 Aggregation: sequencing

We finally consider an aspect of intertemporal poverty which may be thought of as one of the most critical issues for intertemporal poverty measurement, but which has not yet been fully explored in the literature; sensitivity of the trajectory measure to the order in which poverty spells take place. This is important if the measure is to capture *chronicity* or persistence of poverty.

Suppose an individual is observed as poor in five out of ten periods under consideration. Should we care when these take place? Is it worse to have all five at the beginning? At the end? In the middle? Furthermore, do we want to penalise periods that are close together (contiguous) as there is less recovery time in between? Or would movement in and out of poverty be equally distressing? There are no obvious answers in the literature; these are normative choices which should be made explicitly.

The simplest approach, reflected in a property satisfied by most of the intertemporal poverty measures suggested in the literature and which is analogous to ANONYMITY in the aggregation over individuals, is to disregard the sequencing entirely.

(27) TIME SYMMETRY – *The trajectory measure evaluates as equivalent trajectories comprising the same set of levels of wellbeing, regardless of their order.*

Calvo and Dercon (2009) break this symmetry by introduction of a discount factor β to weight more heavily more recent episodes of poverty.

Hoy and Zheng (2011), Bossert *et al.* (2012) and Gradin *et al.* (2011) all discuss the issue of sequencing, reflected in the properties they call *chronic poverty*, *single-spell additive decomposability* and *poverty spell duration sensitivity* respectively, whereby being poor two periods in a row is more than twice as bad as being poor in one period. The latter has the more general definition, reflected in the following property whose name has been chosen for brevity.

(28) CONTIGUITY – *When applied to two trajectories comprising the same set of wellbeings in a different sequence, the trajectory measure evaluates as more poor that sequence in which more of the wellbeings below* z *are contiguous.*

The measures proposed by Bossert *et al.* (2012) and Gradin *et al.* (2011) satisfy this property. However, this is achieved at the cost of substantial discontinuities in the trajectory measure that lead to perverse orderings of trajectories which an applied practitioner may wish to avoid. While we would expect CONTIGUITY to conflict with INTERTEMPORAL TRANSFER it is not obvious that it should not be consistent with CONTINUITY; it is a matter for further research to find a measure which satisfies both of these properties.

9.3.8 Concluding remarks

A consensus on desirable properties for an intertemporal poverty measure has not yet been established by researchers or policy makers. While further research remains to be done, conflicting properties may be argued to be desirable in different contexts, so it is likely that no consensus will ever be established. Rather, the applied practitioner should choose a measure with properties that are appropriate for the context. The discussion of properties above should help the practitioner to make an informed choice amongst them; in the subsequent section we outline the properties satisfied by several measures suggested in the recent literature.

9.4 Intertemporal poverty measures and their properties

We now review several poverty measures proposed in the recent literature that capture the dimension of time. We identify the normative properties that each possesses, some of which have been explicitly discussed by the authors, others of which have been left implicit.

Space precludes a fully comprehensive review, so we have chosen several measures which represent the variety of different functional forms and properties in the literature. These are the measures proposed by Jalan and Ravallion (2000), Foster (2009), Gradin *et al.* (2011), Foster and Santos (Chapter 8, this volume) and Porter and Quinn (2008). We note that while some of these measures were proposed as 'chronic poverty' measures, they are not sensitive to duration or persistence of poverty as such, and so capture other facets of intertemporal poverty than chronicity. They are still interesting and relevant for application but care should be taken regarding context of application and interpretation of results. To facilitate the informed choice of an appropriate measure, we discuss which measures may be appropriate for certain types of poverty analysis. This takes into account the choices between normative properties outlined in Section 9.3, but also includes some empirical issues, such as the availability of different welfare indicators (consumption, income and other dimensions of wellbeing), and how reliable these data can be considered.

9.4.1 The measures

We outline a number of intertemporal poverty measures suggested in the recent literature, and compare their properties. All of the measures in the literature are consistent with the general functional form (9.6) and so satisfy ANONYMITY, SUBSET CONSISTENCY and POPULATION SIZE NEUTRALITY, aggregating over time before over people. As our focus is on the intertemporal properties of the measures, we focus on the trajectory measures p and do not discuss the social aggregation here. As discussed in Section 9.3.2, we do not follow the social aggregation suggested by the authors but normalise all of the measures to P_2 (9.8) so that they all satisfy CW-FOCUS, CW-STRICT MONOTONICITY and CW-STRICT TRANSFER and direct comparisons may be made between evaluations of poverty using the different measures.

Several measures are direct attempts to extend the FGT measures (Foster *et al.*, 1984); we discuss these first.

Jalan and Ravallion's (2000) 'chronic poverty' measure p_{JRC} is the squared poverty gap (FGT-2)[11] applied to each individual's *mean* wellbeing. That is, we first take the simple average of the wellbeing measure over time, to create one observation per individual, and then apply the FGT formula as in the static case.

$$p_{\mathrm{JRC}}(\mathbf{x}_i) = \max\left[0, \frac{1}{T}\sum_{t=1}^{T}\left(1-\frac{x_{it}}{z}\right)\right]^2. \tag{9.9}$$

This measure satisfies ISOQUANT IDENTIFICATION, GENERAL FOCUS, STRICT MONOTONICITY, CONTINUITY, NON-DECREASING COMPENSATION and TIME SYMMETRY. The initial averaging step means that it represents perfect compensation of wellbeing between time periods, satisfying neither INTERTEMPORAL TRANSFER nor DURATION SENSITIVITY. Further, it does not satisfy STRONG FOCUS, INCREASING COMPENSATION or CONTIGUITY.

Foster's (2009) measure is also an extension of the FGT measure for the intertemporal case. As discussed above, Foster incorporates a 'poverty line' τ in the time dimension for the identification step, so that a household is only identified as poor if that household is below the wellbeing poverty line z in a proportion of periods greater than or equal to τ.[12] Calvo and Dercon (2009) compare the Foster and Jalan–Ravallion measures and include generalisations of both.

$$p_{\mathrm{F}}(\mathbf{x}_i) = \frac{1}{T}\sum_{t=1}^{T}\left(1-\frac{x_{it}}{z}\right)^{\alpha}\mathbb{I}(x_{it} \le z)\mathbb{I}\left(\sum_{t=1}^{T}\mathbb{I}(x_{it} \le z) \ge \tau T\right) \tag{9.10}$$

This measure satisfies DURATION-CUTOFF IDENTIFICATION, GENERAL FOCUS, WEAK MONOTONICITY, STRONG FOCUS, RESTRICTED STRICT MONOTONICITY (if $\alpha>0$) and TIME SYMMETRY but not STRICT MONOTONICITY, CONTINUITY, NON-DECREASING COMPENSATION or CONTIGUITY. Its discontinuities mean that it does not satisfy INTERTEMPORAL TRANSFER or DURATION SENSITIVITY although it does satisfy each of these for certain poor trajectories.

Gradin *et al.* (2011) take a similar approach to that of Foster, in that the measure is an intertemporal sum of FGT per-period poverty measures. However, they do not incorporate the duration cutoff for identification. This means that anyone with any poverty at all is included in the group of intertemporally poor. A feature of the measure (that was pioneered by Bossert *et al.* (2012) and generalised in this measure) is to penalise contiguous periods of poverty by weighting the original components of the summation formula so that longer periods in poverty are weighted more heavily.

$$p_{\mathrm{G}}(\mathbf{x}_i) = \frac{1}{T}\sum_{t=1}^{T}\left(1-\frac{x_{it}}{z}\right)^{\alpha} w_{it} \tag{9.11}$$

where

$$w_{it} = \left(\frac{s_{it}}{T}\right)^{\beta}. \tag{9.12}$$

So for example, a single period in poverty enters with a weighting of $(1/T)^{\beta}$; both periods in a two period spell would be weighted by $(2/T)^{\beta}$ as in equation (9.12) above. The Gradin *et al.* measure satisfies WEAK IDENTIFICATION,

General Focus, Weak Monotonicity, Strong Focus, Restricted Strict Monotonicity (if $\alpha > 0$) and Contiguity but not Strict Monotonicity, Continuity, Non-Decreasing Compensation or Time Symmetry. Its discontinuities mean that it does not satisfy Intertemporal Transfer or Duration Sensitivity although it does satisfy each of these for certain poor trajectories. The authors note that Foster (2009) is a special case of their measure if $\beta = 0$ and $\tau = 1/T$.

Foster and Santos (Chapter 8, this volume) propose a measure that generalises the approach of Jalan and Ravallion's chronic poverty measure, but relaxes perfect compensation and allows a choice parameter for the (constant) elasticity of compensation of wellbeing between time periods. To achieve this, the authors construct a measure which is based on the *generalised* mean of the wellbeing indicator (in contrast to the arithmetic mean of Jalan and Ravallion). It is a transformation of μ^i_β below.[13]

$$\mu^i_\beta(\mathbf{x}_i) = \begin{cases} \left[\dfrac{1}{T}\sum_{t=1}^{T}(x_{it})^\beta\right]^{1/\beta} & \text{for } \beta \neq 0, \\ \prod_{t=1}^{T}(x_{it})^{1/T} & \text{for } \beta = 0. \end{cases} \tag{9.13}$$

This measure satisfies Isoquant Identification, General Focus, Strict Monotonicity, Continuity, Intertemporal Transfer, Non-Decreasing Compensation and Time Symmetry. It does not satisfy Strong Focus, Duration Sensitivity, Increasing Compensation or Contiguity.

We have proposed (Porter and Quinn, 2008) a family of poverty measures with the property of increasing compensation, constructed by taking linear combinations of different CES functions. The identification step is again similar in spirit to Jalan and Ravallion and Foster and Santos, as we first aggregate wellbeing for each individual before identifying the poor trajectories. The general trajectory measure is:

$$p_{\mathrm{PQ}}(\mathbf{x}_i) = \max\left[0, \frac{1}{(k+1)T}\sum_{t=1}^{T}\left(\left(\frac{z}{x_{it}}\right)^k + \ln\left(\frac{z}{x_{it}}\right) - 1\right)\right] \tag{9.14}$$

for a parameter $k \geq 0$. The degree of intertemporal compensation decreases as the flexible parameter k increases. Intuitively, the measure allows almost perfect compensation of wellbeing between periods if they are both relatively rich periods. However, the MRC decreases if one or both periods is very poor, reflecting the idea that very low wellbeing cannot be easily compensated by higher wellbeing in another period (an intuitively appealing property in the context of welfare levels that drop close to starvation, as discussed earlier). Relating to other measures, the summand is effectively a linear combination of different constant-elasticity functions suggested by Foster and Santos. The first term in the summand permits little compensation between periods while the log term permits a moderate degree of compensation. The low-compensation term dominates for low wellbeings while the higher-compensation term dominates for higher wellbeings, hence the degree of compensation increases as the wellbeing levels increase; increasing k strengthens the resistance to compensation at low wellbeing levels. In the empirical section we apply $k=2$. In the limiting case when $k=0$, the measure has constant elasticity of compensation of one (wellbeings in two periods are neither complements nor substitutes) and is equivalent to the Foster–Santos measure above when $\beta=0$, which is the geometric mean.

This measure satisfies Isoquant Identification, General Focus, Strict Monotonicity, Continuity, Intertemporal Transfer, Increasing Compensation and Time Symmetry. It does not satisfy Strong Focus, Duration Sensitivity or Contiguity.

9.4.2 Comparison of the measures

The salient properties of the measures outlined above are summarised in Table 9.1.

Table 9.1 Properties of intertemporal poverty measure trajectory orderings

Trajectory measure	*Jalan–Ravallion*	*Foster*	*Gradin et al.*	*Foster–Santos*	*Porter–Quinn*
Identification properties					
Weak identification	✗	✗	✓	✗	✗
Duration-cutoff identification	✗	✓	✗	✗	✗
Isoquant Identification	✓	✗	✗	✓	✓
Depth sensitivity properties					
General focus	✓	✓	✓	✓	✓
Weak monotonicity	✓	✓	✓	✓	✓
Strict monotonicity	✓	✗	✗	✓	✓
Strong focus	✗	✓	✓	✗	✗
Restricted strict monotonicity	✓	✓	✓	✓	✓
Continuity	✓	✗	✗	✓	✓
Transfer properties					
Intertemporal transfer	✗	*	*	✓	✓
Duration sensitivity	✗	*	*	✗	✗
Non-decreasing compensation	✓	✗	✗	✓	✓
Increasing compensation	✗	✗	✗	✗	✓
Other temporal properties					
Time symmetry	✓	✓	✗	✓	✓
Contiguity	✗	✗	✓	✗	✗

Notes

The table shows properties of the trajectory ordering represented by the trajectory measures p as outlined in section 9.4. Formulae for the measures may also be found in this section. * denotes that the measure has the property for some but not all poor trajectories. If any of these trajectory measures is incorporated in an intertemporal poverty measure of the form (9.7) with g strictly convex below z, for example P_2, then the intertemporal poverty measure additionally has the properties Anonymity, Subset Consistency, Population Size Neutrality, CW-Focus, CW-Strict Monotonicity and CW-Strict Transfer.

All of the measures satisfy WEAK MONOTONICITY and GENERAL FOCUS. All also satisfy RESTRICTED STRICT MONOTONICITY (in the case of the FGT-extension measures provided $\alpha > 0$) and those which do not satisfy STRONG FOCUS also satisfy STRICT MONOTONICITY. Foster's (2009) measure and Gradin *et al.*'s (2011) measure (and Bossert *et al.*, 2012 which the Gradin measure generalises) do not satisfy CONTINUITY, in the Foster case due to the double focus step inherent to the measure and for the others due to the weighting that depends on contiguity of poverty. These latter measures satisfy CONTIGUITY which is of course in direct conflict with TIME SYMMETRY.

INTERTEMPORAL TRANSFER does not hold for Jalan and Ravallion's (2000) measure, nor for the measures that satisfy STRONG FOCUS if we are considering transfers in which one period's wellbeing crosses the poverty line. As outlined in the discussion above, any measure that satisfies STRONG FOCUS does not satisfy NON-DECREASING COMPENSATION, and we can see that Porter–Quinn is the only measure that satisfies INCREASING COMPENSATION. DURATION SENSITIVITY only holds for the measures that are based on headcount calculations and therefore also do not satisfy STRICT MONOTONICITY. Some of these issues are discussed as special cases in Calvo and Dercon (2009).

We can see then that the poverty analyst already has a number of options when choosing how to measure poverty over time. One property we have not yet discussed as it is not a formal property, was noted by Foster (2009) – the measure should be easy to understand. Applied policy makers may wish to choose a measure that is quite easy to explain to the general public, which may rule out some of the more complex formulae above (for example Porter–Quinn, Foster–Santos and Gradin *et al.*). On the other hand, if one is worried that the data are somewhat unreliable, it may be unwise to adopt a poverty measure that is highly sensitive to measurement error, which would be the case for measures that are discontinuous. As noted earlier on, we should also think about the type of data that are available to us – in the case of consumption, we may wish to assume that any intertemporal substitution has already taken place by households themselves, and so we should choose a measure that restricts intertemporal transfer. We do not take a strong line on any of these choices in this chapter, as we believe that there are many reasons to measure poverty, the important thing is to begin a discussion which makes clear the normative and practical decisions needed to do the best job given the data and constraints.

9.5 Empirical illustration

In this section we apply the various measures outlined above to a panel dataset from a developing country, to illustrate their similarities and differences. The Ethiopian Rural Household Survey (ERHS) contains data on just over 1100 households in 15 villages, observed at six points in time over a ten year period, 1994–2004. The timing of the rounds is not regular, with fieldwork in 1994, 1995, 1997, 1999 and 2004 (we use data from only one of the two rounds conducted in 1994). We use information about household consumption to compute the indicator of wellbeing. This information is from detailed diaries that the households were asked to keep in the two weeks prior to the survey, including all food that was home grown, bought at market or received as a gift or benefit from government. The poverty line is village-specific and represents the amount needed to consume 2000 calories per day plus some basic non-food items (such as firewood to cook). It is thus an extremely austere poverty line, around one-third of the commonly used 'dollar a day' line. In each round we also deflate value of consumption and the poverty line by a village specific food price index that was collected at the community level and construct a measure of consumption per adult equivalent. It is this deflated, equivalised measure that we use as the individual-period wellbeing indicator. This means that strictly we are evaluating poverty at the household rather than individual level; the data do not allow us to take account of issues of intra-household distribution. As the wellbeing indicator is based on consumption rather than income data, we argue that measures which allow a high degree of intertemporal substitution such as p_{JRC} will not be appropriate. For more details on this survey and the calculation of the by now quite widely used consumption basket, please see Dercon and Krishnan (1998).

Several authors have analysed wellbeing in the ERHS, including most recently Dercon and Porter (2011) and Dercon *et al.* (2012). Comparing cross sections, poverty has fallen in the study villages between 1994 and 2004, with the headcount reducing from just under 40 per cent to just over 20 per cent. Looking at households over time, there is a lot of movement in and out of poverty; only around one-third of households have not experienced poverty in any of the five rounds that we analyse. However, only 2.5 per cent of households had consumption below the poverty line in every visit over the ten year period. Hence we are faced with exactly the kind of problem that was outlined in theory above. Some households have longer periods in poverty, but do not fall very much below the poverty line, while others have fewer episodes of poverty but those are very severe.

We apply all of the measures reviewed in Section 9.4 to evaluate intertemporal poverty in the study villages, having made specific choices of the flexible parameters.[14] For the three extended FGT measures p_{JRC}, p_F and p_G, we take $\alpha=2$.[15] For p_G we take $\beta=1$, the special case in which poverty spells are weighted in direct proportion to their duration and which corresponds closely to the measure suggested by Bossert *et al.* (2012). For the time-dimension poverty line in p_F we take $\tau=0.6$ so that a trajectory is identified as poor if it includes at least three periods below the poverty line. For p_{FS} we take $\beta=0$ reflecting neither complementarity nor substitutability between wellbeings in different periods while for p_{PQ} we take $k=2$ reflecting a significant degree of complementarity for low levels of wellbeing. Each of these different trajectory measures we incorporate into an intertemporal measure of the form (9.8), reflecting a P_2 or 'poverty-gap-squared' approach to social aggregation. This means that the contrasting empirical results that we obtain with the various measures may be ascribed solely to differences in the intertemporal properties of the measures and not to differences in the aggregation over individuals.

Our main results are given in Table 9.2. The first column of results gives the 'headcount' for each measure, that is the proportion of households whose trajectory is identified as poor according to that measure. Each of these households contribute to the calculation of total poverty while those not identified as poor do not. The highest by a considerable degree (64.0 per cent) is Gradin *et al.*'s measure, as any household below the poverty line in any period is identified as poor, there is no further cutoff. The measures that allow less intertemporal substitution (Foster–Santos, Porter–Quinn) have higher headcounts (29.7 and 31.4 per cent respectively) than that of Jalan and Ravallion (13.4 per cent) which allows perfect substitution, so that periods of higher consumption may compensate for some periods in poverty. The headcount for Foster's measure is also rather low (15.4 per cent), reflecting the stringent identification criterion of poverty in at least three of the five periods. Of the measures evaluated, in identification this perhaps most closely captures the concept of *chronicity* of poverty.

The values of the poverty measures form the second column of results in Table 9.2. The values all appear rather low due to the 'poverty-gap-squared' approach to aggregation; all of the measures evaluate trajectories of constant wellbeing exactly as Foster *et al.*'s P_2 measure. Again, we see significant differences between the evaluation of poverty with the different measures, with the highest measure (Porter–Quinn) over five times as high as the lowest measure (Jalan–Ravallion). This may be ascribed to the substantially different approaches to intertemporal substitution embodied in these measures, as discussed above. Porter and Quinn's measure is sensitive to extremely low wellbeing (while Jalan and Ravallion's is not, unless it is sustained) but equally it will be more sensitive to measurement error. Interestingly, the ordering of the different measures is quite different from the ordering of their headcounts. For example, Gradin *et al.*'s measure, despite having the highest headcount, gives the second smallest evaluation of poverty. This is because it heavily discounts non-contiguous periods of poverty while many of the households in the study experience fluctuating levels of wellbeing, transitioning in and out of poverty. Finally, in the third column of Table 9.2 we note the standard deviation of individual households' contributions to the measures.

Table 9.2 Evaluation of poverty measures

Poverty measure	*Headcount*	*Measure*	*Std Dev.*
Jalan–Ravallion ($\alpha=2$)	0.134	0.010	0.044
Foster ($\alpha=2$)	0.154	0.034	0.081
Gradin *et al.* ($\alpha=2$, $\beta=1$)	0.640	0.026	0.058
Foster–Santos ($\beta=0$)	0.297	0.041	0.103
Porter–Quinn ($k=2$)	0.314	0.055	0.129

Notes
Measures defined in text; $n=1179$.

It is clear that there are implications of the choice of measure for anti-poverty policy targeting. Different measures identify different households as poor; there are some households that would not be included if interventions were directed by the measures with lower headcounts. Of course, the poverty analyst is likely also to take public finance constraints into consideration when making choices about anti-poverty policy targeting.

Table 9.3 shows the correlations of individual households' contributions to the different measures. We see that there is a high correlation between Foster's measure and Gradin *et al.*'s. They are constructed in a similar way, being extensions of the cross-sectional FGT measure P_2; although they treat fluctuating trajectories differently, for the chosen parameter values they are exactly equivalent for consistently poor trajectories. If we increased the duration cutoff or the contiguity parameter we would expect to see a lower degree of correlation. Foster–Santos and Porter–Quinn are very highly correlated, due to their similar intertemporal substitution properties around most of the distribution. The Porter–Quinn and Jalan–Ravallion measures have the lowest correlation. This is consistent with our discussion about intertemporal transfer and substitutability above.

Further work could explore these correlations further, examine sensitivities of each measure to changes in their flexible parameters, and profile the intertemporal poor to see differences in rankings, for example over regions or household types.

9.6 Conclusion

This chapter has covered an important yet fledgling topic in the economics literature: how to measure poverty in an intertemporal context. Our aim was to cover some conceptual and practical issues in order to clarify the assumptions and tradeoffs that need to be made when measuring poverty over time. While several measures have been suggested in the recent literature, no one measure has every desirable property that we identify, not least because some of these properties are in conflict. It is the applied poverty analyst who must choose carefully between measures, based on their desired properties, as well as the available data and policy environment.

Table 9.3 Cross-correlation table of poverty measures

Poverty measure	*Jalan–Ravallion*	*Foster*	*Gradin et al.*	*Foster–Santos*	*Porter–Quinn*
Jalan–Ravallion (α=2)	1.000	–	–	–	–
Foster (α=2)	0.716	1.000	–	–	–
Gradin *et al.* (α=2, β=1)	0.833	0.906	1.000	–	–
Foster–Santos (β=0)	0.687	0.914	0.884	1.000	
Porter–Quinn (k=2)	0.608	0.866	0.840	0.983	1.000

Notes
Measures defined in text; n=1179. All correlations are significant at 1%.

There remains further theoretical work to be done. For example, nobody has yet designed a measure that captures chronicity or duration sensitivity and that is also continuous. There is no inherent conflict between these properties and such a measure would be a useful contribution to the applied poverty analyst's toolkit.

Notes

1 Note that some authors, for example Bossert *et al*. (2012) in fact take as their starting point an individual-period indicator of *poverty*, without specifying how the indicator is defined. Such an indicator must always be computed from underlying observations of one or more indicators of the individual's wellbeing, and so fits into our framework with no loss of generality.
2 Formally, a complete, transitive binary relation.
3 Not every possible ordering may be represented by a real-valued measure; a classic counter-example is a lexicographic ordering (Debreu, 1954) which is discontinuous everywhere. However, it is highly unlikely that a poverty analyst would wish to order profiles in a way which is not representable by a real-valued measure. Porter and Quinn (2012) discuss the required restricted discontinuity property in more detail.
4 This is the case for all continuous monotone trajectory orderings and for many discontinuous orderings including all of those discussed in Section 9.4.
5 Hoy and Zheng (2011) apply it more directly, proposing measures of chronic poverty that are separable both over time and across people. This imposes strong restrictions on the properties of the measure that we feel go beyond what may be normatively motivated. Many of the intertemporal poverty measures proposed in the literature are not separable over time.
6 www.chronicpoverty.org/ contains an archive of ten years of research into chronic poverty.
7 See Heckman and Borjas (1980) and Jackman and Layard (1991) for further information.
8 Porter and Quinn (2012) observe that WEAK MONOTONICITY in conjunction with CW-FOCUS imply the *weak focus* property discussed above.
9 A mean preserving transfer between periods.
10 This is analogous to the MRS between commodities in consumer theory.
11 We take $\alpha=2$ for all of the extended FGT measures in this section, though for all of them the parameter remains flexible.
12 Foster also permits a flexible power parameter α; for comparability, we shall take $\alpha=2$; in our empirical section we take $\tau=0.6$.
13 See their chapter for more details on their measure.
14 A more comprehensive study would conduct sensitivity analysis on all of the flexible parameters.
15 $\alpha=2$ is already explicitly assumed by Jalan and Ravallion.

References

Bane, M. J., and D. T. Ellwood (1986): 'Slipping Into and Out of Poverty: The Dynamics of Spells', *Journal of Human Resources*, 21(1), 1–23.

Blackorby, C., W. Bossert and D. Donaldson (2005): *Population Issues in Social Choice Theory, Welfare Economics and Ethics*, Econometric Society Monographs. Cambridge University Press, Cambridge.

Bossert, W., S. Chakravarty and C. D'Ambrosio (2012): 'Poverty and Time', *Journal of Economic Inequality*, 10(2), 145–162.

Calvo, C., and S. Dercon (2009): 'Chronic Poverty and All That', in *Poverty Dynamics*, ed. by T. Addison, D. Hulme and R. Kanbur, pp. 29–58. Oxford University Press, Oxford.

Calvo, C., and F. Fernandez (2012): 'Measurement Errors and Multidimensional Poverty', Discussion paper, Oxford Poverty and Human Development Initiative.

Chakravarty, S. R. (1983): 'A New Index of Poverty', *Mathematical Social Sciences*, 6(3), 307–313.

Debreu, G. (1954): 'Representation of a Preference Ordering by a Numerical Function', in *Decision Processes*, ed. by R. M. Thrall, C. H. Coombs and R. L. Davis, pp. 159–165. John Wiley and Sons, New York.

Dercon, S., J. Hoddinott and T. Woldehanna (2012): 'Growth and Chronic Poverty: Evidence from Rural Communities in Ethiopia', *Journal of Development Studies*, 48(2), 238–253.

Dercon, S., and P. Krishnan (1998): 'Changes in Poverty in Rural Ethiopia 1989–1995: Measurement, Robustness Tests and Decomposition', Discussion Paper WPS/98-7, Centre for the Study of African Economies.

Dercon, S., and C. Porter (2011): 'A Poor Life? Chronic Poverty and Downward Mobility in Rural Ethiopia', in *Why Poverty Persists*, ed. by B. Baulch, pp. 65–96. Edward Elgar, Cheltenham.

Foster, J., J. Greer and E. Thorbecke (1984): 'A Class of Decomposable Poverty Measures', *Econometrica*, 52(3), 761–766.

Foster, J. E. (2009): 'A Class of Chronic Poverty Measures', in *Poverty Dynamics: Interdisciplinary Perspectives*, ed. by T. Addison, D. Hulme and R. Kanbur, pp. 59–76. Oxford University Press, Oxford.

Foster, J. E., and A. F. Shorrocks (1991): 'Subgroup Consistent Poverty Indices', *Econometrica*, 59(3), 687–709.

Gradin, C., C. Del Rio and O. Canto (2011): 'Measuring Poverty Accounting for Time', *Review of Income and Wealth*, 58(2), 330–354.

Heckman, J. J., and G. J. Borjas (1980): 'Does Unemployment Cause Future Unemployment? Definitions, Questions and Answers from a Continuous Time Model of Heterogeneity and State Dependence', *Economica*, 47(187), 247–283.

Hoy, M., and B. Zheng (2011): 'Measuring Lifetime Poverty', *Journal of Economic Theory*, 146(6), 2544–2562.

Jackman, R., and R. Layard (1991): 'Does Long-Term Unemployment Reduce a Person's Chance of a Job? A Time-Series Test', *Economica*, 58(229), 93–106.

Jalan, J., and M. Ravallion (2000): 'Is Transient Poverty Different? Evidence for Rural China', *Journal of Development Studies*, 36(6), 82.

Mendola, D., A. Busetta and A. M. Milito (2011): 'Combining the Intensity and Sequencing of the Poverty Experience: A Class of Longitudinal Poverty Indices', *Journal of the Royal Statistical Society: Series A (Statistics in Society)*, 174(4), 953–973.

Narayan-Parker, D., R. Patel, K. Schaff, A. Rademacher and S. Koch-Schulte (2000): *Voices of the Poor: Can Anyone Hear Us?* vol. 1 of *Voices of the Poor*. Oxford University Press for the World Bank, Oxford.

Porter, C., and N. N. Quinn (2008): 'Intertemporal Poverty Measurement: Tradeoffs and Policy Options', Centre for the Study of African Economies, Working Paper Series 2008-21.

Porter, C., and N. N. Quinn (2012): 'Normative Choices and Tradeoffs when Measuring Poverty over Time', Discussion Paper 56, OPHI Working Paper Series.

Ravallion, M. (1994): *Poverty Comparisons*. Harwood Academic Publishers, Switzerland.

Sen, A. K. (1976): 'Poverty: An Ordinal Approach to Measurement', *Econometrica*, 44(2), 219–231.

Stevens, A. H. (1999): 'Climbing Out of Poverty, Falling Back In: Measuring the Persistence of Poverty over Multiple Spells', *Journal of Human Resources*, 34(3), 557–588.

10 Measuring levels and trends in absolute poverty in the world

Open questions and possible alternatives

Stephan Klasen[1]

10.1 Introduction

At the Millennium Summit in 2000, the world community agreed on eight Millennium Development Goals (MDGs) to ensure poverty reduction and sustainable development for all. The first MDG calls for the eradication of extreme poverty and hunger. The targets agreed are somewhat more modest and refer to halving, between 1990 and 2015, the share of the population living on less than $1 a day, and halving the proportion of the population who suffer from hunger. For these two targets, three indicators were chosen. For the first target, the World Bank has been charged to regularly produce the relevant indicator, i.e. the share of the population that is living on less than a $1 a day. For the second target, there are two indicators. The first is to be monitored by UNICEF and WHO and refers to halving the share of children under five years of age who are underweight, and the second is to be monitored by the FAO and refers to halving the share of the population who are below minimum recommended levels of dietary intake.

While Klasen (2008) and de Haen *et al.* (2011) discuss measurement issues surrounding hunger, the focus here will be on the first target and indicator, the share of the population living below $1 a day. The focus on this issue is particularly pertinent, as the World Bank has published extensive revisions to this indicator (Chen and Ravallion 2010). In fact, the revision of these numbers has been based on the derivation of a new international poverty line against which to measure levels and trends in poverty, as well as a recalculation of poverty levels and trends for all countries going back to 1981. Thus both the baseline for the first MDG target in 1990 has changed, as well as the levels in each subsequent year, thereby also affecting the rate of progress towards this MDG. In this chapter, I will critically review the way the World Bank measures this extreme poverty indicator, present key facts and figures on trends in extreme poverty using data from before and after the revision, and then focus on a critical appraisal of the recent revisions. While there is little reason to believe that these revisions seriously distort *trends* in extreme poverty in the developing world, I will argue that the uncertainty about levels in *absolute* extreme poverty is very high and that the recent revisions have done little to reduce this uncertainty and might have indeed increased it. In fact, given the difficulties to measure poverty using an international poverty line, I will suggest instead that it might be preferable to abandon the efforts to construct such an international poverty line and focus instead on creating consistent and comparable national poverty lines using a common set of methods that could in turn be used to estimate levels and trends in absolute extreme poverty in the world. The chapter is organized as follows: in Section 10.2 I will discuss the $1 a day poverty indicator, its conceptual underpinning and its empirical derivation. The following Section 10.3 will then critically review the revisions that were introduced in 2005 and present key facts and figures that have been affected by this revision. In Section 10.4 I will discuss the implications of these criticisms. In the penultimate Section 10.5, I will discuss advantages and disadvantages of an alternative procedure that has been proposed in the literature which has some promise but also faces significant obstacles. The final Section 10.6 concludes.

10.2 The World Bank's international poverty measure

Measuring poverty consistently in a single country is clearly a challenging task. Among the questions to be asked are the domain in which poverty is to be measured, whether a poverty line separating the poor from the non-poor is invariant across space and time, whether one should consider just the incidence or also the depth of poverty, and whether poverty should be measured at the individual or household level.[2] These are all complex questions that merit detailed discussions as well as high-quality comparable household survey data. For a poverty indicator that attempts to measure levels and trends in poverty in a comparable manner across all developing countries, matters are even more complicated as the inter-country comparability of poverty lines as well as the underlying data will be critical issues. Data availability and comparability issues will necessarily involve simplifications and short-cuts. In fact, until 1990 it was not possible to generate such comparative poverty figures as the coverage of household surveys in developing countries was simply too sparse. In the 1990 World Development Report (World Bank 1990) the World Bank made a first attempt to measure poverty in a comparable way using an international poverty line and measuring poverty for the year 1985. This was based on an international poverty line of the purchasing power equivalent of $1 per capita in 1985 prices. There have been two major updates of this poverty line, once in the World Development Report 2000/01, where the poverty line was shifted to $1.08 in 1993 prices and recently again in Ravallion *et al.* (2009), when it was shifted again to $1.25 in 2005 prices. The one proposed in 2000 became the basis for the first target of MDG1 at the Millennium Summit. The methods for establishing the poverty lines have largely remained the same in these three versions which will be described below.

Before turning to this point, it is useful to point out a number of implicit choices and simplifications that are inherent in this approach to the measurement of poverty. First, the focus is entirely on the income dimension of poverty. Whether such income poverty is correlated with other forms of deprivation or a multidimensional view of poverty consistent with, for example, Sen's capability approach is not considered here (e.g. Sen 1985; Klasen 2000). While this is clearly a narrow view, it is defensible in the context of the MDGs where other forms of deprivation are captured in the other MDGs as well as the hunger target of MDG1.

Second, the international poverty line is invariant in space and time[3] and thus constitutes an *absolute* poverty line that tries to capture the share of people who are in extreme poverty where basic physical survival and health is at risk. Interestingly, recently Ravallion and Chen (2011) proposed a *weakly relative* version of international poverty where, after a certain level of average incomes, the poverty line rises (underproportionately) with mean incomes. Thus this is an issue that can be (and has been) addressed, but currently the focus remains on the absolute version of the poverty line.

Third, poverty depth is not considered in the target for MDG1. Considering the depth of poverty would indeed be preferable, but somewhat harder to communicate and also makes greater demands on the precision of the data.[4]

Fourth, the figures are per capita figures and do not account for differences in household size and composition which is likely to affect the needs of households as well as their ability to economize on resources. This will have the consequence that poverty in regions with large households and many children (such as many countries in Sub-Saharan Africa) is overstated relative to regions where household sizes and the number of children are small (such as China or South-East Asia).[5]

Last, inequality in intrahousehold resource allocation is not considered in these measures. If the per capita income of a household falls below the poverty line, everyone in the household is considered poor, even though some members might have better resource access than others. As a result, this approach is ill suited to examine the differential in poverty by gender or age group and it might indeed affect accurate poverty measurement (see Haddad and Kanbur 1990; Klasen 2007).[6]

While these are all shortcomings of this approach, some of these choices appear defensible in the context of the MDGs where there was a need for a straightforward comparable poverty indicator that would particularly capture levels and trends in extreme income poverty.

Bearing these methodological choices in mind, the big remaining questions are how this international poverty line is actually derived and how it is then used to measure poverty in each developing country so that poverty levels and trends can then be aggregated and compared. This is described in detail in Ravallion *et al.* (2009) and will be summarized here. Let me first turn to the construction of the international poverty line. In all three versions presented by the World Bank (1990, 2000 and 2008), the starting point was always the national poverty lines of a large sample of developing countries, expressed in their national currencies. In order to render them comparable, the results of the so-called International Price Comparison Project (ICP) were used to turn these national poverty lines into international prices (expressed in international $). The ICP rounds, which take place every three to ten years, compare prices of a large basket of goods and services in many different countries to generate exchange rates that appropriately reflect purchasing power differences between countries (see below and Ward (2009)). These so-called purchasing power parity (PPP) exchange rates are used for the translation of national poverty lines into international $ in the hope that this approach will adequately reflect purchasing power differences and thus make these poverty lines comparable. For the 1990 exercise the 1985 ICP was used; for the 2000 revision, the 1993 ICP was used, and the latest poverty estimates are based on the 2005 ICP.

In a second step the poverty lines are plotted against (the log of) per capita incomes and it is regularly found that among low income countries, the poverty lines, turned into international $, are very similar. The average of these poverty lines then is used as the international poverty line, which turned out to be $1.02 in 1985 prices in the 1990 *World Development Report*, $1.08 in 1993 prices in the 2000/01 *World Development Report*, and $1.25 in 2005 prices in the recent revision.

To measure poverty using these poverty lines, the following three steps are then undertaken. First, the international poverty line is turned into a poverty line in national currencies at the benchmark year using the PPP exchange rates from the particular ICP round (1985, 1993 and 2005, respectively). Second, this poverty line is adjusted using national inflation rates to generate poverty lines in national currencies backwards and forwards in time for all years since 1990 (or even since 1981). Third, the share of the population living below this poverty line is then determined using national household income or expenditure surveys.

It is important to emphasize that in each three rounds of calculation (1990, 2000 and 2008), poverty rates were recalculated not only for the most recent years, but for *all* years since the beginning of measurement of poverty at the global level (where the first data point is 1981). Thus we have three sets of poverty estimates for 1981, one based on the 1985 ICP round published in 1990, one for the 1993 ICP round published in 2000 and another one based on the 2005 ICP round published in 2008. The resulting numbers for the same year are in some cases dramatically different and it is not obvious to say which estimate is the most accurate one, an issue that will be discussed in more detail below.

10.3 Poverty levels and trends using the 2005 ICP round

In late 2007, the World Bank published the new PPP exchange rates from the 2005 ICP survey (World Bank 2007). Not only do they represent a more up-to-date set of price comparisons, but this round of the ICP was more comprehensive than all previous rounds, and particularly included China for the first time. Subsequently, it then published new poverty estimates for developing countries going back to 1981 based on these figures. Table 10.1 compares the share of the population below $1.08 a day in 1993 prices by region and from 1990 to 2005 with the new figures which are based on the $1.25 poverty line in 2005 prices. As can be readily seen from the table, there are dramatic differences in levels of poverty in many regions of the developing world using these two methodologies. Changes are particularly extreme in East Asia and the Pacific, but also quite large in Sub-Saharan Africa, the Middle East and South Asia. In China, using the 2005 ICP suggests that 60 per cent of the population lived in extreme poverty 1990, compared to 'only' 33 per cent using the previous ICP. This largely drives the results for all of East Asia. In India and in Sub-Saharan Africa, extreme poverty is now believed to have been much higher in 1990 than was previously thought. This, of course, has direct implications for the first target of MDG1 where the halving of poverty uses 1990 as a baseline. When the goal was agreed in 2000, based on the 1993 ICP, it implied that the share of the extremely poor should fall from about 29 per cent of the population in developing countries in 1990 to half that, i.e. 14.5 per cent in 2015. Using the 2005 ICP for this goal implies that now extreme poverty, expressed in 2005 international prices, needs to fall from nearly 42 per cent in 1990 to 21 per cent in 2015. Thus halving poverty is now based on a much larger share (and correspondingly, number) of poor people.

But also the most recent observations show large discrepancies. Poverty in 2005 is believed to be about ten percentage points higher in South Asia and Sub-Saharan Africa using the 2005 ICP than the 1993 ICP suggested. In all of the developing world, the 2005 ICP suggests that some 26 per cent of the population suffered from extreme poverty in 2005, compared to only 19 per cent when the 1993 ICP is used.

This not only affects poverty rates, but correspondingly substantially changes the absolute number of people who live in extreme poverty. While the 1993 ICP round suggested that 1.5 billion people lived in extreme poverty in 1990 and this figure dropped to 930 million in 2005; the 2005 ICP implies that the number of extremely poor was 1.9 billion in 1990 and about 1.4 billion in 2005 (Chen and Ravallion 2010).

In contrast, the differences in poverty *trends* since 1990 are remarkably small. The recent revision does not change the direction of poverty trends in any region or for the developing world as a whole; poverty rates globally are about 37–38 per cent lower in 2005 than in 1990, regardless of the data used. Also the size of poverty reduction in the different regions changes remarkably little. Poverty reduction in China, and East Asia as a whole remains very rapid, poverty reduction in South Asia also remains sizable and it continues to be the case that poverty reduction in Latin America, Eastern Europe and Central Asia, and Sub-Saharan Africa was very small, and mostly concentrated in the period after 2000. The new figures thus suggest little change to the assessment of overall progress in the first target of MDG: the developing world as a whole seems to be on track to halving extreme poverty. Also, both sets of estimates suggest that this development is largely driven by over-achievement in poverty reduction in East Asia, good progress in South Asia, less progress in the Middle East and Latin America, and hardly any progress in Sub-Saharan Africa which is highly unlikely to reach the goal.

It is not surprising that the revisions only have a large impact on poverty levels, but not on trends in poverty reduction. All that has changed for poverty

Table 10.1 Share of population suffering from extreme poverty, 1993 and 2005 ICP rounds

Region	*1990*		*1996*		*2002*		*2005*	
Poverty line	1.25 (2005)	1.08 (1993)	1.25 (2005)	1.08 (1993)	1.25 (2005)	1.08 (1993)	1.25 (2005)	1.08 (1993)
East Asia and Pacific	56.0	29.8	37.1	16.1	29.6	12.3	17.9	9.1
Of which China	60.2	33.0	36.4	17.4	28.4	13.8	15.9	9.9
Eastern Europe and Central Asia	1.5	0.5	4.5	4.4	5.6	1.3	5.0	0.9
Latin America and Caribbean	10.7	10.2	11.5	8.9	10.1	9.1	8.2	8.6
Middle East and North Africa	5.4	2.3	5.3	1.7	4.7	1.7	4.6	1.5
South Asia	51.1	43.0	46.9	36.1	43.8	33.6	40.3	30.8
Of which India	51.3	44.3	46.6	39.9	43.9	36.0	41.6	34.3
Sub-Saharan Africa	54.9	46.7	57.5	47.7	52.7	42.6	50.4	41.1
Total	41.6	28.7	34.8	22.7	31.0	20.1	25.7	18.1
Total excluding China	35.2	27.1	34.2	24.5	31.9	22.2	28.7	20.7

Sources: Chen and Ravallion (2010), Chen and Ravallion (2007).

Note

The first column refers to the share of the population below the new poverty line of $1.25 a day, the second one to the old poverty line. The differences in the figures are, to a very small degree, also due to changes in survey data; also note that the $1.08 figures in the 2005 column refer to 2004.

trends at the national level is that the poverty line has been changed at the benchmark year and then the same national inflation rates have been used to adjust the poverty line for each year between 1990 and now. Thus there has been a consistent shift of the poverty line across all years (either upwards, as in the case of most developing countries, or downwards, as in the case of very few developing countries). Trends in poverty using a consistent poverty line are mostly driven by changes in average incomes and the income distribution in national currencies rather than the location of the poverty line, and in this respect nothing has changed.[7]

Despite little change in poverty trends, the big question remains what to make of these large revisions in poverty levels in many developing countries. Similarly, the general approach of periodically revising these international poverty lines to generate a whole new time series of poverty estimates should be scrutinized further. This is taken up in the next section.

10.4 Problems and open questions associated with the recent poverty revisions

There are two different sets of questions associated with this approach to measuring extreme poverty in the developing world. The first set asks whether conceptually the current approach to measuring and periodically revising poverty in the developing world using this procedure is a promising way to accurately reflect levels and trends in absolute poverty. This is a subject on which there has been some debate in the past (e.g. Reddy and Pogge 2009; Ravallion 2009); the focus here will not be to repeat this debate but reflect on the conceptual issues in light of the recent drastic revisions. The second set of questions involves potential empirical problems with the two ICP rounds in 1993 and 2005 that differ so much in their conclusions on relative price levels, and thus on poverty levels. This is discussed in more detail by Ward (2009), but I want to weigh in on some of these issues. I will deal with these two issues in turn.

On the conceptual side, I want to address three difficult issues. The first is whether the PPP exchange rates generated by the ICP process are appropriate for the comparisons of the purchasing power of the poor. This is an issue central to the critique of Reddy and Pogge (2009) and, in principle, it would indeed be preferable to only consider those items in the construction of PPP exchange rates that are of relevance to the poor. Deaton and Dupriez (2008) actually constructed consumption baskets close to the poverty line and use those baskets to compare prices. This is conceptually superior, but data-intensive and only possible for a small set of countries. According to results presented in Ravallion *et al.* (2009) and Deaton (2010), the empirical differences between these two sets of results in terms of setting a poverty line are rather small. This will not necessarily mean that this issue has no importance, however, as there is a second related point one must bear in mind. The overall PPPs that are used for the translation of the international poverty line into national currencies can be significantly affected by goods and services not heavily consumed by the poor and this can in turn affect levels of poverty. For example, Heston (2008) argues that in the 2005 ICP the biggest reason for the drastic downward revision of PPP adjusted per capita income (and thus the large upward adjustment of poverty levels) was due to changes in the assumptions about the productivity of government services. While this is an important issue to consider when comparing GDP levels, it is less clear that it is a relevant factor to consider when assessing poverty as the poor are arguably less affected by this change (see Reddy 2008). Similarly, Ward (2009) argues that the ICP 2005 finding that prices in China were much higher than previously thought is based on baskets of goods that are internationally comparable, but largely out of reach of the poor so that also here the question is whether this change should really impinge on poverty measurement. The choices made in the 2005 ICP round to deal with these two difficult conceptual issues suggest that income levels are underestimated and poverty is overestimated, an issue to which I return below.

The second, equally difficult question relates to the intertemporal use of a single cross-country price comparison. As explained above, the poverty estimates from 1981 to 2005 are now based on the 2005 PPP exchange rates which are now used, in combination with national inflation rates, to derive poverty lines for each year. As has been pointed out by many scholars working on these price comparisons, the ICP rounds only provide valid comparisons of purchasing power adjusted GDPs at one point in time, the benchmark year. Thus the use of the 2005 price comparisons to compare poverty levels in 2000, 1995, let alone 1981, is not really an appropriate use of these comparisons. The reason is that the 2005 comparison is based on the global production and demand of goods available in 2005 and global demand and availability of goods changes over time which would affect price comparisons. While it appears rather innocuous to use the 2005 ICP to assess poverty levels in 2003–2006, it seems quite problematic to use these prices for comparisons of poverty levels in 1990, let alone 1981 when the world economy, the structure of world demand and the availability of goods was quite different. This problem will keep getting worse. The next ICP round is planned for 2011 and it seems highly dubious to base comparisons of poverty levels in 1990 or earlier on the assessment of global prices in 2011.

The obvious alternative to this approach would be to use all of the benchmark comparisons years (i.e. 1985, 1993, 1996 and 2005) in an assessment of purchasing power parity exchange rates for different time periods. This is the approach that so far has been taken by the various versions of the Penn World Tables provided by Summers, Heston, and Aten which provide PPP-adjusted GDP per capita levels.[8] There is a vigorous debate (in the context of the Penn World Tables, based on the different ICP rounds) on whether this approach is appropriate as this approach uses information from national accounts evaluated at national prices and mixes them with the international prices at the benchmark years (see Johnson *et al*. 2009).[9]

The very significant problem with applying this approach to poverty measurement is that it will lead to strong deviations of trends in poverty from what we know from national statistics.[10] The extreme case would be China again. If we believed the 1993 ICP and the 2005 ICP to be true reflections of Chinese prices relative to world prices, then poverty would have fallen in China between 1993 and 2005 by only ten percentage points (from 28 to 16 per cent), and from 1996 to 2005 poverty would not have hardly been reduced at all (falling only from 18 to 16 per cent, see Table 10.1) which is entirely inconsistent with the massive growth and considerable poverty reduction using national data which show much faster poverty reduction. A second problem with such an approach would be that the earlier ICP rounds were much more incomplete, omitting large countries such as China in all previous rounds and India in several rounds, so that it is unclear what estimates to use for these countries. A third problem is that there are reasons to believe that the earlier rounds were of lower quality than previous ones so that one might reasonably want to distrust some of their results (see Heston 2008; Ravallion *et al.* 2009).

So one is faced with a rather unpalatable choice between either relying on a single cross-country price comparison in 2005 and pretend it tells us a lot about how relative prices were 15 or even 25 years ago across the world or use the different price comparisons and have to contend with the inconsistencies, the coverage and the differential quality.

The third conceptual problem deals with updating the international poverty line itself. The important point here is that the update not only included a new set of PPPs that affected the poverty line but also a new and much expanded country sample that was used to estimate the new international poverty line. Indeed, as has been argued by Deaton (2010) and shown by Greb *et al.* (2011), the change in the country sample is mostly responsible for the drastic upward revision of absolute levels of poverty in the world.[11] While again there are some good arguments to use a larger sample of countries to estimate the poverty line, this has introduced another change which is essentially arbitrary and could be changed again as soon as more data points on national poverty lines become available.

These conceptual problems appear to be really quite serious and they will get worse over time. Imagine that we are in 2015 where there has yet again been another ICP round and we will first have to generate a new international poverty line yet again (with another expanded sample of countries), then produce an entirely new time series of poverty rates going back all the way to 1990 to determine what the new levels of poverty are and whether indeed we have met the goal.

Apart from these conceptual problems, there are also some measurement problems to deal with in this particular ICP round and the biases it might have produced. There are reasons to believe that in general the quality of the ICP is higher than in previous rounds. The country coverage was wider, there were more consistency checks, and the data gathering and analysis was more thorough (see, e.g. Heston 2008; Ravallion *et al.* 2009). At the same time, two particular empirical problems arose. The first relates to the choice of goods whose prices are being compared (Ward 2008; Heston 2008). There is the inherent tension between international comparability and national representativeness of goods.

Ideally the goods whose prices are to be compared should be highly comparable internationally, but also representative of purchasing habits of a population. In rich countries this is likely to be the case (e.g. one can compare the prices of identical shirts of the same popular brand that are both available and widely used). When we compare prices between a rich and poor country, the tension will arise. We will probably be able to find identical goods in both countries but in the poor country the good is likely to be consumed by a rich and import-oriented elite and thus not representative of the population (e.g. an international brand-name shirt will be bought only by a select group in China). One could alternatively replace the good in a poor country with a cheaper, domestically produced substitute which would be more representative of the population, but inherently less comparable to the good in the rich country. There is no easy way to resolve this tension and choose the 'correct' answer. The claim by Ward (2009) is that in the 2005 ICP one tended to emphasize international comparability over national representativeness, and this led to some high reported prices in countries such as China, India or parts of Sub-Saharan Africa, and thus the associated higher poverty levels. This would essentially mean that poverty is overstated in these countries in the 2005 ICP round as the 'comparable' national goods are actually more affordable, leading to higher purchasing power and lower poverty.

As a result, we are left with a great deal of uncertainty about the levels of extreme poverty and it appears to me that the conceptual and empirical problems are so severe that we can place relatively little confidence in the reported *levels* of poverty using these international poverty lines. To be sure the reported *trends* in global poverty are much less questionable where the changes implied by each revision have been always relatively modest due to the reasons described above.

10.5 Is there an alternative approach to measure global poverty?

Given these problems, the question arises whether there is an alternative approach to measuring extreme poverty at the global level. In principle there is and it is an approach that is already widely used, including by the World Bank, though not usually for measuring poverty at the global level. The approach, which has been suggested by Reddy (2008) and applied as an example in Reddy *et al*. (2008) consists of creating *national* poverty lines using a procedure that is internationally consistent so that then poverty measured in this consistent way could be aggregated across countries. The most common way to generate consistent poverty lines is to link them in some form for a nutritional requirement. There are principally two common ways suggested in the literature to this (see Ravallion 1994, 1998): the food energy intake method and the cost of basic needs method. Briefly, the former asks the question what incomes are empirically needed to allow households to have a specified number of calories per capita (or adult equivalent). This is done by running a regression of caloric intake on incomes (or expenditures) to identify the required expenditures to meet a certain caloric norm. India's poverty line is essentially based on this approach and is based on the incomes that in 1973/74 were sufficient to purchase an adequate diet in rural and urban areas (see Subramanian 2005; Reddy 2007). These poverty lines can then be updated over time by either some consumer prices index (ideally using a basket that reflects purchasing habits of the poor), as done in India, or simply the exercise can be redone in each (survey) year, as apparently was done in Bangladesh in the 1980s where a new poverty line was generated using the expenditure–food intake relationship (Wodon 1997).

The cost of basic needs (CBN) method, now used predominantly by most developing countries as well as the World Bank, is closely related but proceeds somewhat differently. It first chooses a reference group of probably poor people (e.g. the bottom third of the income distribution), examines the level and type of food expenditures to generate a food basket that determines the shares of food types in that basket. The (food) poverty line is then the amount of food expenditures needed so that this basket will provide a pre-defined caloric content (i.e. the food expenditures in each group are proportionately scaled up or down across the entire basket until they deliver this caloric norm). Allowance for non-food items is then made by either taking the average non-food share of those households whose food expenditure equals the food poverty line (upper limit) or whose total expenditure equals the food poverty line (lower limit).[12] Updating of the poverty line can be done in three ways, either by simply using a consumer price index (or one relevant for the poor), or by using the specific prices for the items in the food basket (and either keeping the non-food share fixed or using a new survey to allow it to vary), or by redoing the entire exercise using a new household survey. Most often, the second method is used, i.e. updating the prices of the food basket and (while most often) keeping the non-food share fixed.

Using these nationally set poverty lines and poverty rates using either of these methods, one would examine levels and trends in poverty, country by country, and then simply add up the number of poor people across countries, without reference to an international poverty line. To the extent that these approaches are indeed fully comparable across countries (and time) and all measure how many people have insufficient incomes to consume enough food, one would this way generate a global poverty estimate of the poor. This estimate would obviate the rather complex conceptual and empirical problems inherent in the current PPP-based international poverty lines.

This approach is, in principle, rather straightforward, possible with available household survey data, and this method is being used in many developing countries already to analyse poverty nationally.

Unfortunately, there are several serious conceptual and empirical issues that would need to be tackled. First of all, consistency across space and time will require that one consistently uses one of the two approaches outlined above and also uses consistent choices when actually implementing them (e.g. on updating and on the reference group for the food basket). With enough international coordination, achieving consistency in principle is possible. But it is not obvious on what approach this international poverty measurement should converge.

There are serious problems with each of the approaches taken. The problems with the food–energy intake method are nicely illustrated by Wodon (1997) in Bangladesh. In urban areas, much higher expenditures are apparently required to achieve the caloric norm than in rural areas. Does this merely reflect higher prices in urban areas? Or greater need or preferences for non-food items? To the extent it is tastes, should that be reflected in a poverty line? Updating can create further problems as the comparison between India and Bangladesh illustrates. As shown by Wodon, redoing the food–energy intake method leads to a *falling* poverty line between 1985 and 1989 as the amount of income needed to reach the caloric norm has fallen. Wodon convincingly shows that this is related to the fact that falling incomes lead to a substitution to cheaper calories and thus the falling poverty line is actually a result of households *reacting* to *higher* poverty; this reaction should scarcely be seen as representing falling poverty. Conversely, not updating the food–energy intake poverty line can also lead to serious problems as well, as the Indian case demonstrates. The poverty line developed in 1983 (and confirmed in 1993) was sufficient to purchase 2400 calories per adult in 1973/74. By 1999/2000, (rural) people at the poverty line were only purchasing less than 1900 calories (see Patnaik 2004; Subramanian 2005; Reddy 2007), suggesting that, in some sense, the Indian poverty line is at a level where households are no longer adequately nourished. This would generate a case for updating which would lead to a much higher poverty line in India today, and consequently much higher poverty levels (and much less poverty reduction in recent decades).

But also the CBN method has serious questions and shortcomings. First, it is not obvious who is supposed to be the reference group for the poverty food basket in a country (those close to the poverty line or all below it?), which will affect results. More importantly for international comparisons of poverty lines is the question whether these food baskets (as well as the non-food requirements at the poverty line) should be determined separately for each country, or established internationally. In their illustration, Reddy *et al.* (2009) determine them separately for each country but this might go against the idea of developing an absolute standard that is comparable across countries. In richer countries, the presumed poor will likely consume more expensive calories and have higher non-food needs than in poorer countries; as a result deriving the poverty line for each country contains an element of a relative poverty line, driven by prevailing consumption patterns.[13] Conversely, if one chose a common food basket, it is unclear whether this could adequately account for specificities of climate, food availability, specific needs, etc.

Moreover, updating presents similar problems. If one simply updates using prices from the food basket, the problem identified in India would still largely hold. The reason the people at the poverty line are in 2000 only consuming fewer than 1900 calories is only to a small extent due to the fact that cereal prices rose faster than the price index used for updating the poverty line (Subramanian 2005; Reddy 2007). It is more related to the fact that households apparently switched their consumption habits, turning to more expensive calories and more non-food items. Conversely, redoing the poverty line with each new survey puts into question the inter-temporal comparability of the poverty estimates which each time are based on a different basket of goods. In particular it is unclear whether the driver of the changed baskets are income effects (positive or negative) which might make this again somewhat of a relative (rather than absolute) poverty standard. For international comparisons, the problem will naturally arise that some countries would then frequently update their poverty line, while others do it more rarely (as surveys are done more infrequently) and it is not always clear then which survey estimates should be compared. A benchmark year approach, as done in the ICP, might be one way to deal with this.

There are also a host of empirical issues to consider. They include problems of comparability of questionnaires of household surveys, different extent of measurement error in food expenditures that would affect the construction of national poverty lines, the increasing (and internationally highly variable) difference between mean consumption in household surveys and consumption as reported in the national accounts are the most important ones. Also here, the devil is in the detail as has been shown in Deaton and Kozel (2005) who survey the great debates on poverty in India where these issues are the centre of the debate. In particular, the issue of measurement error and incomparability of survey instruments (and survey implementation) across countries and time are the most serious empirical drawbacks of moving to such an approach to measuring global poverty.

Clearly, this alternative approach of using consistent national poverty lines to measure poverty at the global level would also generate a host of serious conceptual and empirical issues. As a first step, it would be important to investigate the feasibility of this approach in much more detail. Also, it is clear that implementation of such an approach would require a totally different set-up of poverty analysis than is currently in place. In particular, it would either require extensive international coordination and standardization on poverty measurement approaches, which is currently largely absent, or would require an international agency, such as the World Bank, to take all these household surveys and build up comparable national poverty lines for as many countries as possible. Both paths are not entirely infeasible. The first route has been chosen in the creation of the System of National Accounts where all countries of the world agreed to specific rules for the calculation of national income, GDP and other aggregate statistics. The size of the effort was and remains huge, and something rather similar might be needed to implement internationally coordinated poverty measurement. And the second route the World Bank has already gone down quite far in its work on global poverty as well as its work on national poverty. But currently its work on national poverty is highly decentralized and the critical conceptual and empirical issues discussed above have not been addressed.

In short, it would not appear to be entirely infeasible to try out such an alternative approach. If implemented successfully, such an approach would obviate the need for the periodic drastic revisions of poverty levels and would also present a more accurate picture of poverty trends over time. To be sure, the conclusion on poverty trends would, as discussed above, likely not be substantially different from those based on the current method favoured by the World Bank, but they would be based on a better foundation and give us a better grounding on comparative poverty levels.

10.6 Conclusion

The rather drastic revisions of global poverty have revealed significant problems in the measurement of levels and trends in global poverty. In this chapter, I have discussed the nature of these problems and highlighted some potential solutions. It appears that the current method of calculating poverty using a regularly revised international poverty line is beset with considerable conceptual and empirical problems. Instead, it is well worth considering an alternative where global poverty is monitored using internationally consistent national poverty lines. It is time that such an approach was actually investigated more thoroughly whether the conceptual and empirical problems inherent in this alternative approach could be handled.

At the same time, it is important to highlight that there are other substantive problems not touched upon in the chapter. First, for most developing countries we do not have any data on poverty so are unable to say much about levels and trends. Second, the vast majority of countries, particularly in Sub-Saharan Africa, but also countries in other regions of the world, are unlikely to reach the first MDG by 2015; this remains a big challenge for national policy makers as well as the international community.

Notes

1 This is a thoroughly revised, updated and expanded version of a paper that appeared in E. Mack, S. Klasen, T. Pogge and M. Schramm (eds) *Absolute Poverty and Global Justice*. London: Ashgate (2009). I would like to thank participants at the St Gallen Meeting of the International Association for Research in Income and Wealth, Francois Bourguignon, Francisco Ferreira, Elke Mack, Thomas Pogge, Martin Ravallion, Sanjay Reddy, Michael Schramm, and Michael Ward (+) for helpful comments and discussion. This chapter is dedicated to his memory.
2 For a discussion, see, for example, Sen (1982), Ravallion (1994) and Klasen (2000).
3 The poverty line is adjusted only for differences in prices across space and time.
4 Most of the background papers by the World Bank team working on these numbers usually also prepare figures that consider the depth and severity of poverty. See, for example, Chen and Ravallion (2010).
5 In principle, this problem could be addressed by using equivalence scales although there is no consensus in the literature on which scales to use.
6 There are also more detailed measurement questions such as the consistency of household surveys between countries and over time, as well as the consistency of income or consumption information in household surveys with the same information in national accounts which appear to suggest lower rates of extreme poverty and faster poverty reduction. See Chen and Ravallion (2010) and Bhalla (2004) for a discussion.
7 They will have a modest impact, however, as now these trends are evaluated at a different national poverty lines which has consistently been shifted upwards or downwards for all years, and we know formally from the work of Bourguignon (2003),

Klasen and Misselhorn (2007), among others, that the location of the poverty line will affect the pace of poverty reduction; this is due to the fact that the density of incomes around the poverty line clearly affects the impact (distribution-neutral) growth will have on poverty reduction: if the density is very high around the poverty line, poverty reduction (measured in percentage points) from growth will be larger. Measured in percentage changes (rather than percentage point changes), poverty reduction will be higher if the density is low and one is in the left tail of the income distribution (where a percentages reduction of poverty of 50 per cent, say from 2 to 1 per cent, is quite easy to achieve).

8 They do not only consider the benchmark years, but also national growth rates and try to render the two pieces of information consistent. For details see the documentation in Heston *et al*. (2006).

9 In addition, the switch of baseline year can have a significant impact both on levels and on growth rates of GDP in developing countries as was also demonstrated by same authors. This, in fact, was one reason to change the methods of generating the Penn World Tables between version 6.2 and 6.3. The latter is more robust to changes in the benchmark year. It is also far from clear that the solution proposed by Johnson *et al*. (2009), to just use the benchmark years for the calculation of growth rates, would be a satisfactory alternative, either conceptually or empirically.

10 The same was true if the Johnson *et al.* (2009) approach was used, where only the benchmark years are actually used; see the China example just discussed.

11 This may appear surprising at first sight since \$1.25 in 2005 prices does not appear to be much higher than \$1.08 in 1993 prices (in fact, in real terms it appears lower). But the higher price levels found in the ICP in many poor countries in 2005 would have lowered their poverty lines (in PPP terms) so that, had one just changed the PPPs but used the same country sample us before, the international poverty line would have actually fallen to \$1.05, with little change in the number of the poor. See Greb *et al.* (2011) for further discussion, including the question of statistical estimation issues in estimating this international poverty line.

12 Reddy *et al*. (2009) in their proposal for comparable poverty lines using such an approach rely on this method in their illustrative analysis.

13 Of course, there might be a good justification for doing so. As Reddy *et al*. (2009) want to create a capability-based poverty line based on the capability 'adequate nourishment', fulfilling that capability might indeed require more resources in a richer country than in a poorer one, making it still 'absolute' in the space of capabilities while relative in the space of resources (see also Sen 1984 on that); in this sense one would come closer to a weakly relative poverty line suggested by Ravallion and Chen (2010) in the income space.

References

Bhalla, S. (2004), 'A comparative analysis of estimates of global inequality and poverty', *CESifo Economic Studies* 50:1, 85–132.

Bourguignon, F. (2003), 'The growth elasticity of poverty reduction', in T. Eicher and S Turnovsky (eds) *Inequality and growth* (Cambridge, MA: MIT Press).

Chen, S. and Ravallion, M. (2007), 'Absolute poverty measures for the developing world, 1981–2004', *Proceedings of the National Academy of Sciences* 104, 16757–16762.

Chen, S. and Ravallion, M. (2010), 'The developing world is poorer than we thought, but no less successful in the fight against poverty', *Quarterly Journal of Economics* 125:4, 1577–1625.

De Haen, H., Klasen, S. and Qaim, M. (2011), 'What do we really know? Metrics for food insecurity and malnutrition', *Food Policy* 36:6, 760–769.

Deaton, A. (2010), 'Price indexes, inequality, and the measurement of world poverty', *American Economic Review* 100:1, 5–34.

Deaton, A. and Dupriez, O. (2008), 'Poverty PPPs around the world: an update and progress report', Mimeographed, Development Data Group, World Bank.

Deaton, A. and Kozel, V. (eds) (2005), *The great Indian poverty debate* (London: Macmillan).

Greb, F., Klasen, S., Pasaribu, S. and Wiesenfarth, M. (2011), 'Dollar a day re-revisited', Courant Research Center Working Paper.

Haddad, L. and Kanbur, R. (1990), 'How serious is the neglect of intrahousehold inequality?' *Economic Journal* 100:402, 866–881.

Heston, A. (2008), 'The 2005 global report on purchasing power parity estimates: a preliminary review', *Economic and Political Weekly* 43:11, 65–69.

Heston, A., Summers, R. and Aten, B. (2006), 'Penn World Table Version 6.2', Center for International Comparisons of Production, Income and Prices at the University of Pennsylvania.

Johnson, S., Larson, W., Papageorgiou, C. and Subramanian, A. (2009), 'Is newer better?' Penn World Table revisions and their impact on growth estimates, Mimeographed, MIT.

Klasen, S. (2000), 'Measuring poverty and deprivation in South Africa', *Review of Income and Wealth* 46:1, 33–58.

Klasen, S. (2007), 'Gender-related indicators of well-being', in M. McGillivray (ed.) *Human well-being: concepts and measurement* (Basingstoke: Palgrave Macmillan).

Klasen, S. (2008), 'Poverty, undernutrition, and child mortality: some inter-regional puzzles and their implications for research and policy', *Journal of Economic Inequality* 6:1, 89–115.

Klasen, S. and Misselhorn, M. (2007), 'Determinants of the growth semi-elasticity of poverty reduction', *Ibero-America Institute for Economic Research Working Paper* No. 176 (Göttingen: Ibero-America Institute).

Patnaik, U. (2004), 'The republic of hunger', Public Lecture on the occasion of the 50th Birthday of Safdar Hashmi, organized by SAHMAT (Safdar Hashmi Memorial Trust) on 10 April 2004, New Delhi.

Ravallion, M. (1994), *Poverty comparisons* (Chur: Harwood Academic Press).

Ravallion, M. (1998), 'Poverty lines in theory and practise', Living Standards Measurement Survey World Paper No. 133, Washington DC: World Bank.

Ravallion, M. (2009), 'A reply to Reddy and Pogge', in S. Anand, P. Segal and J. Stiglitz (eds) *Debates on the measurement of global poverty* (Oxford: Oxford University Press).

Ravallion, M. and Chen, S. (2011), 'Weakly relative poverty', *Review of Economics and Statistics* 93:4, 1251–1261.

Ravallion, M., Chen, S. and Sangraula, P. (2009), 'Dollar a day revisited', *World Bank Economic Review* 23(2): 163–84.

Reddy, S. (2007), 'Poverty estimates: how great is the debate?' *Economic and Political Weekly* 10 February.

Reddy, S. (2008), 'The World Bank's new poverty estimates: digging deeper into a hole'. Available at: www.socialanalyis.org (last accessed 22 January 2009).

Reddy, S. and Pogge, T. (2009), 'How not to count the poor', in S. Anand, P. Segal and J. Stiglitz (eds) *Debates on the measurement of global poverty* (Oxford: Oxford University Press).

Reddy, S., Visaria, S. and Asali, M. (2009), 'Inter-country comparisons of income poverty based on a capability approach', in K. Basu and R. Kanbur (eds) *Arguments for a better world* (Oxford: Oxford University Press, Vol. II).

Sen A.K. (1982), 'Poor relatively speaking', in A. Sen (ed.) *Choice, welfare, measurement* (London: Blackwell).

Sen A.K. (ed.) (1984), *Resources, values and development* (Cambridge, MA: Harvard University Press).

Sen, A.K. (1985), *Commodities and capabilities* (Amsterdam: North-Holland).

Subramanian, S. (2005), 'Unravelling a conceptual muddle', *Economic and Political Weekly* 40:1, 57–66.

Summers, R., Heston, A. and Aten, B. (2006), 'The Penn World Tables Version 6.2', Center for International Comparisons of Production, Income and Prices at the University of Pennsylvania.

Ward, M. (2009), 'Identifying absolute poverty in 2005: the measurement question', in E. Mack, S. Klasen, T. Pogge and M. Schramm (eds) *Absolute poverty and global justice* (London: Ashgate).

Wodon, Q. (1997), 'Food energy intake and cost of basic needs: measuring poverty in Bangladesh', *Journal of Development Studies* 34:2, 66–101.

World Bank (1990, 2000/01), *World Development Report* (New York: Oxford University Press).

World Bank (2007, 2008), *Global monitoring report* (New York: Oxford University Press).

Part III

Small area estimation methods

11 Small area methodology in poverty mapping

An introductory overview

Ray Chambers and Monica Pratesi

11.1 Introduction

The need for reliable estimates of poverty, inequality and life condition indicators for communities across the globe has increased substantially in recent years, due in large part to the more detailed information needs of policy makers charged with developing strategies for poverty alleviation. In particular, there has been a steep increase in the demand for more detailed information about the geographic distribution of these indicators. As a consequence, poverty maps have become a powerful tool for designing better policies and interventions, and many national statistical agencies are now developing, evaluating and implementing poverty estimation methodologies. An example is the major investment that the European Commission has made in funding the research projects SAMPLE (Small Area Methods for Poverty and Living condition Estimates – www.sample-project.eu) and AMELI (Advanced Methodology for European Laeken Indicators – www.unitrier.de).

At its heart, poverty mapping is about combining survey data that measures poverty incidence with auxiliary information about the population of interest. In effect, basic data derived from answers given to survey questions about income and consumption are combined with data about other, correlated, variables. These variables are commonly obtained from other surveys, from population censuses or from administrative registers. However, in all cases it is essential that they refer to the same domains or population units.

Auxiliary information can also consist of geo-coded data about the spatial distribution of these domains and units, obtained via geographic information systems. For example, the data can be obtained from digital maps that cover the domains of interest and so allow for the calculation of the centroids of these domains, their borders, perimeter, areas and the distances between them. All these attributes are helpful in the analysis of social-economic data relating to these domains since these often display spatial structure, i.e. they are correlated with the so-called geography of the landscape.

New developments in communications technology are now leading to the measurement and storage of huge amounts of spatio-temporal auxiliary data relating to households and individuals, e.g. the use of point of sale data on the amount and location of household transactions and the spatial tracking of the movements of individuals by means of GPS systems linked to their mobile phones and other communication devices. These data sources are sometimes collectively referred to as Big Data. Their use in poverty measurement is still somewhat limited, so they are not considered here.

The application of small area methods to poverty assessment typically depends on the data collection strategies adopted by national statistical agencies. Often the domains of interest are defined at a geographical level, e.g. administrative areas corresponding to local levels of governance. Unfortunately, agency budget constraints mean that this level of geographic coverage is impossible for major national surveys. Instead these surveys are designed to obtain reliable estimates at higher levels of spatial aggregation, such as regions. Thus, alternative methodologies are necessary before reliable estimates of poverty indicators can be used to construct poverty maps on which to base political decisions aimed at reducing vulnerabilities and difficult living conditions. These alternate methodologies treat the larger regions as planned domains while the smaller areas of interest, such as provinces and municipalities, are considered to be unplanned domains, with typically small, or even zero, sample sizes.

In general, overcoming the problem of small unplanned domains requires that one either (i) increases the sample size of the survey in the specific domains of interest (sometimes referred to as top up sampling) or (ii) resorts to small area estimation techniques. Oversampling in specific domains is usually a very costly alternative and may not be necessary, in the sense that the accuracy of the estimates calculated using standard methods and based on the enhanced sample may not be any better than the accuracy of the corresponding estimates calculated using the original sample, but employing small area estimation methods (see Giusti *et al.*, 2012). As a consequence we now see increasing demand from both official and private institutions for statistical estimates of poverty and living conditions for smaller unplanned domains based on the application of small area estimation methodologies.

11.2 Poverty indicators

The so-called Laeken indicators are a core set of statistical indicators on poverty and social exclusion agreed by the European Council in December 2001 in the Brussels suburb of Laeken, Belgium. They include measures of the incidence of poverty and of the intensity of poverty. They are an important tool for identifying poverty and inequality when making comparisons between areas of interest.

In the chapters making up this Part of the book the authors focus mainly on the estimation of the incidence of poverty using the *Head Count Ratio* (also known as at-risk-of-poverty-rate) and on the intensity of poverty using the *Poverty Gap.* These are special cases of the generalized measures of poverty introduced by Foster, Greer and Thorbecke (1984), hereafter FGT. Denoting the poverty line by t, the class of FGT poverty measures for a small area d are indexed by $\alpha \geq 0$ and are defined as:

$$F_{ad} = \frac{1}{N_d}\sum_{j=1}^{N_d}\left(\frac{t-y_{jd}}{t}\right)^{\alpha} \mathrm{I}(y_{jd} \le t). \tag{11.1}$$

The poverty line t is a level of income that defines the state of poverty. Units, i.e. individuals or households, with income below t are considered poor. Here y is a measure of income for unit j, N_d is the number of units in area d, I is the indicator function (equal to 1 when $y_{jd} \le t$ and 0 otherwise) and α is a "sensitivity" parameter. When $\alpha=0$, F_{ad} is the *Head Count Ratio* or HCR indicator whereas when $\alpha=1$, F_{ad} is the *Poverty Gap* or PG indicator.

The HCR indicator is a widely used measure of poverty and is easily seen to be the proportion of units in the area with income at or below the poverty line. The popularity of this indicator is due to its ease of construction and interpretation. However, it implicitly assumes that all poor units are in the same situation. Thus, the easiest way of reducing the HCR for an area is by targeting benefits to units just below the poverty line because they are cheapest to move across the line. Poverty alleviation policies based on the HCR could therefore be sub-optimal. For this reason one also calculates an estimate of PG. This indicator can be interpreted as the average shortfall of poor units. It shows how much would have to be transferred to the poor to bring their expenditure up to the poverty line.

The FGT poverty indicator (11.1) depends on the empirical cumulative distribution of income values in area d. Consequently this distribution function represents an important source of information for the living conditions in that area. In particular, many quantities of interest in poverty assessment (e.g. the median income, the income quantiles), as well as poverty indicators like HCR and PG, can be computed from an estimate of this distribution function. It immediately follows that estimation of the cumulative distribution function of income within an area is an important aspect of poverty analysis for the area. In particular, we note that estimates of traditional monetary-based poverty indicators accompanied by estimates of the cumulative distribution functions of the income variables of interest then allow a more detailed analysis of poverty.

Of course, it is also true that these measures are generally considered to be starting points for more in depth studies of poverty and living conditions in an area, with analyses carried out using non-monetary indicators providing a more complete picture of poverty and deprivation (Cheli and Lemmi, 1995). In particular, since poverty is often a relative concept, the set of indicators used to characterize it is generally enlarged to include other indicators for vulnerable groups, i.e. those likely to enter into a state of poverty. For more discussion of this point, see the results of the SAMPLE project.

11.3 Small area methods for the estimation of poverty indicators

The straightforward approach to calculating estimates of FGT poverty indicators for an area of interest is to compute direct estimates of these indicators for the area. These are appropriate when there is a reasonably large (say >50) sample in the area. By a direct estimate we mean here an estimator that depends only on the sample data from the area. For example, let w_{jd} denote the sampling weight of unit j in the sample s_d of units in area d, i.e. j is one of the n_d sampled units from small area d. The direct estimators of the FGT poverty indicators are then of the form

$$\hat{F}_{\alpha d}^{dir} = \frac{1}{\sum_{j \in s_d} w_{jd}} \sum_{j \in s_d} w_{jd} \left(\frac{t - y_{jd}}{t} \right)^{\alpha} I(y_{jd} < t), \quad d = 1, \ldots, D. \tag{11.2}$$

Typically $\sum_{i \in s_d} w_{jd} = N_d$, the population size of small area d, but this is not always the case. When the sample in the area of interest is of limited size, estimators such as (11.2) are of limited use because of their variability. In these cases small area estimation techniques are employed. In order to motivate these techniques, we note that the population mean of y in small area d can be written as

$$m_d = N_d^{-1} \left(\sum_{j \in s_d} y_{jd} + \sum_{k \in r_d} y_{kd} \right), \tag{11.3}$$

where r_d denotes the $N_d - n_d$ non-sampled units in area d. Since the y values for the non-sampled units in r_d are unknown, they need to be predicted.

Random effects models are a popular tool for small area estimation, as they include random area effects to account for between area variation. To illustrate, suppose that the population values y_{jd} follow a linear model defined by a vector $\mathbf{x}_{jd}$ of unit level covariates and area-level random effects γ_d. Then the Empirical Best Linear Unbiased Predictor (EBLUP) of the mean of y in small area d is:

$$\hat{m}_d^{EBLUP} = N_d^{-1} \left[\sum_{j \in s_d} y_{jd} + \sum_{k \in r_d} (\mathbf{x}_{kd}^T \hat{\beta} + \hat{\gamma}_d) \right]$$

where $\hat{\beta}$ and $\hat{\gamma}_d$ are Restricted Maximum Likelihood estimators of β and γ_d. See Rao (2003) for further details on the use of the EBLUP approach in small area estimation.

In many applications, unit level data on y and $\mathbf{x}$ are not available, but direct estimates of the average values $\bar{y}_d$ of y in the small areas, together with the corresponding area average values $\bar{\mathbf{x}}_d$ of $\mathbf{x}$, are available. In this situation one can still adopt a linear model for these direct estimates that includes area-specific random effects, and so calculate an EBLUP. These area level models are often referred to as Fay–Herriot models, after Fay and Herriot (1979), who first introduced them. A standard assumption in many applications based on linear models with random area effects is that the area effects are independent. This is unlikely to be the case where geographic contiguity of areas results in spatially correlated data. Extensions to both the unit level and area level EBLUPs to allow for spatially correlated area effects have been proposed in the literature. In this section, the paper by Salvati *et al.* (2012) investigates how small area estimation based on an area level model that allows for spatial proximity effects via a spatially varying mean function compares with small area estimation based on a Fay–Herriot model with a spatially invariant linear mean function but with spatially correlated area effects.

A widely used approach to poverty mapping is based on the methodology proposed by Elbers, Lanjouw and Lanjouw (2003), which is often referred to as the World Bank (WB) or ELL method and is reviewed in this book by Haslett. This assumes a nested error linear regression model, fitted to clustered survey data from the population of interest, with the random effects in the model corresponding to the clusters used in the survey design. It also assumes the availability of unit level census data from a time period as close as possible to that of the survey. The response variable y for the regression model, which is not available in the census, is typically the logarithm of a welfare variable, e.g. income or consumption, and the explanatory variables $\mathbf{x}$ in the model are assumed to be available both in the survey and in the census datasets. Once the model has been estimated using the survey data, the fitted model parameters are combined with the census micro-data to produce multiple imputed values of the welfare variable for all members of the population by making independent draws from the fitted model. For each such draw, values of the poverty indicators of interest for the different small areas are calculated. These are averaged over the simulations to produce the final estimates of these quantities, with the simulation variability of these estimates used as an estimate of their uncertainty.

Molina and Rao (2010) have recently proposed a modification of the ELL method that introduces random area effects (rather than random cluster effects) into the linear regression model for the welfare variable, and also generates the simulated values of y by making independent draws from the conditional distribution of this variable given its observed sample values and the associated (census) values of $\mathbf{x}$. Chapter 13 by these authors in this part of the book contains a description of their method. As with the ELL method, given the simulated values of y at each draw, the values of the poverty indicators of interest for the resulting simulated population are calculated, and the final estimates for these indicators obtained by averaging their draw specific values. These authors refer to this approach as Empirical Best since it is easy to see that the resulting estimate of (11.1) is an estimate of the conditional expectation of this quantity given the sample data and the census micro-data. A bootstrap approach is then used to estimate the mean squared error of this estimator.

Both the ELL method and the modification proposed by Molina and Rao (2010) require multiple microsimulations of a target population, and so can be very computer intensive in practical applications. A fast approximation suggested in Ferretti and Molina (2012) is to first resample from the census values of $\mathbf{x}$ and to then simulate the corresponding sample values of y. These microsimulated sample values are used as inputs in calculating direct estimates of the poverty indicators of interest, see (11.2), which are then averaged over the simulations in the usual way. Note however that these fast approximations are no longer Empirical Best, since they are not estimators of the conditional expectations of the population values of the poverty indicators.

An alternative approach to small area estimation that relaxes the parametric assumptions of the random effects models that underpin the EBLUP, ELL and EB approaches was recently proposed by Chambers and Tzavidis (2006) and Tzavidis *et al.* (2010). This approach models the M-quantiles of y given $\mathbf{x}$, rather than the expected value (as would be the case with a standard regression model), and then characterizes the differences between the small areas using different M-quantile orders, and hence different fitted models. As described in Chapter 15 by Tzavidis *et al.* in this part of the book, this approach can be combined with the concept of smearing (Duan 1983) to estimate the value of (11.1).

Smearing-based prediction was introduced by Chambers and Dunstan (1986) as a method of predicting a finite population distribution function. Its application to predicting (11.1) can be summarized as follows. Suppose that our sample values of y in area d can be written as $y_{jd}=\hat{\mu}_d(\mathbf{x}_{jd})+e_{jd}$ where $\hat{\mu}_d$ denotes a fitted mean function specific to area d and the e_{jd} are the corresponding sample residuals for area d. Further, let

$$z_{jd}=\left(\frac{t-y_{jd}}{t}\right)^{\alpha}\mathrm{I}(y_{jd}\le t).$$

Then (11.1) is just the area d mean of z_{jd}, and the same decomposition as set out in (11.3) applies. Consequently, we can estimate the value of (11.1) by predicting its unknown non-sample component. This leads to the smearing predictor of (11.1),

$$\hat{F}_{\alpha d}^{smear}=N_d^{-1}\left(\sum_{j\in s_d}z_{jd}+\sum_{k\in r_d}\hat{z}_{kd}^{smear}\right). \tag{11.4}$$

In this expression the smearing-based predictor of a particular non-sample value z_{kd} is given by

$$\hat{z}_{kd}^{smear}=n_d^{-1}\sum_{j\in s_d}\left(\frac{t-\hat{y}_{kjd}}{t}\right)^{\alpha}\mathrm{I}(\hat{y}_{kjd}\le t), \tag{11.4}$$

where $\hat{y}_{kjd}=\hat{\mu}_d(\mathbf{x}_{kd})+e_{jd}$. Note that the smearing approach requires specification of $\hat{\mu}_d(\mathbf{x})$. This estimate is typically based on the fitted M-quantile model for y in area d, but can be based on any predictive model for y, included a regression model with random area effects. The mean squared error of this estimator can be estimated using a bootstrap approach. See Chapter 15 by Tzavidis *et al.* for details.

11.4 Comments on the use of small area methods in poverty assessment

Given the limited resources typically available for poverty assessment surveys, using the data collected in these surveys to study relative differences in poverty across the study population (poverty mapping) inevitably requires the application of small area estimation methods. Since the choice of an appropriate small area estimation method is something that is highly relevant to the world beyond academe, their proper use requires the application of statistically sound modelling methods. Analysts applying poverty mapping methods provide guidance to decision makers about level and variation of deprivation across a study population, allowing the efficient allocation and monitoring of funding for poverty alleviation. In order to ensure that this guidance is sound, these analysts need to ensure that the modelling that underpins the small area estimation method is fit for purpose.

For this reason we recommend that the nature of the data used to measure poverty, and the adequacy of the statistical model used to represent variability in these data, is given careful consideration at the very beginning of the poverty mapping exercise. This is important since all the small area estimation methods discussed in this section rely on the predictive power of the model for the poverty measure of interest. This requires a wide selection of covariates, with known values that are available at small area or finer levels, which are both related to the poverty measure and measured along with it, and so can be used to 'explain' the variability in this measure. A secondary, but still important, consideration is then the capacity of the method used to provide valid estimates of the accuracy of the resulting small area estimates.

All the small area estimation methods described in this part of the book model the relationship between the variable underpinning the poverty measure and the covariates using the survey data. Where they differ is in how the fitted model is then used to make predictions for the distribution of the poverty measure in the small areas of interest. The ELL method and its variants do this by using the fitted model to microsimulate the distribution of the poverty variable in each small area. This has the distinct practical advantage that no linking of survey and population units is required. On the other hand, it requires that the covariates used in fitting the model using the survey data are defined in exactly the same way in the census data, and there is a negligible temporal disconnect between these two data sources. On the other hand, the EB method of Molina and Rao (2010) assumes that the survey and census are linked, so the same covariates are used in both, and then microsimulates only the unsurveyed population. This has definite benefits when the sample in a small area is a reasonable proportion of its population. As the chapter by Haslett points out, however, this situation is unlikely to hold in practice. In any case, it is clear that both approaches require careful model specification searches, both in terms of the scale of measurement of the poverty variable (typically logarithmic) and choice of potential covariates, as well as in deciding how the random effects in the model should be defined. The inclusion of geo-coded data is theoretically possible using both approaches, but depends on the level at which this information is available. At a minimum, one could consider correlated area effects. However, this would require changes to the way these effects are simulated under both approaches and would be unlikely to improve the predictive power of the model unless the survey data included values from most, if not all, of the small areas.

The use of spatial information is likely to be most productive when the available model covariates are weak. In this case, spatial information can substantially strengthen prediction for non-sampled areas – provided there is significant spatial correlation. In particular, Chapter 14 by Salvati *et al.* in this part of the book demonstrates how area level models can include spatial information basically into two ways, either via a non-linear spatial trend in the mean structure of the model, or via the inclusion of spatial correlation in the distribution of the random area effects. These ideas extend to unit level random effect and M-quantile models, see Opsomer *et al.* (2008) and Salvati *et al.* (2012).

Income and expenditure distributions are often long-tailed, containing many outliers. In such cases, the Gaussian assumptions that underpin random effects models need to be considered carefully, and both the ELL and EB methods use a logarithmic transformation of the outcome variable in order to make them more credible (see the chapters by Haslett, and Molina and Rao in this part of the book). However, identifying such a 'robustifying' transformation can be difficult, and use of this approach requires a back-transformation from the transform scale model predictions. This can lead to problems since modelling errors associated with the second order structure of the model translate into back-transformation bias in the final estimates – see Chapter 17 of Chambers and Clark (2012). An alternative approach to controlling the effect of outliers is to use outlier robust estimation. Outlier robust small area estimation under a linear mixed model is discussed in Sinha and Rao (2009), but this approach can be numerically unstable. In contrast, quantiles are much less susceptible to long-tailed distributions and outliers, and in this context we note that the M-quantile approach to poverty estimation does not require strong distributional assumptions and, because it is automatically robust to the presence of outlying observations (but not necessarily outlying area effects), it can be applied directly to raw income data (see the Chapter 15 by Tzavidis *et al.*).

As noted earlier, unbiased estimation of the uncertainty of the estimated small area poverty measures is important, even when the small area estimators are themselves unbiased. Estimated standard errors that are biased downwards can lead to distinctions being made between estimated poverty levels at small area level that are not in reality significantly different, while estimated standard errors that are biased upwards can lead to real diversity between areas being hidden. All contributors to this part of the book therefore provide methods for estimating the mean squared errors of the small area estimators that they discuss. However, this is a research area where considerable work still needs to be done. Analytical expressions of mean squared error of small area estimators are generally complex and so virtually all the small area methods discussed in the rest of this section resort to empirical methods like the bootstrap to assess measures of uncertainty like mean squared error. Correctly applied, bootstrap methods allow a good approximation to true uncertainty, but their application often requires experienced staff and computational effort. In particular, the estimation of the correctly calibrated confidence intervals for small area quantities remains an important challenge. Another challenge is integrating this uncertainty into poverty maps, which at present tend to focus on presentation of the geographic variation of the small area point estimates. Such 'uncertainty augmented' poverty maps have the potential to offer a much better tool for evidence-based policy making.

Small area estimation methods work well when there is good auxiliary, or covariate, data to support the modelling process. These methods depend on models that link the poverty variable of interest with covariate information in a way that can be used to predict the unobserved values of this variable for the non-sampled population units. For example, in order to estimate the mean value of equivalized household income (or other poverty indicators) in the areas of interest, EU-SILC (European Survey on Income and Living Conditions) data are used to estimate statistical models which are then combined with census data to predict equivalized household income for non-sampled households in the EU-SILC target population. An immediate consequence is that any model that we use for this purpose can only depend on variables that are common to both EU-SILC and census data. Furthermore, the type of model used can depend on the level of aggregation of the census data. If we have access to unit level census data then we can use unit level models. If only census aggregates are available, then we have a choice between area level modelling and unit level modelling but with area level covariates. At the time of writing, there is no precise theory to guide this choice, with initial results reported in Namazi-Rad and Steel (2011) indicating that it is not a trivial problem.

Modern information technology resources allow unprecedented access to the data held in administrative registers. These data sources can cover the entire population of interest, and often contain variables that can be employed as explanators in statistical modelling. For example, social security and taxation registers contain data on claimants of various social benefits, data on dwelling categories for local taxation etc. Administrative data are not collected for statistical purposes, however, so care must be taken when utilizing them to ensure comparability of concepts, definitions, target population, collection mode and the purpose for which these data have been collected. In addition, unless the units in the target possess unique identifiers, data linkage problems arise when attempting to 'connect' the register and the survey data sets. The use of probabilistic data linkage is possible, but brings with it bias issues caused by both incorrect links as well as missing links (survey units that cannot be linked to the register). Initiatives on combining data to support small area statistics are in place in many developed economies, but similar initiatives are much less likely to be in place in developing countries where the need for poverty maps is more pressing.

Finally, we note the interesting developments currently taking place with regard to the small area estimation of non-monetary measures of poverty. Poverty is a multidimensional concept and can be defined in a number of ways. The small area estimation theory that is the focus of this part of the book uses a cost-of-basic-needs (CBN) approach since this is most popular, especially in developing countries. Calculated poverty lines represent the level of per capita expenditure or income needed to meet the basic needs of the members of a household. Because prices vary among geographical areas, these poverty lines can be calculated separately for different regions for which income, expenditure and price information is available. Alternatively, household per capita expenditure can be adjusted using regional price indices to give real per capita expenditure in which case a single poverty line can be applied across the country. However, as described in other parts of this book, poverty indicators – both monetary and non-monetary – can also be defined as fuzzy indicators. The concept of belonging to the set of 'poors' – as defined in the more convenient way for the situation under study – is often better measured using fuzzy set theory, since poverty is a dynamic condition, and often the result of a gradual process of deprivation and vulnerability. Fuzzy poverty indicators (Pratesi *et al.*, 2010) are then useful additions to the suite of traditional poverty indicators, and the small area estimation theory described in this part of the book can also be applied to them in order to give a more complete picture of the poverty situation in the areas of interest.

References

Chambers, R.L. and Clark, R.G. (2012). *An Introduction to Model-Based Survey Sampling with Applications*. Oxford University Press: Oxford.

Chambers, R.L. and Dunstan, R. (1986). Estimating distribution functions from survey data. *Biometrika 73*, 597–604.

Chambers, R.L. and Tzavidis, N. (2006). M-quantile models for small area estimation. *Biometrika 93*, 255–268.

Cheli, B. and Lemmi, A. (1995). A totally fuzzy and relative approach to the multidimensional analysis of poverty. *Economic Notes 24*, 115–134.

Duan, N. (1983). Smearing estimate: a nonparametric retransformation method. *Journal of the American Statistical Association 78*, 605–610.

Elbers, C., Lanjouw, J.O. and Lanjouw, P. (2003). Micro-level estimation of poverty and inequality. *Econometrica 71*, 355–364.

Fay, R. and Herriot, R. (1979). Estimation of income from small places: an application of James-Stein procedures to census data. *Journal of the American Statistical Association 74*, 269–277.

Ferretti, C. and Molina, I. (2012). Fast EB method for estimating complex poverty indicators in large populations. *Journal of the Indian Society of Agricultural Statistics 66*, 105–120.

Foster, J., Greer, J. and Thorbecke, E. (1984). A class of decomposable poverty measures. *Econometrica 52*, 761–766.

Giusti, C., Marchetti, S., Pratesi, M. and Salvati, N. (2012). Robust small area estimation and oversampling in the estimation of poverty indicators. *Survey Research Methods 6*, 155–163.

Molina, I. and Rao, J.N.K. (2010). Small area estimation of poverty indicators. *Canadian Journal of Statistics 38*, 369–385.

Namazi-Rad, M.-R. and Steel, D. (2011). What level of statistical model should we use in small domain estimation? Working Paper 02-11, Centre for Statistical and Survey Methodology, University of Wollongong.

Opsomer, J.D., Claeskens, G., Ranalli, M.G., Kauermann, G. and Breidt, F.J. (2008). Nonparametric small area estimation using penalized spline regression. *Journal of the Royal Statistical Society, Series B 70*, 265–286.

Pratesi, M., Giusti, C., Marchetti, S., Salvati, N., Tzavidis, N., Molina, I., Durbán, M., Grané, A., Marín, J.M., Veiga, M.H., Morales, D., Esteban, M.D., Sánchez, A., Santamaría, L., Marhuenda, Y., Pérez, A., Pagliarella, M.C., Rao, J.N.K. and Ferretti, C. (2010). SAMPLE deliverables 12 and 16: final small area estimation developments and simulations results. Available at: www.sample-project.eu/it/theproject/ deliverables-docs.html.

Rao, J.N.K. (2003). *Small Area Estimation.* John Wiley: New York.

Salvati, N., Pratesi, M., Tzavidis, N. and Chambers, R. (2012). Small area estimation via M-quantile geographically weighted regression. *Test 21*, 1–28.

Sinha, S.K. and Rao, J.N.K. (2009). Robust small area estimation. *Canadian Journal of Statistics 37*, 381–399.

Tzavidis, N., Marchetti, S. and Chambers, R. (2010). Robust prediction of small area means and distributions. *Australian and New Zealand Journal of Statistics 52*, 167–186.

12 Small area estimation of poverty using the ELL/PovMap method, and its alternatives

Stephen Haslett

12.1 Introduction

The standard design-based methods for analyzing sample survey data use no models at analysis stage. For surveys that are self-weighting, mean estimates (and mean estimates within subpopulations, including those based on areas) are simply the average of the relevant sample observations. For more complex surveys, scaling or weighting of observations is required to better reflect the structure of the population (or subpopulation) being sampled; the unique weight to ensure that estimates are unbiased is the inverse of its selection probability for each observation. These ideas can be extended to estimation of totals, and of other parameters of interest. The classic texts on this topic include Hansen *et al.* (1953) and Cochran (1977). References that also include more recent developments in sampling theory are Lohr (2010) and Fuller (2009); sample surveys for developing and transition countries are outlined and discussed in United Nations (2005). The underlying idea of probability-based or design-based sampling was developed in the seminal paper of Neyman (1934). These methods are characterized by them using, for estimates in a domain or subpopulation of interest, only the sample data collected within that domain. Such estimates are called direct estimates. There have also been extensions from design-based to model-based estimation, even for direct estimates. See Valliant *et al.* (2000) for example.

Generally, the design-based methods that produce direct estimates have seen considerable development over the past seventy years or more, and can work well. However, as anyone who has been asked to design a sample survey will be aware, estimates using these randomization methods are often requested or required at a finer level than is possible with the resources available. In the extreme, estimates can be requested even for areas for which no sample has been collected at all.

Historically, within government statistical agencies internationally there has been resistance to any use of statistical modeling. This reluctance reflects the need for sound government statistics not only to be unbiased, but also to be seen to be unbiased, so that the risk of introducing bias via applying statistical models to survey data was for a considerable period seen as unacceptable.

Nevertheless, even within the wider statistical community, sound modeling methods that might provide more accurate estimates than were possible using direct estimates attracted considerable research interest. The essential reference on small area estimation that developed from this research is Rao (2003), where the focus is on fitting models to survey data using auxiliary information from the same survey. Rao's principal emphasis, reflecting the needs and concerns of most major government statistical agencies at the time, is to provide usable estimates that are for somewhat but not markedly smaller areas than are possible using direct estimation.

Among geographers and economists, there has also been interest in getting better estimates via models, and the literature of both disciplines has developed in parallel to, but almost completely independently of, the statistical research that is summarized in Rao (2003). The geographers developed what they call spatial microsimulation techniques (e.g,. Ballas *et al.* 2003, via Birkin and Clarke 1995 and Bramley and Smart 1996) which are structurally equivalent to the method that Elbers *et al.* (2003) published in the econometric literature. Not surprisingly, there are common origins, e.g., Hentschel *et al.* (2000) make reference to Bramley and Smart (1996).

These techniques are intended to provide finer level estimates than those of Rao (2003). There are however some differences even between the geographers and the econometricians. The most important of these are that geographers use implicit models while the econometricians' emphasis is on explicit models, and that the geographers are often interested in time series projections to produce scenarios for comparing policy implications. See Haslett *et al.* (2010b) for a full discussion of the underlying parallels, the benefits of explicit models, and data and modeling requirements for useful projection.

Up until 2003, the structurally similar econometric and geographic literature was different in a major way from the statistical research literature: they made use of census (or pseudo-census) data in addition to modeling sampled data. Given a correct model (and this is a central issue to which we will return below) incorporating census data gives additional accuracy.

Molina and Rao (2010) represents a rapprochement between the various approaches. It uses sound methods for fitting models to survey data, as well as incorporating census data in a way that is different from ELL when there are a reasonable number of sampled units in small areas, but which reduces to essentially the type of synthetic estimator used by ELL when there are a larger number of small areas and smaller or zero sample sizes in each. Of course, like ELL, when Molina and Rao (2010) is used for a large number of small areas it requires more careful model fitting than is necessary when the number of small areas is relatively small. Further details are given in Section 12.4.

All these techniques (i.e. those of econometricians, geographers, and statisticians) can be classified as small area estimation.

In summary, the core aim of small area estimation is to improve the accuracy of direct estimates or to allow reliable estimates to be calculated at a finer level, by use of statistical models. Supplementary data sources may be used but, while this can have considerable advantage, it is not essential. The core data used for small area estimation usually comes from sample surveys.

12.2 Small area estimation of poverty

Small area estimation has many applications. When applied to poverty related variables, it is an important analytical tool in targeting to a finer level the delivery of food and other aid in developing countries. There are a wide variety of small area estimation methods, but the principal method used for poverty estimation (and hence poverty mapping) at small area level has been Elbers, Lanjouw, and Lanjouw (2003). The ELL method has also been extended to nutrition measures such as kilocalorie consumption, and underweight, stunting, and wasting in children under five years of age.

ELL uses both survey data and census data, and a model based on the survey data to make household level predictions for all the census observations. These are then added or averaged within predefined (and usually administratively based) small areas. When the model is correct, using the ELL method can give considerable accuracy improvements over direct estimation and survey-only based small area estimation methods, but when it is not correct the estimated standard errors can be severe underestimates. This chapter explores the ELL method and its underlying assumptions, briefly considers the advantages and disadvantages of the World Bank's PovMap software (Zhao 2006) that implements ELL (including its most recent version PovMap2 – Zhao and Lanjouw 2009), and outlines useful diagnostics for and variations of alternatives to ELL.

Poverty can be defined in a number of ways. The most common uses the cost-of-basic-needs (CBN) approach, in which calculated poverty lines represent the level of per capita expenditure needed to meet the basic needs of the members of a household, usually after making a small allowance for non-food consumption. A food poverty line is first established, based on the cost of basic food requirements. Then a minimal non-food allowance is added, using an amount equal to the typical non-food expenditure of households whose food expenditure is close or equal to the food poverty line. Because prices vary among geographical areas, poverty lines can be calculated separately for different regions for which price information is available, or instead household per capita expenditure can be adjusted using regional price indices to give real per capita expenditure in which case a single poverty line can be applied across the country.

Poverty incidence is defined as the proportion of individuals living in a given area who are in households with an average per capita expenditure below the poverty line. *Poverty gap* is the average distance below the poverty line (which by definition is zero for those individuals on or above the line). It represents resources needed to bring all poor individuals up to the poverty line. *Poverty severity* measures the average squared distance below the line, by giving more weight to the very poor. These three measures can be placed in a common mathematical framework, the so-called FGT measures (Foster, Greer, and Thorbecke, 1984):

$$P_{\alpha} = \frac{1}{N}\sum_{i=1}^{N}\left(\frac{z - E_i}{z}\right)^{\alpha} \cdot I(E_i < z) \tag{12.1}$$

where N is the population size of the area, E_i is the expenditure of the ith individual, z is the poverty line, and $I(E_i<z)$ is an indicator function (which is equal to one when expenditure is below the poverty line, and zero otherwise). Poverty incidence, gap, and severity then correspond to $\alpha=0$, 1, and 2 respectively.

Using alternative or additional measures extends the original research of ELL. For example, we can estimate incidence, gap, and severity of under-nourishment at the small area level. Such figures are used by the UN World Food Programme (WFP) for example, in conjunction with the poverty estimates for prioritizing food and related development assistance and support.

The World Health Organization (WHO) and WFP, along with other international aid agencies, also use three measures of child malnutrition based on a child's height, weight, and age. Stunting or low height-for-age is defined as height at least two standard deviations below the median height for a specified reference population. Underweight or low weight-for-age is similarly defined. Wasting is based on standardized weight-for-height. The data used as a reference standard in the definitions was established by the National Center for Health Statistics/Centers for Disease Control in the USA in 1975 (Hamill *et al.* 1979). Implicit in the use of a single international reference standard is the assumption that variations in height and weight for children below five years are largely environmental rather than genetic. Since 1975 there have been arguments put forward for more localized standards, although these by construction complicate international comparisons of stunting, underweight, and wasting.

Usually, the emphasis in child malnutrition measurement is nutrition status of children under the age of sixty months (i.e. below five years). Within a particular area, stunting is defined as the proportion of such children with a standardized height-for-age value below –2: children below –3 are considered "severely stunted". Similarly underweight is the proportion with a standardized weight-for-age value below –2, and severe underweight below –3. Stunting can be regarded as evidence of chronic malnutrition. Underweight reflects both chronic malnutrition and acute malnutrition, and is usually a current condition resulting from inadequate food intake, past episodes of under-nutrition, or poor health conditions. Wasting is the proportion with a standardized weight-for-height value under –2, and severe wasting below –3. Wasting can be an indicator of acute malnutrition, i.e. of children who have eaten adequately in the past, but are currently not doing so.

12.3 The ELL methodology

12.3.1 The regression framework

This section provides a brief overview of small area estimation and the Elbers *et al.* (2003) ELL method.

We consider a target variable, denoted by Y, for which we seek estimates for a number of small subpopulations. These subpopulations usually correspond to small geographical areas, but can instead represent different subgroups that may be collocated (in which case the technique is sometimes called small domain estimation).

In the original ELL method for poverty measures, Y is log-transformed per capita expenditure. For extensions to the under-nourishment measures, log kilocalorie intake per person or per adult equivalent is used instead. For stunting, underweight and wasting in children, Y is standardized height-for-age, weight-for-age, and weight-for-height respectively. Provided there are at least some sample data available for each small subpopulation, direct estimates of Y for these subpopulations can be derived from the sample survey data, for which Y has been measured directly on the final-stage sampled units (e.g. households or eligible children). Because sample sizes within even the sampled subpopulations are typically very small, these direct estimates are however generally not reliable. The core idea of small area estimation is that auxiliary information, denoted X, which is available from the survey and may also be available from other sources such as a census even for unsampled parts of the population, can be used to improve the estimates, giving lower standard errors than are possible using only direct estimates.

In the ELL method, but not in those small area methods covered in Rao (2003), X represents additional variables that have been measured for the *whole* population, either by a census or via a GIS database. (For the Rao 2003 methods X is generally available only on the sampled units but, unlike ELL, the range of statistical models can be nonlinear.)

For ELL, a linear regression-type relationship between Y and X namely:

$$Y = X\beta + u \tag{12.2}$$

is estimated from the survey data only, using both the available target variable and the auxiliary variables. (For the Rao 2003 methods, this model can be more general and may be nonlinear as for example in logistic regression that instead uses proportions.) In (12.2) β represents the regression coefficients determining the effect of the X variables on Y, and u is a random error term that represents the part of Y that cannot be explained using the auxiliary information. If we can assume that this same relationship applies to the whole population, it can be used to predict Y for all units for which we have measured X but not necessarily Y. Even though these predictions contain prediction error that may be substantial at household or child level, when amalgamated over the subpopulations of interest small area estimates based on these predicted Y values will often have smaller standard errors than the direct estimates, even given this uncertainty in the predicted values, because they are based on much larger samples. The idea is to "borrow strength" from the considerably greater coverage of the census data (which since it includes all the population may be several orders of magnitude larger than the sample size for the survey).

12.3.2 Clustering

Standard methods of fitting regression models to data cannot be used however, not only because the survey data need to be weighted (so that standard errors of the fitted parameters using standard statistical software can be incorrect), but because the survey data are collected from a sample that is usually both stratified and clustered. Thus for complex survey designs, the units on which measurements have been made are not selected independently. The clustering is not only a design property of the sample, but also of the population which is grouped naturally into geographical clusters that often contain units with similar characteristics, since households tend to cluster together into villages or other small geographic or administrative units that are themselves relatively homogeneous. Households close together tend to be more alike than households that are distant from one another, which is why the selected households are not independent. When there is such structure in the population, and there are clusters in the sample (often called primary sampling units, or psu in design terms), the regression model above can be more explicitly written as:

$$Y_{ij} = X_{ij}\beta + c_i + e_{ij} \qquad (12.3)$$

where Y_{ij} represents the measurement on the *j*th unit in the *i*th psu, c_i the error term held in common by the *i*th psu, and e_{ij} the household level error within the psu. The relative importance of the two sources of error can be measured by their respective variances σ_c^2 and σ_e^2. For this model and survey data, there are estimation issues linked to unequal selection probabilities and to non-independence of the units sampled. Ghosh and Rao (1994) give an overview of how to obtain small area estimates, together with standard errors, for this model when the data have been collected using a complex sampling scheme.

Where ELL is extended, and individual level data is available, as it is for stunting, underweight, and wasting in children under five years of age, an additional error term at child level within household needs to be added. In the general explanation below we focus on (12.3) to establish general principles for separating the variation at "higher" and "lower" levels. When there are three error terms rather than two, the three terms form a sequence in which the psu or cluster remains the highest aggregation level, household is intermediate, and individual level variation is at the finest or lowest level.

The auxiliary variables X_{ij} can be useful primarily in explaining the cluster level variation, or the household level variation. The more variation that is explained at a particular level, the smaller the associated error variance, σ_c^2 or σ_e^2.

The estimate for a particular small area will typically be the average of the predicted *Y*s in that area. Because the standard error of a mean gets smaller as the sample size gets bigger, the contribution to the overall standard error of the variation at each level, household and cluster, depends on the sample size at that level. The number of households in a small area will typically be much larger than the number of clusters, so to get small standard errors for estimates at small area level, it is important that, at the higher level, the unexplained cluster level variance σ_c^2 should be small. Two important diagnostics of the model-fitting stage, in which the relationship between Y and X is estimated for the survey data, are the R^2 measuring how much of the variability in Y is explained by X, and the variance ratio $\sigma_c^2 / (\sigma_c^2 + \sigma_e^2)$ measuring how much of the unexplained variation is at the cluster level. Of these a low variance ratio is the more important because R^2 is strongly dependent on whether the model is fitted to household data (in which case R^2 tends to be higher) or to individual data (in which case R^2 tends to be lower). Note that although σ_c^2 and σ_e^2 are parameters, even for the same survey and choice of Y, they are different for different models with different regressors. If the model has three levels, for good small area estimation we want as much of the overall variance in the variance components to be at the lowest level (or perhaps the middle level) in the hierarchy rather than at cluster level, because the population contains fewer clusters than people. Incorporating GIS data and cluster level means into X as contextual variables can be particularly useful in lowering the variance ratio.

As already noted, the other important aspect of clustering is its effect on parameter estimation for the model. For social and socioeconomic surveys, the survey data modeled have usually been collected using a complex survey design, which uses weighting, stratification, and clustering. Accounting adequately for the survey design's complexity requires use of specialized statistical routines and software (Skinner *et al.* 1989; Chambers and Skinner 2003; Lehtonen and Pakhinen 2004; Longford 2005) so as to get consistent estimates for the regression coefficient vector β and its variance V_β. Suitable software packages include Stata, Sudaan, SPSS Complex Surveys, and the survey package in R. There are also limited specialized routines in SAS. The correct fitting procedures for regression models given complex survey data is an important aspect of small area estimation, and one for which the original ELL paper requires some technical modification – for details see Haslett *et al.* (2010a), and for an econometric perspective see also Tarozzi and Deaton (2009).

12.3.3 The ELL method

The ELL methodology was designed specifically for the small area estimation of poverty incidence, gap, and severity, based on per capita household expenditure. The target variable Y in (12.2) is log-transformed expenditure. The logarithm is used to make the highly right-skewed distribution of untransformed expenditure more symmetrical. The measurements on Y and X are assumed to be available from a survey, and X is also assumed available for all of the relevant population from a census that has been administered contemporaneous with the survey.

The first step is to identify, from a set of candidate variables, a set of auxiliary variables X that are in the survey and are also available for the whole population from the census. It is important that all the candidates should be in both survey and census and be defined and measured in a consistent way in both these data sources. In practice, this requires detailed and time consuming checking and cross-referencing of questionnaires, field manuals, and the statistical properties of the variables themselves. The model (12.3) is then estimated for the survey data, by using appropriate software and incorporating stratification, clustering, and weights via inverse sampling probabilities. From (12.2) and (12.3), the residuals (or estimated model errors) $\hat{u}_{ij}$ from this analysis are used to define the cluster level residuals $\hat{c}_i = \hat{u}_{i\cdot}$, where the dot denotes averaging over j, and the household level residuals $\hat{e}_{ij} = \hat{u}_{ij} - \hat{c}_i$.

The cluster level effects c_i are assumed all to come from the same distribution, but the household level effects e_{ij} may be heteroskedastic. The ELL method models this by allowing the variance σ_e^2 to depend on a subset Z of the auxiliary variables:

$$g(\sigma_e^2) = Z\alpha + r \tag{12.4}$$

where $g(.)$ is an appropriately chosen link function, α represents the effect of Z on the variance, and r is a random error term. The general form usually involves a logistic-type link function, fitted using the squared household level residuals:

$$\ln\left(\frac{\hat{e}_{ij}^2}{A - \hat{e}_{ij}^2}\right) = Z_{ij}\alpha + r_{ij} \tag{12.5}$$

The fitted variances $\hat{\sigma}_{e,ij}^2$ can be calculated from this model and used to produce standardized household level residuals $\hat{e}_{ij}^* = \hat{e}_{ij} / \hat{\sigma}_{e,ij}$. These are then mean-centered to sum to zero, either across the whole survey data set or separately within each cluster. Although there is a focus on heteroskedasticity in the econometric literature, standardizing the household level residuals using (12.4) and (12.5), as ELL requires, is in practice of limited importance for two reasons. First, experience in Bangladesh, Nepal, and Cambodia suggests the R^2 for fitting (12.5) is usually well below 5 percent, and second (as discussed above) it is not the modeling household level variance but of the cluster level variance (and even more importantly of an additional component, area level variance, if it is not negligible) that is the more critical component in fitting (12.3).

In the small area estimation methods outlined in Rao (2003), the estimated model (12.3) (or its extensions) is applied to the known X values in the sample only to produce predicted Y values, which are then averaged over each small area to produce a point estimate, the standard error of which is inferred from appropriate asymptotic theory.

For poverty mapping using ELL, our interest is not always directly in Y, which for ELL is log-transformed expenditure modeled via a linear model, but in several nonlinear functions of Y (namely poverty incidence, gap, and severity). The ELL method obtains unbiased estimates and standard errors for incidence, gap, and severity by using a bootstrap procedure, and the census data (not just the survey data) as described below.

12.3.4 Bootstrapping

Bootstrapping is a collection of statistical procedures that use computer-generated random numbers to resample from the original sample to simulate the distribution of an estimator (Efron and Tibshirani 1993). In the case of poverty mapping, to bootstrap we construct not only one predicted value:

$$\hat{Y}_{ij} = X_{ij}\hat{\beta} \tag{12.6}$$

(where $\hat{\beta}$ represents the estimated coefficients from fitting the model) but a large number, B, of alternative predicted values:

$$Y_{ij}^{b} = X_{ij}\beta^{b} + c_{i}^{b} + e_{ij}^{b}\,,\ b = 1,\ldots B \tag{12.7}$$

in such a way that we can take account of their variability, and thus estimate the accuracy of the predictor (12.6), or aggregations of it over the small areas.

The statistical analysis of the model chosen for Y provides information about how best to include variability in the predicted values. For example, because $\hat{\beta}$ is an unbiased estimator of β with variance V_{β}, for each $b=1,\ldots, B$ we can draw β^{b} independently from a multivariate normal distribution with mean $\hat{\beta}$ and variance matrix V_{β}. Putting aside (as ELL do) the need to test for area level random effects, the cluster level effects c_{i}^{b} can be drawn from the empirical distribution of c_i, i.e. taken randomly *with* replacement from the set of cluster level residuals $\hat{c}_i$; this is necessary because the actual cluster level residuals are known only for the clusters in the sample not for all the clusters in the census. (We could, as Molina and Rao (2010) suggest, use the actual residual for small areas and clusters that are in the sample, but for the clusters at least, because the fraction of clusters sampled is very low, the benefit is very small.)

If we want to take account of unequal variances (heteroskedasticity) in the household level residuals, we draw α^{b} from a multivariate normal distribution with mean $\hat{\alpha}$ and variance matrix V_{α}, then combine it with Z_{ij} via (12.4) to give a predicted variance $\sigma_{e,ij}^{b}$ and use this to adjust the household level effect, so that:

$$e_{ij}^{b} = e_{ij}^{*b} \times \sigma_{e,ij}^{b}$$

where e_{ij}^{*b} is a random draw from the empirical distribution of e_{ij}^{*}, either for the whole data set or within the cluster chosen for h_i. Otherwise, if no allowance is to be made for heteroskedasticity of the household level residuals, setting $e_{ij}^{b} = e_{ij}^{*b}$ suffices.

Each complete set of bootstrap values Y_{ij}^{b}, for a fixed value of b, will yield a set of small area estimates. In the case of poverty estimates, for each household exponentiating Y to give predicted expenditure $E_{ij}=\exp(Y_{ij})$, and then applying equation (12.1) gives the FGT measures. For poverty gap and severity, this is *not* equivalent to summing the Y_{ij} in each small area and then exponentiating.

To summarize, for ELL for each of the B predictions for each census household, the poverty incidence, gap, and severity are calculated using (12.1), then aggregated to small area level. The variation in these B small area estimates then provides an estimate of the standard error of the small area estimate in each small area conditional on the model being correct.

12.3.5 Interpretation of standard errors

The estimated standard error of each small area estimate reflects the uncertainty in that estimate, provided the model is correct. A useful rule of thumb, given tables of small area estimates and their standard errors, is to take two estimated standard errors on each side of the point estimate to represent the range of values within which we would expect the true value to be 95 percent of the time. To compare areas, the standard errors need to be combined (by squaring each, adding, and finding the square root). If the difference, plus or minus the estimated standard error of the difference, gives an interval that contains zero, then to a first approximation, the two estimates are not significantly different at the 5 percent level.

The small area estimates from different small areas are however correlated to some extent, so the more accurate alternative, provided the bootstrap estimates are available, is to consider the percentage of times the bootstrap estimate from one of two specified small areas exceeds the other.

When two or more estimates from different small areas are being compared, for example when deciding on priority areas for receiving development assistance, these methods permit testing of whether observed differences are indicative of real differences in poverty between the small areas. Using them serves to remind users of poverty maps that the information in them represents estimates which may not always be precise.

The size of the standard errors for ELL poverty estimates depends on a number of factors. Putting aside for the moment the need to test for area level random effects, the poorer the fit of the model (12.3) in terms of large σ_c^2 (or to a lesser extent σ_e^2), or a large $\sigma_c^2 / (\sigma_c^2 + \sigma_e^2)$ ratio, or (again to a lesser extent) low R^2, the more variation in the target variable will be unexplained and the greater the standard errors of the small area estimates will be. The population size, both in terms of number of households and number of clusters in each small area, is also important. Generally, for ELL standard errors decrease in proportion to the inverse square root of the population size in each small area, so that it is necessary to find a balance between having many small areas, and the estimates from them being acceptably accurate. Usually the levels at which aggregation are possible are determined by the hierarchical structure of national administrative boundaries, so that considerable effort at finding suitable models is required to ensure sufficiently accurate small area estimates at what is often a prespecified preferred level of administrative aggregation. Nevertheless, if we decide to create a poverty map such as Figure 12.1 from our small area estimates, at a level for which the standard errors are generally acceptable, there may still be some, smaller, areas for which the standard errors are larger than we would like.

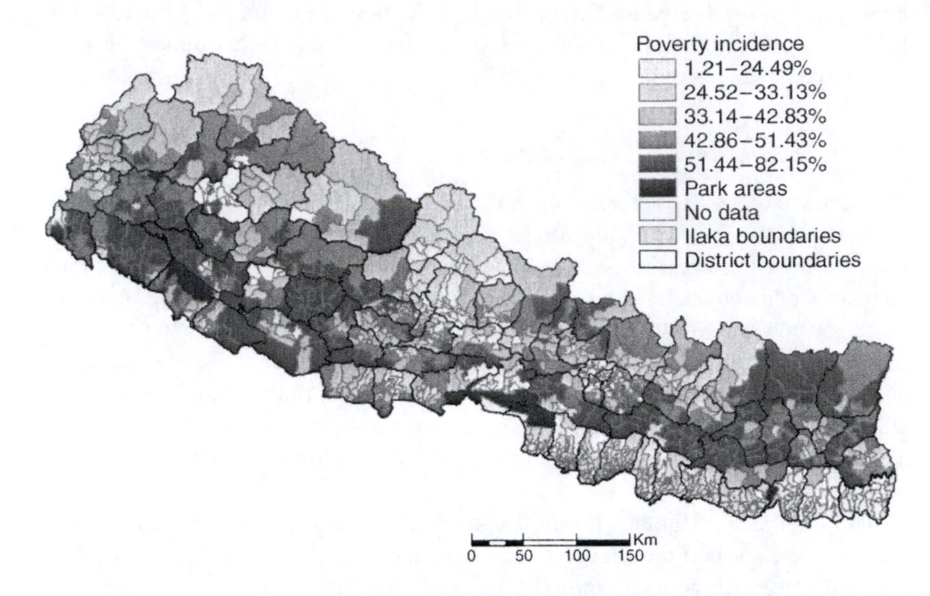

Figure 12.1 Poverty incidence (P0 – proportion of people below the poverty line) in Nepal 2001–2003 at the subdistrict (*ilaka*) level, based on small area estimation techniques.

The sample size used in fitting the model is also important. The bootstrapping methodology includes the variability in the estimated regression coefficients $\hat{\alpha}$ and $\hat{\beta}$. If the sample size is small, these estimates will be very uncertain and the standard errors of the small area estimates will be correspondingly large. This problem is also affected by the number of explanatory variables included in the auxiliary information, X and Z. We can always increase the apparent explanatory power of the model (i.e. increase R^2 from the survey data) by increasing the number of X variables, or by dividing the population into distinct subpopulations and fitting separate models in each as ELL recommend, but the increased uncertainty in the estimated coefficients may actually result in an overall *loss* of precision when the model is used to predict values for the census data. We must take care not to "over-fit" the model.

There is a small uncertainty in the estimates, and indeed the standard errors, due to the bootstrapping methodology, because it uses a finite number of bootstrap estimates to approximate the distribution of the estimator. This is generally a small effect. In any event, the bootstrap error can be decreased, at the expense of computing time, by increasing the number of bootstrap simulations B. An often used rule is to set $B = 100$.

Finally, and perhaps most importantly, the integrity of the estimates and standard errors depends on the model fitted using (12.3) being correct, both in terms of the fit being good and unbiased and because the model must apply to the census population in the same way that it applies to the sample. To have the same model apply to survey and census relies on good matching of survey and census variables used in X to provide valid auxiliary information. Care is also necessary to avoid spurious relationships or artefacts which appear statistically, to be true in the sample but do not hold in the population. Such problems can be caused by choosing variables indiscriminately from a very large set of possibilities (such as all available variables and their interactions to high order), or by fitting too many variables. Doing so can lead to models with higher R^2 and small area estimates with apparently small standard errors, but these estimated standard errors would be underestimates, perhaps severely so, a problem that is exacerbated if area level random effects are not tested for and fitted where required. For this reason, field verification as the final step in poverty mapping is extremely important, especially for small areas that appear to have anomalous estimates not in line with local knowledge.

The requirement for variables to match between survey and census is one reason that special care is needed if survey and census are not from the same time period. Changes between periods can simply be changes in level of the X variables, which (within bounds) the model can accommodate. But such changes can also be structural changes, because the interpretation of particular variables has changed and the model is no longer appropriate. Determining whether changes are structural or not is a very difficult and somewhat subjective task. Both types of change have the potential to add to standard errors of estimates, and in some cases to produce biased small area estimates. Even without such bias, interpretation can be affected – for example, differences of several years in dates of survey and census make it difficult to know to which period or date even otherwise sound small area estimates apply. See also the related comments on updating small area estimates in Section 12.4.

12.3.6 Software

As mentioned in Section 12.3.2, accounting adequately for the complexity of the survey design when fitting models to survey data via (12.3) requires use of specialized statistical routines and software so as to get consistent estimates for the regression coefficient vector β and its variance V_{β}, and suitable software packages include Stata, Sudaan, SPSS Complex Surveys, the survey package in R, and the limited specialized routines in SAS. Having fitted the model, use of the bootstrap, and estimation of small area estimates and their standard errors, requires additional programming.

PovMap (Zhao 2006) and PovMap2 (Zhao and Lanjouw 2009) provide alternative off-the-shelf software for all components of fitting ELL models to poverty data. This can simplify implementation, and is the reason the World Bank often runs short courses for government statistical office staff and other interested people in the use of this software.

There are however several provisos and cautions, all technical, some of them focused on the implementation the ELL methodology given a particular survey and census, others on the software itself.

Considering implementation first, no matter what software is used there are preliminary and associated tasks. These include examining questionnaires and field manuals, talking to field staff about implementation issues, discussing with local experts the translation of questionnaires into English from local languages (and how the questions actually asked of respondents relate to the translation), ensuring categories within questions in different questionnaires are equivalent (or can be collapsed to be equivalent), and checking whether the equivalence of questions and categories (even if questions appear identical) is supported by what should be equivalent statistics and scores from survey and census. These cannot simply be reduced to computer-based tasks.

Often, if there are multiple target variables, Y (e.g., malnutrition as well as poverty measures), there will also be multiple surveys and hence more than two survey questionnaires to be explored in detail and compared with the census questionnaire before modeling of survey data can formally begin.

During model fitting, it is also necessary to try many possible models and candidate variables that might be included in X in (12.3), estimate and re-estimate variance components (including small area level effects) and check the basis of those calculations (e.g., how many clusters have been sampled in each small area), extend the analysis to fitting preliminary small area estimates, go back to the modeling phase and explore further survey-based models, and their associated small area estimates, check small area estimates for accuracy to determine the smallest acceptable area at which reliable estimates can be provided. Throughout this process, care must be taken to consider spatial effects by testing for inclusion of regional effects and interactions in the model. This is a better alternative than ignoring such effects in the model, or – at the other extreme – subsetting data and fitting separate models. Care is also needed not to develop many candidate variables for X by using many high level interactions, not to overfit, and to include (or at least check for) an additional variance component at small area as well as cluster and household level effects.

Whatever software is used, these phases involve months of research, even if there are not logistical delays in getting final datasets. Completing all phases adequately, and knowing which issues need special attention for particular data also requires a high level of statistical expertise. This means that fitting small area models well usually requires a very high level of technical skill and years of experience, not simply a short course of weeks or days. Without this knowledge and experience, even when using off the shelf, blackbox software such as PovMap or PovMap2, there is a high risk that the small area estimates derived will be misleading or wrong. There is an even higher risk that estimates of their accuracy (via estimated standard errors) will be severe underestimates, and that this will engender an ill-founded sense of security in the results.

ELL-type methods are more sensitive than those outlined in Rao (2003) to such considerations. The principal reasons are that ELL-type methods use census data (which requires matching), incorporate a wider range of variables into the model than the Rao (2003) methods do (making model choice more difficult and more critical), and that ELL-type methods by using census data provide small area estimates at a much finer level (for example, Molina and Rao (2010) consider around thirty small areas for Spain, while Haslett and Jones (2006) consider about 1000 for Nepal).

Second, there are also internal issues in the structure of the PovMap software itself. Most of these are highly technical, but even for users they still have influence on the utility of the software. Detailed discussion of PovMap algorithms for fitting (12.3) can be found in Haslett *et al.* (2010a). For a different econometric perspective using instrumental variables, see Tarozzi and Deaton (2009). The user manual for PovMap2 (Zhao and Lanjouw 2009) focuses, as it should, on explaining how to use the software, and there is no other detailed technical documentation available for the PovMap2 software, so whether the technical issues in PovMap have been corrected in PovMap2 remains unclear.

Nevertheless, as a balance to these necessary provisos, the results when small area estimation is done well (whether PovMap or other software is used) remain extremely useful, and the World Bank's and Peter Lanjouw's continued involvement in supporting small area estimation of poverty and raising its profile should be much commended.

12.4 Extensions to ELL

Elbers *et al.* (2003) has become a focus in the literature but, in rudimentary form at least, was preceded by Hentschel *et al.* (2000), Fofack (2000), Bigman *et al.* (2000), Elbers *et al.* (2001), Alderman *et al.* (2002), and Elbers *et al.* (2002). In comparison with Rao (2003), the novel idea in Elbers *et al.* (2003) and to a lesser extent in the geographical literature (e.g., Williamson *et al.* 1998) is incorporating census predictions into small area estimation. In many respects, the other part of the ELL procedure, fitting regression-type models to survey data, already had a long history (e.g., Fuller 1975; Scott and Holt 1982). Regression for survey techniques had also been extended to small area estimation based on survey data (e.g., Ghosh and Rao 1994). By 2003 the statistical literature on the topic was extensive and it is this material about using such survey-based models for small area estimation that is summarized in Rao (2003).

The extensions to ELL fall into a number of categories: extensions of coverage and presentation, clarification of technique and data requirements, and extensions in concept.

Extensions of coverage are perhaps simplest to summarize. These include using ELL's incorporation of census data technique, applied to other poverty related measures such as those already mentioned (under-nourishment, child malnutrition) and others (e.g., Gini coefficients). There are also extensions to small area estimation of unemployment (e.g., Haslett *et al.* 2008b) and small domain estimation of expenditure for different ethnic groups (e.g., Haslett *et al.* 2008). However, although these two papers extend ELL to time series of small area estimates plus using administrative databases of variables related to the target variables, and to subpopulations not based on areas, at core it is still use of unit level census-type data that separates them from the techniques covered in Rao (2003).

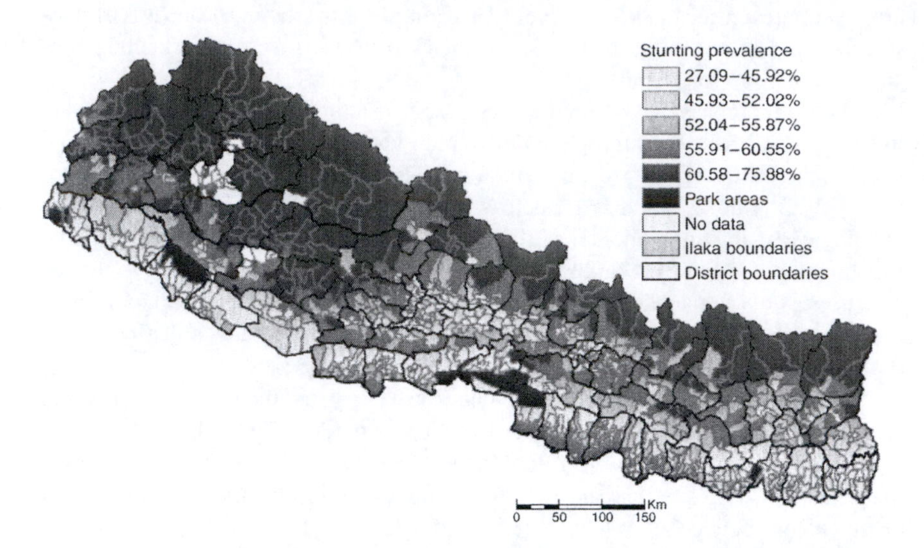

Figure 12.2 Stunting prevalence (S2 – i.e., >2 standard deviations below height-for-age) in Nepal 2001–2003 at the subdistrict (*ilaka*) level, based on small area estimation techniques.

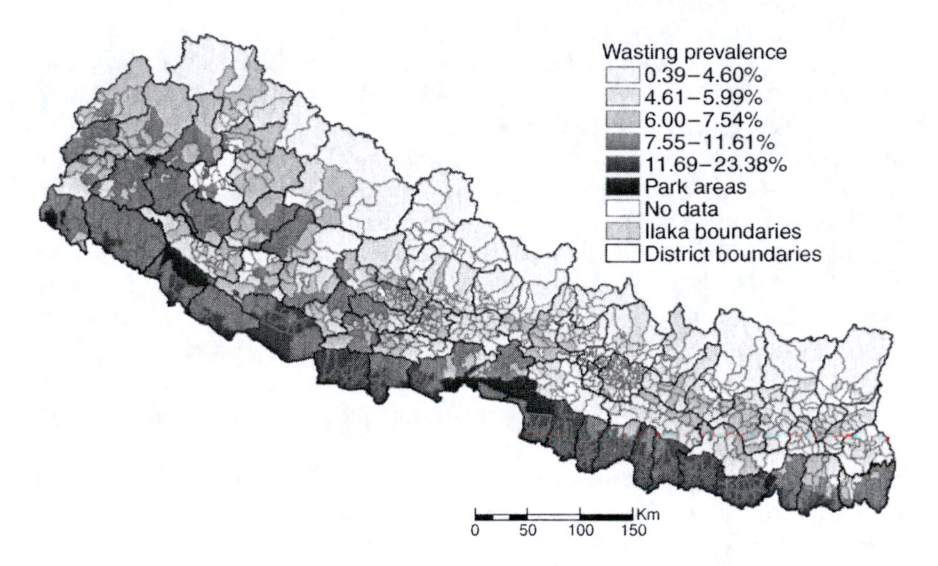

Figure 12.3 Wasting prevalence (W2 – i.e., >2 standard deviations below weight-for-age) in Nepal 2001–2003 at the subdistrict (*ilaka*) level, based on small area estimation techniques.

Another coverage or rather presentation issue is the use of poverty maps versus poverty estimates. Maps, such as Figures 12.1–12.3 for Nepal have major pictorial advantages over tables of estimates and their standard errors. Maps provide a useful overall impression and connect estimates from areas that are close together in a way that is difficult to do in a table, especially for someone unfamiliar with local names and geography. They allow comparison of different aspects of an underlying phenomenon, and to see differences in location of different indicators (for example, that stunting in children under five years of age in Nepal is mainly in the mountains, but underweight mainly in the Terai or plains). They are the reason that small area estimation of poverty is frequently called poverty mapping. But maps have disadvantages too. They present estimates in banded categories or ranges that are identically colored or shaded for each category, so the actual value of poverty in a given small area cannot be discerned. They also do not include standard error estimates, and (consequently) establishing which areas have significantly different poverty from a map of poverty estimates alone is not possible. Maps then are not sufficiently detailed for aid allocation. There are also questions of coloring, some preferring each map to be in many colors, others preferring shades of the same color. Particular caution is needed when a map is multi-colored as it can be difficult to gauge the extent of poverty if there is no clear rationale for color choice. Sometimes, to circumvent this issue, the colors of the spectrum (red, orange, yellow, green, blue, indigo, and violet) are used to provide a natural ordering. Nevertheless, maps of many colors do have more visual impact. Using shades of the same color on each map, but a different color on each map is another option. This may lack the impact of a kaleidoscope but, as well as allowing an impression without detailed use of a key, maps remain interpretable even when photocopied in black and white. This benefit is less important for softcopy maps.

Maps are however able to provide one useful diagnostic that tables of estimates and standard errors cannot. One interesting aspect of ELL-type techniques, and indeed of many small area estimation techniques, is that the spatial relationship between small areas is not exploited when fitting models, even though it would be possible to smooth the estimates spatially by using such techniques as conditional autoregressive (CAR) models, for example (see Cressie 1993 for details.) However such additional smoothing, while it tends to produce estimates in adjacent small areas that are similar, can by doing so disguise inadequate small area models. In practice, what can be more informative is to map estimates without spatial smoothing to see whether adjacent small areas that are not separated by natural boundaries (such as rivers or mountains) have similar estimates. Collections of adjacent small areas (especially areas with similar geographical features) that have identical or similar colors or shades tend to indicate a good model, while a mosaic of colors or shades is usually strong grounds for further model checking and refitting.

Other diagnostics can be useful too. Some such as variance ratios and to a lesser extent the proportion of variance explained by the survey-based model (R^2) have been indicated earlier. There is of course also the significance level for parameter estimates associated with candidate variables for the regression, but as for R^2 (which provides a model fit statistic at a level somewhat finer than that of the amalgamated small area estimates themselves) some more careful consideration is warranted, since it is possible for variables to be highly significant in the regression but essentially irrelevant to good small area estimation. An example is child's age in small area estimation of stunting, underweight, and wasting which, although often highly significant in the regression itself, varies so little in its distribution for different small areas that its final effect on the small area estimates themselves is usually negligible.

Extensions to ELL that have involved clarification of technique and data requirements include:

a Discussion of the central role of questionnaires, field manuals, and local knowledge including languages when matching survey and census data, in addition to checking statistical equivalence such as possible in PovMap2. See, for example, Haslett and Jones (2005, 2006).

b Variations on regression modeling methods, summarized in Haslett *et al.* (2010a) and discussed from an alternative econometric viewpoint in Tarozzi and Deaton (2010).

c Improvements to accuracy of estimated standard errors when using ELL and improvements, in order to limit the possibility of standard error underestimation. See Molina and Rao (2010) and Haslett and Jones (2010).

Extensions in concept are varied. The most obvious is to fit different types of statistical model to the survey data. For poverty incidence, modeling incidence directly through a logistic regression model fitted to the probability of a household (and its occupants) being in poverty using the survey data is a possibility. There is a range of reasons that ELL focused on log of expenditure, however:

a Fitting logistic models will not give small area estimates for poverty gap and poverty severity. Modeling log expenditure can give small area estimates of both, in addition to poverty incidence. The core problem is that the distributional form of any model for poverty gap and poverty severity needs to contain a cutoff (i.e., setting of gap and severity to zero for all households that are not in poverty). This leads to zero inflation, an aspect that usually needs to be modeled separately.

b Fitting mixed models even for the logistic can be more problematic than fitting linear mixed models of the form used by ELL for log expenditure. This would be a more minor technical issue if the survey design allowed easy estimation of variance components, but even for log-expenditure models there may for example be few clusters sampled in each small area, and this makes estimation more complicated.

c In the middle of the range of possible poverty values (say 30 to 70 percent) the logistic regression is reasonably well modeled by a linear regression since in this range both the logistic function and its variance is approximately constant.

Empirical Bayes and composite methods have also been considered (see Longford (2005) part II and Longford (2007) for composite estimation details, and Haslett and Jones (2010) for a comparison of the two methods with ELL). In both these alternative types of model the small areas are regarded as random rather than fixed, and the models use small area level rather than household level data. Some versatility is lost because the estimated standard errors of the small area estimates are then unable to reflect the distribution of expenditure within small area, and the use of aggregate rather than household level information means estimates have larger estimated standard errors than ELL.

The paper by Molina and Rao (2010) has already been mentioned in Section 12.1 as providing a synthesis of ELL and the spatial microsimulation on the one hand, and the methods used by statisticians for small area estimation up until 2003 on the other. Similarly to Haslett and Jones (2010), but with the addition of census predictions similar to ELL, Molina and Rao (2010) derive nonlinear small area estimates using Empirical Bayes and a nested error model. They focus on poverty indicators, but their methodology is applicable even to general non-linear parameters. They use a parametric bootstrap to estimate the mean squared error of their Empirical Bayes estimators, and estimate poverty incidence and poverty gap by gender in the fifty Spanish provinces together with estimated mean squared errors. Molina and Rao (2010) claim, quite rightly, that their method is better than ELL in situations where there is a reasonable sample size in each small area, because then their estimator incorporates the additional information in the direct estimate as well as a synthetic component via the model. The small number of provinces studied and used by Molina and Rao (2010) as small areas relative to ELL is critical to this "bestness" however. When the number of small areas is larger by an order of magnitude or more for a similar population size, as it is in most applications of small area estimation in developing and transition countries, the optimality of Molina and Rao's method no longer holds because their small area estimator reduces essentially to that of ELL with an added variance component at small area level.

This brings us back to the issue of the extent of aggregation used to form small area estimates that was first mentioned in Section 12.1. At low levels of aggregation, small area estimates have two main sources of error – that which comes via the error in estimating the regression parameters and that which comes from the variance components. The latter can be controlled by increasing the size of the small areas by greater aggregation, but the error in the regression coefficients remains whatever the extent of the aggregation. At some point in this aggregation sequence, the variance of the small area estimates from the estimated regression coefficients is approximately equal to the sampling error of the direct estimates. At that stage, almost nothing is gained by using small area estimation techniques rather than direct estimation. Note that situations in which Molina and Rao's method has advantages over ELL is nearer this break-even point than those situations where the two methods are essentially equivalent.

In the range of aggregation for which ELL and Molina and Rao are essentially equivalent, no matter which of the two methods is used, the critical issues are matching of survey and census data, choice of regressor variables (Miller 2002), and limiting possible use of high order interactions in models, using regional effects and selected regional interactions rather than fitting different models to data subsets, using sound regression techniques for survey data, and taking care to estimate all variance components well during model fitting. Models that use more regression parameters than are used or needed in the models discussed in Rao (2003) are also favored, because such models reduce the size of the variance components and make the need for a small area level variance component less likely, i.e., there is benefit from a well-based, comprehensive model that uses more significant variables because it limits the size of the variance components. Indeed the advantages can be even more pronounced – in a number of poverty studies (in Bangladesh, Nepal, Philippines, and Cambodia, for example), although difficult to estimate well, the size of the variance component at small area level (given there is a cluster level variance component in the model) has been negligible.

One question often asked is how to apply ELL when survey and census are not contemporaneous. The answer is "with caution", as the time period to which such estimates apply is not clear. Lanjouw and van der Wiede (2006) suggest focusing on structural variables that do not change their relationship to poverty over time, and using only these in small area models. However this begs the question of which variables are structural. An alternative that sidesteps this problem is that of Isidro (2010) and Isidro *et al.* (2010) which first fit a model using contemporaneous survey and census, then update this using an extension of structure preserving estimation called ESPREE by using a set of margins from the new survey, and the average pseudo-census derived from the initial model predictions. Standard errors can also be estimated by using a set of updated margins from a new survey that make allowance for the sampling error, and a set of multiway tables constructed from the range of bootstrapped pseudo-censuses which are the result of the initial model predictions being applied to the census.

The final aspect of extensions to ELL is more general and brings us back to the initial stages of any small area estimation project. Small area estimation requires survey data but, in the main, surveys are designed with other purposes in mind. This, for example, is why variance components are often so difficult to estimate well. Recent research on this topic includes Münnich and Burgard (2012) and Haslett (2012), and these papers along with the earlier research of Marker (2001) and Singh *et al.* (1994) should be consulted if a survey that will be used for small area estimation is being designed, even if small area estimation is one purpose among many.

References

Alderman, H., Babita, M., Demombynes, G., Makhata, N., and Ozler, B. (2002) How low can you go? Combining census and survey data for mapping poverty in South Africa, *Journal of African Economics*, 11, 2, 169–200.

Ballas, D., Clarke, G.P., and Turton, I. (2003) A spatial microsimulation model for social policy micro-spatial analysis, in B. Boots, A. Okabe & R. Thomas (eds) *Modelling Geographical Systems: Statistical and Computational Applications*, Kluwer, Dordrecht, 143–168.

Bigman, D., Dercon, S., Guillaume, D., and Lambotte, M. (2000) Community targeting for poverty reduction in Burkina Faso, *World Bank Economic Review*, 14, 1, 167–193.

Birkin, M. and Clarke, G.P. (1995) Using microsimulation methods to synthesize census data, in S. Openshaw (ed.) *Census Users' Handbook*, GeoInformation International, London, 363–387.

Bramley, G. and Smart, G. (1996) Modelling local income distributions in Britain, *Regional Studies*, 30, 3, 239–255.

Chambers, R.L. and Skinner, C.J. (2003) *Analysis of Survey Data*, John Wiley and Sons, New York.

Cochran, W.G. (1977) *Sampling Techniques*, 3rd edition, John Wiley and Sons, New York.

Cressie, N. (1993) *Statistics for Spatial Data*, revised edition, John Wiley and Sons, New York.

Efron, B. and Tibshirani, R.J. (1993) *An Introduction to the Bootstrap*, Chapman and Hall, New York, London.

Elbers, C., Lanjouw, J.O., and Lanjouw, P. (2002) Micro-level estimation of welfare, Policy Research Working Paper 2911, Development Research Group, World Bank, Washington DC.

Elbers, C., Lanjouw, J.O., and Lanjouw, P. (2003) Micro-level estimation of poverty and inequality, *Econometrica*, 71, 1, 355–364.

Elbers, C., Lanjouw, J.O., Lanjouw, P., and Leite, P.G. (2001) Poverty and inequality in Brazil: new estimates from combined PPV-PNAD data, World Bank. www.rimisp.org/FCKeditor/UserFiles/File/documentos/docs/pdf/elbers_et_al_poverty_and_inequality_in_brazil.pdf.

Fofack, H. (2000) Combining light monitoring surveys with integrated surveys to improve targeting for poverty reduction, *World Bank Economic Review*, 14, 1, 195–219.

Foster, J.E., Greer, J., and Thorbeck, E. (1984) A class of decomposable poverty measures, *Econometrica*, 52, 3, 761–766.

Fuller, W.A. (1975) Regression analysis for sample surveys, *Sankhya* C 37, 117–132.

Fuller, W.A. (2009) *Sampling Statistics*, John Wiley and Sons, New York.

Ghosh, M. and Rao, J.N.K. (1994) Small area estimation: an appraisal, *Statistical Science*, 9, 1, 55–93.

Hamill, P.V.V., Dridz,T.A., Johnson, C.Z., Reed, R.B., Roche, A.F., and Moore, W.M. (1979) Physical growth: National Center for Health Statistics percentile, *American Journal of Clinical Nutrition*, 32, 3, 607–621.

Hansen, M.H., Hurwitz, W.N., and Madow, W.G. (1953) *Sample Survey Methods and Theory*: *Vol I & II*, John Wiley and Sons, New York.

Haslett, S. (2012) Practical guidelines for design and analysis of sample surveys for small area estimation, Special Issue on Small Area Estimation, *Journal of the Indian Society of Agricultural Statistics*, 66, 1, 203–212. www.isas.org.in/jisas/jsp/onlinelatestissue.jsp.

Haslett, S., Isidro, M., and Jones, G. (2010a) Comparison of survey regression techniques in the context of small area estimation of poverty, *Survey Methodology*, 36, 2, 157–170.

Haslett, S. and Jones, G. (2005) Small area estimation using surveys and censuses: some practical and statistical issues. *Statistics in Transition*, 7, 3, 541–555.

Haslett, S. and Jones, G. (2006) Small area estimation of poverty, caloric intake and malnutrition in Nepal. Nepal Central Bureau of Statistics/World Food Programme, United Nations/World Bank, Kathmandu, September, 184 pages, ISBN: 999337018-5.

Haslett, S. and Jones, G. (2010) Small area estimation of poverty: the aid industry standard and its alternatives, *Australian and New Zealand Journal of Statistics*, 52, 4, 341–362.

Haslett, S., Jones, G., and Enright, J. (2008a) Small-domain estimation of Māori expenditure patterns, *Official Statistics Report Series*, 3, Statistics New Zealand, ISSN 1177-5017, 87 pages. www.statisphere.govt.nz/further-resources-and-info/official-statistics-research/series.aspx.

Haslett, S., Jones, G., Noble, A., and Ballas, D. (2010b) More for less? Using existing statistical modeling to combine existing data sources to produce sounder, more detailed, and less expensive, *Official Statistics Report Series*, 5, Statistics New Zealand, ISSN 1177-5017, 75 pages. www.statisphere.govt.nz/further-resources-and-info/official-statistics-research/series.aspx.

Haslett, S., Noble, A., and Zabala, F. (2008b) New approaches to small area estimation of unemployment, *Official Statistics Report Series*, 3, Statistics New Zealand, ISSN 1177-5017, 144 pages. www.statisphere.govt.nz/further-resources-and-info/official-statistics-research/series.aspx.

Hentschel, J., Lanjouw, J.O., Lanjouw, P., and Pogge, J. (2000) Combining census and survey data to trace the spatial dimensions of poverty: a case study in Ecuador, *World Bank Economic Review*, 14, 1, 147–165.

Isidro, M. (2010) Intercensal updating of small area estimates, Unpublished PhD thesis, Massey University, New Zealand.

Isidro, M., Haslett, S., and Jones, G. (2010) Extended structure-preserving estimation method for updating small-area estimates of poverty, *Joint Statistical Meetings, Proceedings of the American Statistical Association 2010*, Vancouver, Canada. www.amstat.org/meetings/jsm/2010/onlineprogram.

Lanjouw, P. and van der Wiede, R. (2006) Determining changes in welfare distributions at the micro-level: updating poverty maps, Powerpoint presentation at the NSCB Workshop for the NSCB/World Bank Intercensal Updating Project, Philippines.

Lehtonen, R. and Pakhinen, E. (2004) *Practical Methods for Design and Analysis of Complex Sample Surveys*, 2nd edition, John Wiley and Sons, New York.

Lohr, S. (2010) *Sampling: Design and Analysis*, 2nd edition, Brooks/Cole, Cengage Learning, Kentucky, USA.

Longford, N.T. (2005) *Missing Data and Small Area Estimation: Modern Analytical Equipment for the Survey Statistician*, Springer-Verlag, New York.

Longford N.T. (2007) On standard errors of model-based small-area estimators, *Survey Methodology*, 33, 69–79.

Marker, D.A. (2001) Producing small area estimates from national surveys: methods for minimizing use of indirect estimators, *Survey Methodology*, 27, 2, 183–188.

Miller, A. (2002) *Subset Selection in Regression*, 2nd edition, Chapman and Hall/CRC, New York, London.

Molina, I. and Rao, J.N.K. (2010) Small area estimation of poverty indicators, *Canadian Journal of Statistics*, 38, 3, 369–385.

Münnich, R.T. and Burgard, J.P. (2012) On the influence of sampling design on small area estimates, *Journal of the Indian Society of Agricultural Statistics*, 66, 1, 145–156. www.isas.org.in/jisas/jsp/onlinelatestissue.jsp.

Neyman, J. (1934) On the two different aspects of the representative method: the method of stratified sampling and the method of purposive selection, *Journal of the Royal Statistical Society*, 97, 558–606.

Rao, J.N.K. (2003) *Small Area Estimation*, John Wiley and Sons, New York.

Scott, A.J. and Holt, D. (1982) The effect of two stage sampling on ordinary least squares methods, *Journal of the American Statistical Association*, 77, 380, 848–854.

Singh, M.P., Gambino, J., and Mantel, H.J. (1994) Issues and strategies for small area data, *Survey Methodology*, 20, 1, 3–22.

Skinner, C.J., Holt, D., and Smith, T.M.F. (1989) *Analysis of Complex Survey Data*, John Wiley and Sons, New York.

Tarozzi, A. and Deaton, A. (2009) Using census and survey data to estimate poverty and inequality for small areas, *Review of Economics and Statistics*, 91, 773–792.

United Nations (2005) *Household Sample Surveys in Developing and Transition Countries*, Department of Economic and Social Affairs, Statistics Division, Studies in Methods Series F No. 96, ST/ESA/STAT/SER.F/96, ISBN 92-1-161481-3, United Nations, New York. http://unstats.un.org/unsd/hhsurveys/pdf/Household_surveys.pdf.

Valliant, R., Dorfman, A.H., and Royall, R.M. (2000) *Finite Population Sampling and Inference: A Prediction Approach*, John Wiley and Sons, New York.

Williamson, P., Birkin, M., and Rees, P. (1998) The estimation of population microdata by using data from small area statistics and samples of anonymised records, *Environment and Planning* A, 30, 5, 785–816.

Zhao, Q. (2006) *User Manual for PovMap*, World Bank. http://siteresources.worldbank.org/INTPGI/Resources/342674-1092157888460/Zhao ManualPovMap.pdf.

Zhou, Q. and Lanjouw, P. (2009) *PovMap2: A User's Guide*, World Bank. http://go.worldbank.org/QG9L6V7P20.

13 Estimation of poverty measures in small areas

Isabel Molina and Jon N.K. Rao

13.1 Introduction

When statistical information is demanded at local level, often the survey sample sizes in the local areas are not large enough to produce reliable area-specific or direct estimates and then small area estimation techniques are required. Most of these techniques are based on models that establish some relation across all areas between the target variable and some available auxiliary variables, and this relation allows us to "borrow strength" from all areas to find estimates in a particular area. So far model-based small area estimation has largely focused on means or totals, using either area level models or unit level models. Empirical best linear unbiased prediction (EBLUP), empirical Bayes or empirical best (EB) and hierarchical Bayes (HB) methods have been extensively used for point estimation and for measuring the variability of the estimators. The book by Rao (2003) gives an extensive account of those methods. Several review papers have also appeared after the publication of Rao's book that cover newer developments in small area estimation, see e.g. Jiang and Lahiri (2006) and Datta (2009).

Area level models use only direct area estimates and area level covariates. Such area level data avoid confidentiality issues due to the aggregation process and are therefore more easily available than unit level data. Widely used area level models are the Fay–Herriot (FH) models, introduced by Fay and Herriot (1979), to estimate mean per capita income in small places in the US. A disadvantage of these models is that sampling variances of direct area estimators are assumed to be known. In practice, these variances are estimated from the within area unit level data for the target variable and then smoothed using some models, but the accuracy of the smoothed estimates is generally not known.

When unit level data are available, models established for the individual units, called unit level models, avoid the loss of information due to aggregation. These models are much more powerful because they incorporate more information in the estimation process, compared to area level models. The first model of this kind was proposed by Battese *et al.* (1998) and will be called hereafter BHF model. This model includes, apart from the usual individual errors, random effects for the areas representing the between area variation that is not explained by auxiliary variables. These models also avoid the assumption of known sampling variances of direct estimators.

Basic poverty indicators are the head count ratio, called here poverty incidence, which is simply the proportion of individuals with welfare under the poverty line, and the poverty gap, measuring the mean relative distance to the poverty line of the individuals with welfare under the poverty line. The FGT class of poverty indicators (Foster *et al.* 1984) contains the previous two as particular cases. This class requires the specification of a poverty line. Other poverty indicators, such as the fuzzy monetary and supplementary poverty indicators (Betti *et al.* 2006), are based on ranking the individuals with respect to their level of poverty or welfare and do not require the definition of a poverty line. Inequality measures include the Gini coefficient, Theil index, the generalized entropy class of measures, which includes the Theil index as a particular case, and Atkinson's inequality measures. For a description of many of these poverty and inequality measures see e.g. Neri *et al.* (2005). Most of these indicators are rather complex non-linear functions of the income or welfare of individuals. This means that the available small area methodology for totals or means cannot be applied in a straightforward way.

Few approaches to find efficient estimators of poverty indicators at local level have been studied. Here the most popular approaches are discussed. The first one is based on the FH model introduced above. This approach has been regularly used by the US Census Bureau, within the Small Area Income and Poverty Estimates (SAIPE) project (www.census.gov/hhes/www/saipe). This project produces estimates for counties, school districts and states of poor school-age children. These estimates are then used for the allocation of federal funds to local jurisdictions. Similarly, in Europe, the FH model was applied under the project EURAREA (www.statistics.gov.uk/eurarea) to estimate linear parameters such as mean income. As already mentioned, the problem with FH models is the difficulty in finding reliable estimators of sampling variances due to the small sample sizes within areas. Another disadvantage of these models is that each particular poverty indicator has a different mathematical expression and therefore may require indicator-specific modelling.

This chapter concentrates on approaches for unit level data that can be applied automatically to practically any area parameter, not only poverty or inequality indicators. These methods require the knowledge of the values of auxiliary variables for each population unit. The possible drawback is that they can be computationally expensive, especially in the case of large populations or when the indicators require rankings or pairwise comparisons of population individuals.

The first known method for small area estimation of general non-linear indicators under a unit level model was developed by Elbers, Lanjouw and Lanjouw (2003) and will be called the ELL method. Using that method, the World Bank has been releasing small area estimates of the FGT measures for several countries. The model combines both census and survey data and produces simulated censuses of the variables of interest using the bootstrap resampling method. Estimates for any desired small areas are produced from the simulated censuses.

The average over simulated censuses of the resulting area estimates is taken as the area estimate and the variance of the estimates is taken as a measure of variability of the estimate.

Another approach, based on the empirical best/Bayes methodology, has been recently introduced by Molina and Rao (2010) under the support of the European project SAMPLE (www.sample-project.eu/). This method is similar in spirit to the ELL method in that it combines survey and census auxiliary data. Nevertheless, it gives the estimator with minimum mean squared error or "best predictor" (more exactly, a Monte Carlo approximation of it). This is done under the assumption that there exists a transformation of the incomes of the individuals, or other welfare variable used to measure poverty, such that the transformed incomes follow the BHF model. Mean squared errors of the EB estimators are approximated by a parametric bootstrap method. EB method provides estimators with notably better efficiency (approximately the "best") because it uses the precious information provided by the sample more extensively by conditioning on the sample responses.

A variation of the EB method, called the fast EB method, has been recently introduced by Ferretti and Molina (2012). This method is computationally much faster and therefore it becomes suitable to estimate computationally complex poverty indicators, such as the fuzzy monetary and fuzzy supplementary indicators, in large populations.

We report the results of model-based and design-based simulation studies on the relative performance of EB, ELL and direct area-specific estimators. Our results show that the EB estimators can be considerably more efficient than the ELL and the direct estimators, and that the ELL estimators can be even less efficient than the direct estimators. Results also indicate that a bootstrap mean squared error (MSE) estimator appropriately tracks the true MSE.

13.2 Best predictor under a finite population

Let $\mathbf{y}$ be a random vector containing the values of a random variable associated with the units of a finite population. Let $\mathbf{y}_s$ be the sub-vector of $\mathbf{y}$ corresponding to the sample elements and $\mathbf{y}_r$ be the sub-vector of out-of-sample elements and consider without loss of generality that the elements of $\mathbf{y}$ are sorted as $\mathbf{y} = (\mathbf{y}_s', \mathbf{y}_r')'$. Now consider a real measurable function $\delta = h(\mathbf{y})$ of the random vector $\mathbf{y}$. The objective is to predict $\delta = h(\mathbf{y})$ using the sample data $\mathbf{y}_s$. Let $\hat{\delta}$ denote a predictor of δ based on $\mathbf{y}_s$. The MSE of $\hat{\delta}$ is defined as

$$MSE(\hat{\delta}) = E_{\mathbf{y}}\{(\hat{\delta} - \delta)^2\}, \tag{13.1}$$

where $E_{\mathbf{y}}$ denotes expectation with respect to the distribution of the population vector $\mathbf{y}$. The best predictor of δ is the function of $\mathbf{y}_s$ that minimizes (13.1). Consider the conditional expectation $\delta^0 = E_{\mathbf{y}_r}(\delta \mid \mathbf{y}_s)$, where the expectation is taken with respect to the distribution of $\mathbf{y}_r$ given $\mathbf{y}_s$. Note that δ^0 is a function of the sample data $\mathbf{y}_s$. Note that we can rewrite (13.1) in terms of δ^0 as

$$MSE(\hat{\delta}) = E_{\mathbf{y}}\{(\hat{\delta} - \delta^0)^2\} + 2E_{\mathbf{y}}\{(\hat{\delta} - \delta^0)(\delta^0 - \delta)\} + E_{\mathbf{y}}\{(\delta^0 - \delta)^2\}.$$

In this expression, the last term does not depend on $\hat{\delta}$. For the second term, observe that

$$E_{\mathbf{y}}\{(\hat{\delta} - \delta^0)(\delta^0 - \delta)\} = E_{\mathbf{y}_s}[E_{\mathbf{y}_r}\{(\hat{\delta} - \delta^0)(\delta^0 - \delta) \mid \mathbf{y}_s\}]$$
$$= E_{\mathbf{y}_s}[(\hat{\delta} - \delta^0)\{\delta^0 - E_{\mathbf{y}_r}(\delta \mid \mathbf{y}_s)\}] = 0$$

Thus, the best predictor of δ is the estimator $\hat{\delta}$ that minimizes $E_{\mathbf{y}}\{(\hat{\delta} - \delta^0)^2\}$. Since this quantity is non-negative and its minimum value is zero, the best predictor of δ is

$$\hat{\delta}^B = \delta^0 = E_{\mathbf{y}_s}\{E_{\mathbf{y}_r}(\delta \mid \mathbf{y}_s). \tag{13.2}$$

Note that the best predictor is unbiased in the sense that $E_{\mathbf{y}}(\hat{\delta}^B - \delta) = 0$ because

$$E_{\mathbf{y}_s}(\hat{\delta}^B) = E_{\mathbf{y}_s}\{E_{\mathbf{y}_r}(\delta \mid \mathbf{y}_s)\} = E_{\mathbf{y}}(\delta)$$

Typically, $\mathbf{y}$ follows a distribution depending on an unknown parameter vector $\boldsymbol{\theta}$. This parameter vector can be estimated using the sample data $\mathbf{y}_s$. Then, the empirical best (EB) predictor of δ, denoted $\hat{\delta}^{EB}$, is equal to (13.2), with the expectation taken with respect to the distribution of $\mathbf{y}_r|\mathbf{y}_s$ with $\boldsymbol{\theta}$ replaced by an estimator $\hat{\boldsymbol{\theta}}$. The EB predictor is not exactly unbiased, but the bias coming from the estimation of the parameter $\boldsymbol{\theta}$ is typically negligible.

13.3 EB estimation of FGT poverty indicators

Consider a population U partitioned in D domains or areas $U_1,\ldots, U_D$ of sizes $N_1,\ldots, N_D$. A sample s_d of size n_d is drawn from domain U_d, $d=1,\ldots, D$. Let E_{dj} be the value of a quantitative welfare measure for j-th individual within d-th domain and z a poverty line defined for the population. In this section, the target is to estimate the FGT poverty indicators for domain d, given by

$$F_{\alpha d} = \frac{1}{N_d}\sum_{j=1}^{N_d}\left(\frac{z - E_{dj}}{z}\right)^{\alpha} I(E_{dj} < z), \qquad \alpha = 0,1,2. \tag{13.3}$$

Following Section 13.2, to find the best predictor of $F_{\alpha d}$ we must express $F_{\alpha d}$ in terms of a domain vector $\mathbf{y}_d$, for which the conditional distribution of the out-of-sample sub-vector $\mathbf{y}_{dr}$ given sample data $\mathbf{y}_{ds}$ is known. Under a multivariate normal distribution, all conditional distributions are also normal. The distribution of the welfare variables E_{dj} is seldom normal due to the typical right-skewness of these kinds of economic variables. However, after some transformation, such as log, the resulting distribution might be approximately normal. Thus, we assume that there is a one-to-one transformation of the welfare variables, $Y_{dj}=T(E_{dj})$, that follows a normal distribution. In particular, we assume that Y_{dj} follows the BHF model, given by

$$Y_{dj} = \mathbf{x}'_{dj}\boldsymbol{\beta} + u_d + e_{dj}, \quad j = 1,\ldots,N_d, \; d = 1,\ldots,D,$$

$$u_d \sim \text{iid}\, N(0,\sigma_u^2), \quad e_{dj} \sim \text{iid}\, N(0,\sigma_e^2), \tag{13.4}$$

where $\mathbf{x}_{dj}$ is a column vector with the values of p explanatory variables, u_d is a random area-specific effect and e_{dj} is the residual error for the same observation. Let $\mathbf{y}_d = (\mathbf{y}'_{ds}, \mathbf{y}'_{dr})'$ be the vector containing the values of the transformed variables Y_{dj} for the sample and out-of-sample units within domain d. Then $F_{\alpha d}$, given by (13.3), can be expressed as a function of $\mathbf{y}_d$,

$$F_{\alpha d} = \frac{1}{N_d}\sum_{j=1}^{N_d}\left(\frac{z - T^{-1}(Y_{dj})}{z}\right)^{\alpha} I\{T^{-1}(Y_{dj}) < z\} =: h_\alpha(\mathbf{y}_d) \quad \alpha = 0,1,2.$$

Thus, the FGT poverty measure of order α is a non-linear function $h_\alpha(\mathbf{y}_d)$ of $\mathbf{y}_d$. Then by Section 13.2, the best predictor of $F_{\alpha d}$ under squared loss is given by

$$\hat{F}_{\alpha d}^{B} = E_{\mathbf{y}_{dr}}\{h_\alpha(\mathbf{y}_d) \mid \mathbf{y}_{ds}\} = \int_R h_\alpha(\mathbf{y}_d) f(\mathbf{y}_{dr} \mid \mathbf{y}_{ds}) d\mathbf{y}_{dr}, \tag{13.5}$$

where $f(\mathbf{y}_{dr}|\mathbf{y}_{ds})$ is the joint density of $\mathbf{y}_{dr}$ given the observed data vector $\mathbf{y}_{ds}$. Due to the complexity of the function $h_\alpha(\cdot)$, the expectation in (13.5) cannot be calculated explicitly. However, this expectation can be approximated by Monte Carlo, by generating L replicates $\{\mathbf{y}_{dr}^{(\ell)}; \ell = 1,\ldots L\}$ of $\mathbf{y}_{dr}$ from the distribution of $\mathbf{y}_{dr}|\mathbf{y}_{ds}$, attaching the sample elements to the generated vectors $\mathbf{y}_d^{(\ell)} = ((\mathbf{y}_{dr}^{(\ell)})', \mathbf{y}'_{ds})'$, calculating the target quantity for each replicate $F_{\alpha d}^{(\ell)} = h_\alpha(\mathbf{y}_d^{(\ell)})$ and averaging over the L replicates, that is,

$$\hat{F}_{\alpha d}^{B} \approx \frac{1}{L}\sum_{\ell=1}^{L} F_{\alpha d}^{(\ell)}. \tag{13.6}$$

Let $\mathbf{X}_d = (\mathbf{x}_{d1},\ldots,\mathbf{x}_{dN_d})'$, $d=1,\ldots, D$, $\mathbf{I}_k$ the identity matrix of dimension $k \times k$ and $\mathbf{1}_k$ a column vector of ones of length k. From (13.4), it holds that $\mathbf{y}_d$, $d=1,\ldots, D$, are independent with

$$\mathbf{y}_d \sim N(\mathbf{X}_d\boldsymbol{\beta}, \mathbf{V}_d),$$

where $\mathbf{V}_d = \sigma_u^2 \mathbf{1}_{N_d}\mathbf{1}'_{N_d} + \sigma_e^2\mathbf{I}_{N_d}$, $d=1,\ldots, D$. Consider the decomposition into sample and out-of-sample elements of $\mathbf{X}_d$ and of the elements of the covariance matrix,

$$\mathbf{X}_d = \begin{pmatrix}\mathbf{X}_{ds}\\ \mathbf{X}_{dr}\end{pmatrix}, \mathbf{V}_d = \begin{pmatrix}\mathbf{V}_{ds} & \mathbf{V}_{dsr}\\ \mathbf{V}_{drs} & \mathbf{V}_{dr}\end{pmatrix}$$

Due to the normality assumption, the distribution of the out-of-sample vector $\mathbf{y}_{dr}$ given the sample data $\mathbf{y}_{ds}$ is also normal and it is given by $\mathbf{y}_{dr}|\mathbf{y}_{ds} \sim N(\boldsymbol{\mu}_{dr|s}, \mathbf{V}_{dr|s})$, where the conditional mean vector and covariance matrix are given by

$$\boldsymbol{\mu}_{dr|s} = \mathbf{X}_{dr}\boldsymbol{\beta} + \sigma_u^2 \mathbf{1}_{N_d - n_d} \mathbf{1}'_{n_d} \mathbf{V}_{ds}^{-1} (\mathbf{y}_{ds} - \mathbf{X}_{ds}\boldsymbol{\beta}), \tag{13.7}$$

$$\mathbf{V}_{dr|s} = \sigma_u^2 (1 - \gamma_d) \mathbf{1}_{N_d - n_d} \mathbf{1}'_{N_d - n_d} + \sigma_e^2 \mathbf{I}_{N_d - n_d}, \tag{13.8}$$

when $\gamma_d = \sigma_u^2(\sigma_u^2 + \sigma_e^2 / n_d)^{-1}$. Note that the application of the Monte Carlo approximation (13.6) involves simulation of D multivariate Normal vectors of sizes $N_d - n_d$, $d = 1, \dots, D$, from the conditional distribution of $\mathbf{y}_{dr}|\mathbf{y}_{ds}$. Then this process has to be repeated a large number of times L, something computationally unfeasible for $N_d - n_d$ large. This can be avoided by noting that the conditional covariance matrix $\mathbf{V}_{dr|s}$, given by (13.8), corresponds to the covariance matrix of a vector $\mathbf{y}_{dr}$ generated by the model

$$\mathbf{y}_{dr} = \boldsymbol{\mu}_{dr|s} + v_d \mathbf{1}_{N_d - n_d} + \boldsymbol{\varepsilon}_{dr} \tag{13.9}$$

with new random effects v_d and errors $\boldsymbol{\varepsilon}_{dr}$ that are independent and satisfy

$$v_d \sim N(0, \sigma_u^2(1 - \gamma_d)) \text{ and } \boldsymbol{\varepsilon}_{dr} \sim N(\mathbf{0}_{N_d - n_d}, \sigma_e^2 \mathbf{I}_{N_d - n_d}).$$

Using model (13.9), instead of generating a multivariate normal vector of size $N_d - n_d$, we need to generate only univariate normal variables $v_d \sim N(0, \sigma_u^2(1 - \gamma_d))$ and $\varepsilon_{dj} \sim N(0, \sigma_e^2)$ independently, for $j \in r_d$, and then obtain the corresponding out-of-sample elements Y_{dj} from (13.9) using as means the corresponding elements of $\boldsymbol{\mu}_{dr|s}$ given by (13.7). In practice, the model parameters $\boldsymbol{\theta} = (\boldsymbol{\beta}', \sigma_u^2, \sigma_e^2)'$ are replaced by consistent estimators $\hat{\boldsymbol{\theta}} = (\hat{\boldsymbol{\beta}}', \hat{\sigma}_u^2, \hat{\sigma}_e^2)'$, such as the maximum likelihood (ML) or restricted ML (REML) estimators, and then the variables Y_{dj} are generated from (13.9) with $\boldsymbol{\theta}$ replaced by $\hat{\boldsymbol{\theta}}$, leading to the empirical best (EB) estimator $\hat{F}_{ad}^{EB}$ of F_{ad}.

13.4 Fast EB method

For very large populations and/or computationally complex indicators such as the fuzzy monetary and supplementary poverty indicators of Betti *et al.* (2006), the EB method described in Section 13.3 might be unfeasible. Calculation of these indicators requires sorting all population elements and, in the EB method, this needs to be repeated for each Monte Carlo replicate $\ell = 1, \dots, L$, which might be too time consuming for large N and large L.

A faster version of the EB method has been recently introduced by Ferretti and Molina (2012). This method is based on replacing in the Monte Carlo approximation (13.6), the true value of the indicator in population ℓ by a design-based estimator based on a sample drawn from this population. In this way, only the model responses corresponding to the sample elements need to be generated, avoiding the generation of the full population of responses and the sorting of all the population elements. Another advantage of this variation of the EB method is that the identification of the sample units in the register from where the auxiliary variables are obtained is not necessary; linking the units in the two files is often not possible.

More concretely, for each Monte Carlo replicate ℓ, the fast EB method takes a sample $s(\ell) \subseteq U$. Although there is freedom concerning the sampling design and sample allocation when choosing this sample and better choices can be made, a straightforward approach is to use the same sampling scheme and the same sample size allocation as in the original sample s. Then the values of the auxiliary variables corresponding to the units drawn in $s(\ell)$ are taken, $\mathbf{x}_{dj}$, $j \in s_d(\ell)$, where $s_d(\ell)$ is the subsample from d-th domain. Using those values, the corresponding responses $Y_{dj}, j \in s_d(\ell)$, are generated for $d=1,\ldots, D$ from (13.9). Let us denote the vector containing those values as $\mathbf{y}_{s(\ell)}$. With the sample data $\mathbf{y}_{s(\ell)}$, calculate an approximately design unbiased estimator of the poverty indicator; in the case of the FGT poverty measure, let us denote it $\hat{F}_{\alpha d}^{DB(\ell)}$. Finally, the fast EB estimator is given by

$$\hat{F}_{\alpha d}^{FEB} \approx \frac{1}{L}\sum_{\ell=1}^{L} \hat{F}_{\alpha d}^{DB(\ell)}.$$

Simulation results not reported here indicate that the fast EB method loses little efficiency as compared with the EB estimator described in Section 13.3 and it allows to overcome computational problems due to large populations or due to more complex poverty indicators.

13.5 Parametric bootstrap for MSE estimation

The parametric bootstrap for finite populations, introduced by González-Manteiga *et al.* (2008), can be used to derive MSE estimators for the EB and fast EB estimators of the area poverty indicators. This method works as follows:

1 Fit the nested-error model (13.4) by ML, REML or Henderson method III (Rao, 2003) to obtain model parameter estimators $\hat{\boldsymbol{\beta}}$, $\hat{\sigma}_u^2$ and $\hat{\sigma}_e^2$.

2 Generate bootstrap domain effects from

$$u_d^* \sim iidN(0,\hat{\sigma}_u^2), \qquad d = 1,\ldots,D.$$

3 Generate, independently of $u_1^*,\ldots,u_D^*$, unit errors.

$$e_{dj}^* \sim iidN(0,\hat{\sigma}_e^2), \qquad j = 1,\ldots,N_d, d = 1,\ldots,D.$$

4 Generate a bootstrap population of responses from the model

$$Y_{dj}^* = \mathbf{x}_{dj}'\hat{\boldsymbol{\beta}} + u_d^* + e_{dj}^*, \qquad j = 1,\ldots,N_d, d = 1,\ldots,D.$$

5 Let $\mathbf{y}_d^* = (Y_{d1}^*, \ldots, Y_{dN_d}^*)'$ denote the vector of generated bootstrap responses for domain d. Calculate target quantities for the bootstrap population

$$F_{\alpha d}^* = h_\alpha(\mathbf{y}_d^*), \quad d = 1, \ldots, D.$$

6 Take the elements Y_{dj}^* with indices contained in the sample s, $\mathbf{y}_s^*$. Fit the model to the bootstrap sample $\mathbf{y}_s^*$ and obtain bootstrap model parameter estimators, denoted $\hat{\sigma}_u^{2*}$, $\hat{\sigma}_e^{2*}$ and $\hat{\boldsymbol{\beta}}^*$.

7 Obtain the bootstrap EB estimator of $F^*_{\alpha d}$ through the Monte Carlo approximation, denoted $\hat{F}_{\alpha d}^{EB*}$, $d = 1, \ldots, D$.

8 Repeat steps (2)–(7) a large number of times B. Let $F_{\alpha d}^*(b)$ be true value and $\hat{F}_{\alpha d}^{EB*}(b)$ the EB estimator obtained in b-th replicate of the bootstrap procedure, $b = 1, \ldots, B$.

9 The bootstrap estimator of the MSE is then given by

$$\text{mse}_B(\hat{F}_{\alpha d}^{EB}) = B^{-1} \sum_{b=1}^{B} \{\hat{F}_{\alpha d}^{EB*}(b) - F_{\alpha d}^*(b)\}^2.$$

13.6 ELL estimation of FGT poverty indicators

The method of Elbers *et al*. (2003), called ELL method, assumes a nested error model on the transformed welfare variables similar to (13.4) but using random cluster effects, where the clusters may be different from the small areas. In fact, the small areas are not even specified in advance. Then ELL estimators of domain poverty indicators $F_{\alpha d}$ are computed by applying a method similar to the bootstrap procedure described in Section 13.5. To make their method comparable with the EB method described in Section 13.3, here we consider that the clusters are the same as areas. In this case, ELL estimators of domain FGT poverty indicators can be obtained as follows:

1 With the original sample data $\mathbf{y}_s$, fit the BHF model (13.4). Let $\hat{\boldsymbol{\beta}}$, $\hat{\sigma}_u^2$ and $\hat{\sigma}_e^2$ be the estimators of $\boldsymbol{\beta}$, σ_u^2 and σ_e^2 in model (13.4).

2 Generate area/cluster effects $\{u_d^*; d = 1, \ldots, D\}$ by bootstrap. Here we consider the parametric bootstrap, in which cluster effects are generated as $u_d^* \sim \text{iid } N(0, \hat{\sigma}_u^2)$.

3 Generate bootstrap model errors $\{e_{dj}^*; j = 1, \ldots, N_d, d = 1, \ldots, D\}$ independently of the cluster effects. Here we consider parametric bootstrap errors

$$e_{dj}^* \sim \text{iid } N(0, \hat{\sigma}_e^2), \quad j = 1, \ldots, N_d, d = 1, \ldots, D.$$

4 Construct a population vector $\mathbf{y}^* = ((\mathbf{y}_1^*)', \ldots, (\mathbf{y}_D^*)')'$ from the bootstrap model

$$Y_{dj}^* = \mathbf{x}_{dj}' \hat{\boldsymbol{\beta}} + u_d^* + e_{dj}^*, \quad j = 1, \ldots, N_d, d = 1, \ldots, D. \tag{13.10}$$

5 Calculate the true bootstrap domain parameters $F^*_{\alpha d} = h_\alpha(\mathbf{y}^*_d)$, $d=1,\ldots, D$.

6 The ELL estimator of $F_{\alpha d}$ is then given by the bootstrap mean

$$\hat{F}^{ELL}_{\alpha d} = E_*(F^*_{\alpha d}).$$

Finally, the MSE estimator obtained by the ELL method is the bootstrap variance of $F^*_{\alpha d}$, that is,

$$\text{mse}(\hat{F}^{ELL}_{\alpha d}) = E_*\{F^*_{\alpha d} - E_*(F^*_{\alpha d})]\}^2,$$

where E_* denotes the expectation with respect to bootstrap model (13.10) given the sample data. Note that $E_*(F^*_{\alpha d})$ is tracking $E(F_{\alpha d})$ and $E_*\{F^*_{\alpha d} - E_*(F^*_{\alpha d})]\}^2$ is tracking $E\{F_{\alpha d}-E(F_{\alpha d})\}^2$. In practice, ELL estimators are obtained from a Monte Carlo approximation by generating a large number, A, of population vectors $\mathbf{y}^{*(a)} = ((\mathbf{y}_1^{*(a)})',\ldots,(\mathbf{y}_D^{*(a)})')'$, $a=1,\ldots, A$, from model (13.10), calculating the bootstrap domain parameters for each population a in the form $F^{*(a)}_{\alpha d} = h_\alpha(\mathbf{y}^{*(a)}_d)$, $d=1,\ldots, D$, and later averaging over the A populations; that is, taking

$$\hat{F}^{ELL}_{\alpha d} \approx \frac{1}{A}\sum_{a=1}^{A} F^{*(a)}_{\alpha d} \text{ and } mse(\hat{F}^{ELL}_{\alpha d}) \approx \frac{1}{A}\sum_{a=1}^{A}(F^{*(a)}_{\alpha d} - \hat{F}^{ELL}_{\alpha d})^2.$$

Note that, in contrast to the EB method described in Section 13.3, the population vectors $\mathbf{y}^{*(a)}$ in ELL method are generated from the marginal distribution of the model responses instead from the conditional distribution given sample data $\mathbf{y}_s$, and they do not contain the observed sample data. Thus, if the model parameters were known, the ELL method as described here would not be using the sample data at all.

13.7 Model-based simulation experiment

This section describes a simulation study to analyse the performance of the EB method to estimate area poverty incidences and poverty gaps. Populations of size $N=20{,}000$ and composed of $D=80$ areas with $N_d=50$ elements in each area $d=1,\ldots, D$, were generated from model (13.4). The transformation $T(\cdot)$ defined in Section 13.3 is taken as $T(x)=\log(x)$, that is, the welfare variables E_{dj} are the exponentials of the model responses Y_{dj}.

As auxiliary variables in the model, we considered two binary variables X_1 and X_2 apart from the intercept. These binary variables were simulated from Bernoulli distributions with probabilities $p_{1d}=0.3+0.5d/D$ and $p_{2d}=0.2$, $d=1,\ldots, D$. Here, p_{1d} is a linear function of the area index for X_1 and p_{2d} is a constant. We consider sample indices s_d with within area sample sizes $n_d=50$ drawn independently in each area d by simple random sampling without replacement, $d=1,\ldots, D$. Variables X_1 and X_2 for the population units and sample indices were held fixed over all Monte Carlo simulations. The regression coefficients were taken as

$\boldsymbol{\beta}=(3,0.03,\ -0.04)'$. Using these values, the mean welfare E_{dj} increases when moving from the case ($X_1=0, X_2=0$) to ($X_1=1, X_2=0$), but decreases from ($X_1=0, X_2=0$) to ($X_1=0, X_2=1$). Hence, the highest income level is reached for $X_1=1$ and $X_2=0$. Note also that the probabilities p_{1d} of $X_1=1$ increase with the area index but p_{2d} of $X_2=1$ are constant, so that the last areas will have more individuals with larger Y_{dj} (less poor) and then the FGT poverty measures will decrease with the area index.

The random area effects variance was taken as $\sigma_u^2=(0.15)^2$ and the model error variance, $\sigma_e^2=(0.5)^2$. The poverty line was fixed at $z=12$, which is roughly 0.6 times the median of the welfare variables E_{dj} for a population generated as mentioned above. Hence, the poverty incidence for the simulated populations is approximately 16 percent.

We generated $I=10{,}000$ population vectors $\mathbf{y}^{(i)}=((\mathbf{y}_1^{(i)})',\ldots,(\mathbf{y}_D^{(i)})')'$ from the true model described above. For each population i, true values of poverty indicators, direct, EB and ELL estimators were computed as described below:

a The true area poverty incidences and gaps for each area $d=1,\ldots,D$ and each population i were obtained as

$$F_{ad}^{(i)}=\frac{1}{N_d}\sum_{j=1}^{N_d}\left(\frac{z-E_{dj}^{(i)}}{z}\right)^{\alpha} I(E_{dj}^{(i)}<z), E_{dj}^{(i)}=\exp(Y_{dj}^{(i)}).$$

b Using the sample part of the i-th population vector, $\mathbf{y}_s^{(i)}=((\mathbf{y}_{1s}^{(i)})',\ldots,(\mathbf{y}_{Ds}^{(i)})')'$, direct estimators of $F_{ad}^{(i)}$ were calculated as

$$\hat{F}_{ad}^{(i)}=\frac{1}{n_d}\sum_{j\in s_d}\left(\frac{z-E_{dj}^{(i)}}{z}\right)^{\alpha} I(E_{dj}^{(i)}<z), E_{dj}^{(i)}=\exp(Y_{dj}^{(i)}).$$

c The nested-error model given in (13.4) was fitted to the sample data $(\mathbf{y}_s^{(i)},\mathbf{X}_s)$. Then, substituting the estimated model parameters in (13.7) and (13.8), $L=50$ out-of-sample vectors $\mathbf{y}_r^{(i\ell)}=((\mathbf{y}_{1r}^{(i\ell)})',\ldots,(\mathbf{y}_{Dr}^{(i\ell)})')'$, $\ell=1,\ldots,L$ were generated from the conditional distributions of $\mathbf{y}_{dr}|\mathbf{y}_{ds}$ using (13.9) for $d=1,\ldots,D$. The sample data was attached to the generated out-of-sample data to form the population vector $\mathbf{y}_d^{(i\ell)}=((\mathbf{y}_{ds}^{(i\ell)})',\ldots,(\mathbf{y}_{dr}^{(i\ell)})')'$. Area poverty incidences and gaps were obtained for each population as

$$F_{ad}^{(i\ell)}=\frac{1}{N_d}\sum_{j=1}^{N_d}\left(\frac{z-E_{dj}^{(i\ell)}}{z}\right)^{\alpha} I(E_{dj}^{(i\ell)}<z),\ E_{dj}^{(i\ell)}=\exp(Y_{dj}^{(i\ell)}),\ d=1,\ldots,D.$$

The Monte Carlo approximations to the EB estimators of the poverty incidences and gaps were calculated by averaging over Monte Carlo populations, as

$$\hat{F}_{ad}^{EB(i)}=\frac{1}{L}\sum_{\ell=1}^{L}F_{ad}^{(i\ell)},\ d=1,\ldots,D.$$

d Finally, to implement the ELL estimators of the poverty indicators, we fitted model (13.4) to sample data $\mathbf{y}_s$, obtaining estimators of model parameters. Using those estimated model parameters, $A=50$ populations were generated by means of a parametric bootstrap as described in Section 13.6. The area poverty indicators were computed for each population i, and finally the ELL estimators $\hat{F}_{ad}^{ELL(i)}$ were obtained by averaging over the $A=50$ populations.

After (a)–(d), means over Monte Carlo populations $i=1,\ldots, I$ of true values of poverty indicators were computed as

$$E(F_{ad}) = \frac{1}{I}\sum_{i=1}^{I} F_{ad}^{(i)},\ d=1,\ldots,D$$

Then the biases were computed for EB, direct and ELL estimators as $E(\hat{F}_{ad}^{EB})-E(F_{ad})$, $E(\hat{F}_{ad})-E(F_{ad})$ and $E(\hat{F}_{ad}^{ELL})-E(F_{ad})$ respectively. The MSEs over Monte Carlo populations of the three estimators were also computed as $E(\hat{F}_{ad}^{EB}-F_{ad})^2$, $E(\hat{F}_{ad}-F_{ad})^2$ and $E(\hat{F}_{ad}^{ELL}-F_{ad})^2$.

Figure 13.1 reports the biases and MSEs of the three estimators of the poverty gaps for each area. Figure 13.1a shows that the EB estimator has the smallest absolute biases followed by the ELL estimator. However, in terms of bias, all estimators seem to perform reasonably well. However, in terms of MSE, observe in Figure 13.1b that the EB estimator is significantly more efficient than the ELL and the direct estimators. This figure reveals that, in this simulation study, the ELL estimators are less efficient than the direct estimators. Conclusions for the poverty incidence are similar but plots are not reported here.

Turning to MSE estimation, the parametric bootstrap procedure described in Section 13.5 was implemented with $B=500$ replicates and the results are plotted in Figure 13.2 for the poverty gap. The number of Monte Carlo simulations was $I=500$ and the true values of the MSEs were independently computed with $I=50{,}000$ Monte Carlo replicates. Figure 13.2 shows that the bootstrap MSE estimator tracks the pattern of the true MSE values. Similar results were obtained for the poverty incidence.

Remark 1. Note that we used $L=A=50$ for the EB and ELL methods in the simulation studies. A limited comparison of EB estimators for $L=50$ with the corresponding values for $L=1000$ showed that the choice $L=50$ gives fairly accurate results. In practice, however, when dealing with a given sample data set, it is advisable to use larger values of L such as $L\geq 200$.

13.8 Design-based simulation experiment

This section describes a simulation experiment to study the performance of the estimators under repeated sampling from a fixed population. We generated a fixed population with the same parameters as showed in Section 13.7. Then,

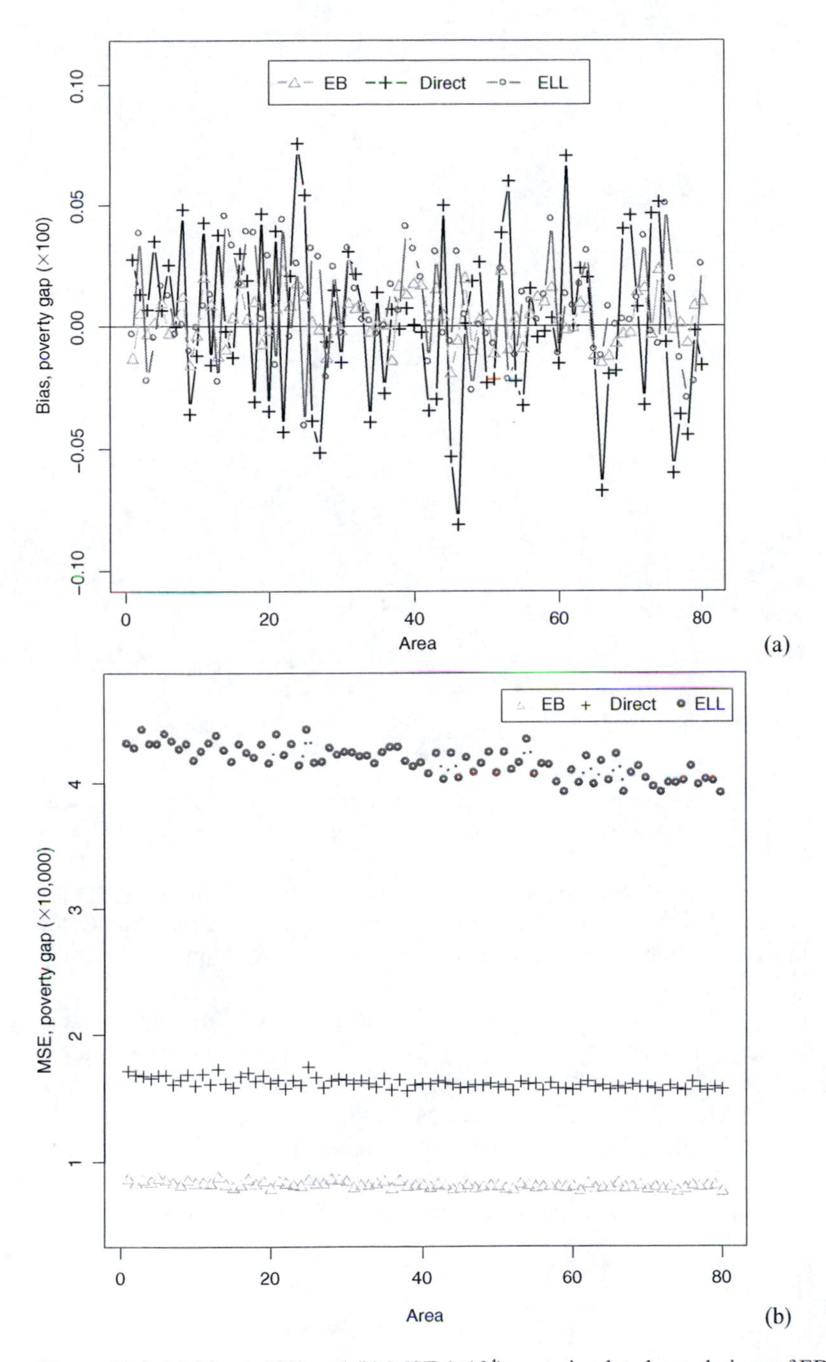

Figure 13.1 (a) Bias (×100) and (b) MSE (×10^4) over simulated populations of EB, direct and ELL estimators of the poverty gap F_{1d} for each area d under model-based simulations.

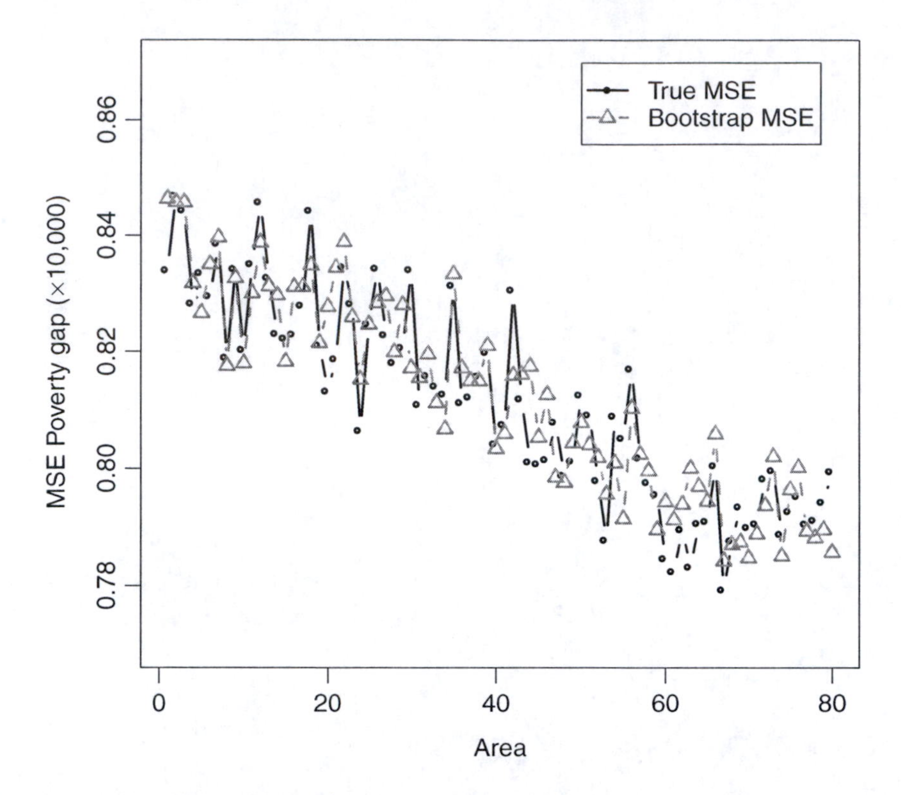

Figure 13.2 True MSE ($\times 10^4$) of EB predictor of poverty gap and average of bootstrap MSE estimates obtained with $B=500$ for each area d.

independently, $I=1000$ samples were drawn from the population using simple random sampling without replacement within each area. For each sample, the three estimators of the poverty incidences and gaps were computed, namely direct, EB and ELL estimators.

Figure 13.3 displays the design bias and the design MSE of the estimators of the poverty gaps for each area. As expected, the direct estimator shows practically zero bias, whereas the EB estimators show small biases, and the largest biases are for ELL estimators. Concerning MSE, the ELL estimators show small MSEs for some of the areas but very large for other areas. In contrast, the MSEs of EB and direct estimators are small for all areas; the MSEs of the EB estimators are smaller than the corresponding ones of the direct estimators for most of the areas.

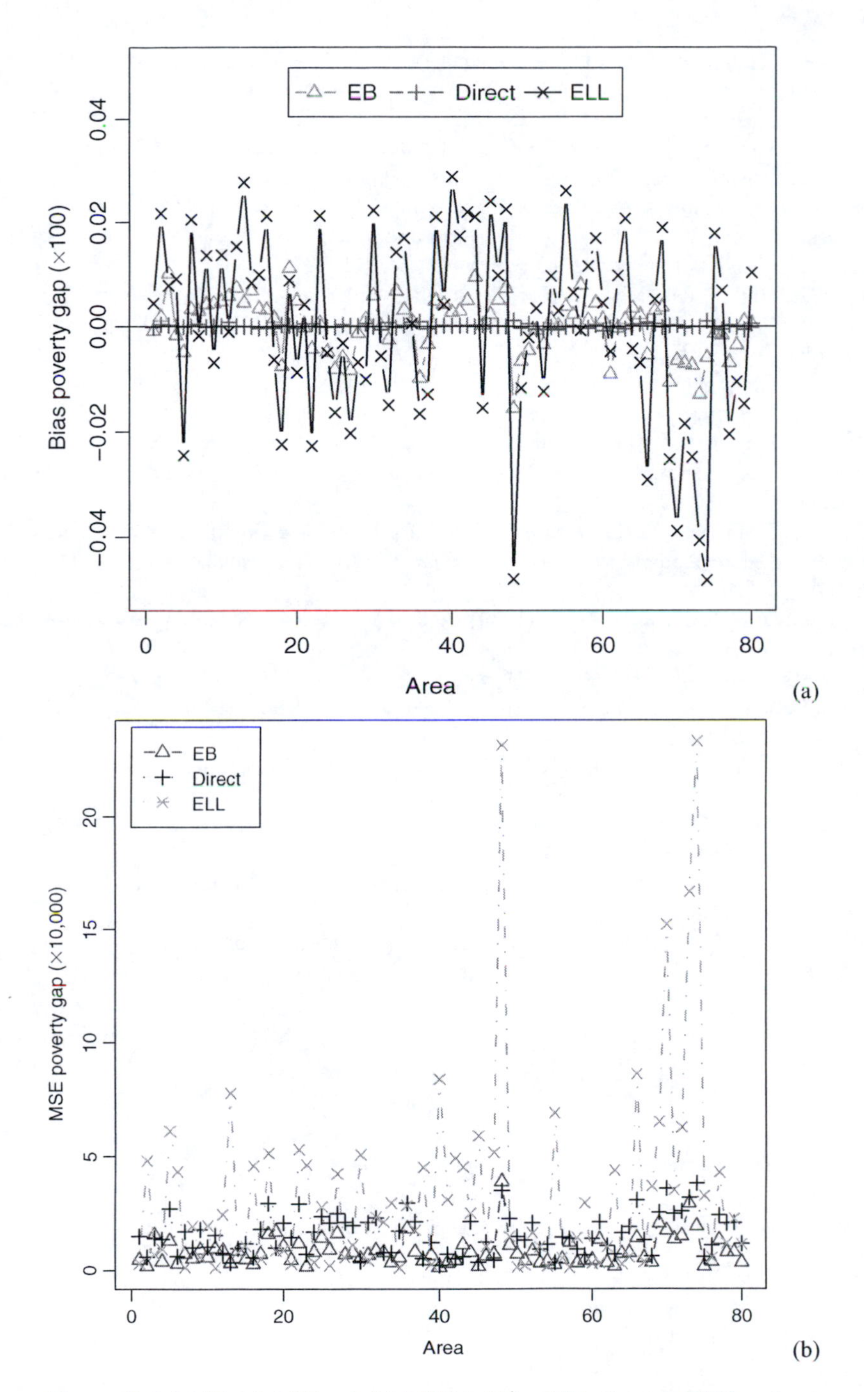

Figure 13.3 (a) Bias (×100) and (b) MSE ($\times 10^4$) of EB, direct and ELL estimators of the poverty gap F_{1d} for each area d under design-based simulations.

References

Battese, G.E., Harter, R.M. and Fuller, W.A. (1988). An Error-Components Model for Prediction of County Crop Areas Using Survey and Satellite Data. *Journal of the American Statistical Association*, 83, 28–36.

Betti, G., Cheli, B., Lemmi, A. and Verma, V. (2006). Multidimensional and Longitudinal Poverty: An Integrated Fuzzy Approach. In Lemmi, A., Betti, G. (eds) *Fuzzy Set Approach to Multidimensional Poverty Measurement*, 111–137, Springer, New York.

Cheli, B. and Lemmi, A. (1995). A Totally Fuzzy and Relative Approach to the Multidimensional Analysis of Poverty. *Economic Notes*, 24, 115–134.

Datta, G.S. (2009) Model-based Approach to Small Area Estimation. In Pfeffermann, D., Rao, C.R. (eds) *Handbook of Statistics: Sample Surveys*, 29B, 251–288, North Holland, Amsterdam.

Elbers, C., Lanjouw, J.O. and Lanjouw, P. (2003). Micro-level Estimation of Poverty and Inequality. *Econometrica*, 71, 355–364.

Fay, R.E. and Herriot, R.A. (1979). Estimation of Income from Small Places: An Application of James-Stein Procedures to Census Data. *Journal of the American Statistical Association*, 74, 269–277.

Ferretti, C. and Molina, I. (2012). Fast EB Method for Estimating Complex Poverty Indicators in Large Populations. *Journal of the Indian Society of Agricultural Statistics*, 66, 105–120.

Foster, J., Greer, J. and Thorbecke, E. (1984). A Class of Decomposable Poverty Measures. *Econometrica*, 52, 761–766.

González-Manteiga, W., Lombardía, M.J., Molina, I., Morales, D. and Santamaría, L. (2008). Bootstrap Mean Squared Error of a Small-area EBLUP. *Journal of Statistical Computation and Simulation*, 78, 443–462.

Jiang, J. and Lahiri, P. (2006). Mixed Model Prediction and Small Area Estimation. *Test*, 15, 1–96.

Molina, I. and Rao, J.N.K. (2010). Small Area Estimation of Poverty Indicators. *Canadian Journal of Statistics*, 38, 369–385.

Neri, L., Ballini, F. and Betti, G. (2005). Poverty and Inequality Mapping in Transition Countries. *Statistics in Transition*, 7, 135–157.

Rao, J.N.K. (2003). *Small Area Estimation*. John Wiley & Sons, Hoboken, NJ.

14 The use of spatial information for the estimation of poverty indicators at the small area level

Nicola Salvati, Caterina Giusti and Monica Pratesi

14.1 Introduction

In this chapter we focus on poverty and social exclusion mapping. Poverty and social exclusion are unevenly distributed both geographically and across social groups. This is particularly true for Italy, which is characterised by a low degree of regional cohesion (European Commission, 2005).

In Italy, the European Survey on Income and Living Conditions (EU-SILC) is conducted yearly by ISTAT to produce estimates of the prevalent living conditions of the population at national and regional (NUTS-2) levels. Regions are planned domains for which EU-SILC estimates are published, while the provinces (LAU1 level) and municipalities (LAU2 level) are unplanned domains. The regional samples are based on a stratified two-stage sample design: in each province the municipalities are the Primary Sampling Units (PSUs), while the households are the Secondary Sampling Units (SSUs). The PSUs are stratified according to administrative region and population size; the SSUs are selected by means of systematic sampling in each PSU.

We note that for areas smaller than the administrative regions, such as administrative provinces (LAU1) or the local labour systems we consider in this chapter, the municipalities sampled may be very few or none.

Direct estimates, that is estimates obtained applying standard weighted design-based estimators, may therefore have large errors at provincial level and may not even be computable at municipality level or for groups of municipalities (for example Local Economic Systems and Local Labour Systems), thereby requiring recourse to small area estimation (SAE) techniques.

The most popular class of models for small area estimation are random effects models, which include random area effects to account for between-area variation beyond that explained by auxiliary variables (Fay and Herriot, 1979; Battese *et al*., 1988; Rao, 2003; Jiang and Lahiri, 2006). Under this class of models the Best Linear Unbiased Predictor (BLUP) is obtained under the assumption of uncorrelated random area effects (Fay and Herriot, 1979). Details about this predictor and its empirical version (EBLUP) for small area parameters (total or mean) can be found in Rao (2003, Ch. 7) and in Jiang and Lahiri (2006). The EBLUP takes advantage of between small area variation, especially when this is not large in relation to the within small area variation (Rao, 2003). In many applications between and within variation are likely to be influenced by the spatial position of small areas and possible further improvement in the EBLUP estimator can be gained by including spatial information. There is extensive literature on the utility of spatial models in the context of SAE (see Cressie, 1991; Pfeffermann, 2002; Singh *et al.*, 2005; Pratesi and Salvati, 2008). It is well known that in many situations the location of the small areas can be so important as to cast doubt on the assumption of spatial independence in the Fay–Herriot model (1979).

Geographic information and geographical modelling can be valuable tools in describing and understanding many phenomena. It is a matter of fact that social economic phenomena have a spatial distribution, which is conditioned by nature and by the actions of man. When we look at a landscape we clearly recognise the effects of this combined action, and we note the evidence of the famous first law of geography: 'everything is related to everything else, but near things are more related than distant things' (Tobler, 1970). This law is also valid for small geographical areas: neighbouring areas are more likely to have similar values of the target parameter than areas which are far from each other. This evidence suggests that an adequate use of geographic information and geographical modelling can help in producing more accurate estimates for small area parameters. If the target is a small area parameter, geographical information constitutes a valid help in order to take advantage of the information available for the related areas.

In this chapter we illustrate different models that use geographical information and may be applied to produce estimates with adequate standard deviations, using data from sample surveys, administrative sources and population censuses. The focus is on area-level Fay–Herriot type models that allow for spatially correlated random area effects (Salvati, 2004; Singh *et al.*, 2005; Saei and Chambers, 2005; Petrucci and Salvati, 2006; Pratesi and Salvati, 2008, 2009). In particular, the idea of extending the Fay–Herriot model to include spatial correlation was first proposed by Cressie (1991), although it was not completely developed for SAE. More recently, Pfeffermann (2002) has shown that the loss of efficiency in SAE can be substantial when correlation between small areas is ignored. Nonetheless, when the spatial correlation is weak, Fay–Herriot models extended to include a correlation parameter do not lead to significant efficiency gains. Recently, the problem of introducing a common autocorrelation parameter among small areas through the Simultaneously Autoregressive (SAR) process has been taken into account and extended to the Fay–Herriot model (Salvati, 2004; Singh *et al.*, 2005; Petrucci *et al.*, 2005; Petrucci and Salvati, 2006; Pratesi and Salvati, 2008, 2009).

An alternative method to take into account the spatial information in SAE has been developed by Giusti *et al.* (2012). The authors propose a semiparametric version of the basic Fay–Herriot model based on P-splines, so that we can also handle situations in which the functional form of the relationship between the variable of interest and the covariates cannot be specified a priori. This is often the case when the data are supposed to be affected by spatial proximity effects. In these cases P-spline bivariate smoothing can easily introduce spatial effects to the area level model. Opsomer *et al.* (2008) proposed a similar small area model based on P-splines, but under the assumption that all the data are available at the unit level, which can be a restriction in some situations.

In this chapter we apply the previous models to data from the EU-SILC 2008 survey to estimate some indicators of poverty and living conditions (mean equivalised income, head count ratio and poverty gap) for each local labour system (LLS) in three Italian regions: Lombardy, Tuscany and Campania. The LLSs are defined as collections of contiguous municipalities which are supposed to form a single labour market, similar to travel-to-work areas used in other countries; according to the official EU nomenclature of local units they are intermediate between LAU1 and LAU2 levels. The choice of these three regions among the 20 existing in Italy is motivated by the geographical disparities that characterise the country. In particular, we believe that these data can be used to investigate the so-called 'north–south' divide, since each of the three regions, located one in the north, one in the centre and one in the south, can be considered as representative of these broad geographic regions of Italy.

We discuss model diagnostics and statistics, summarising the performance of small area estimators in order to select a small area working model and assess the validity of the estimates obtained.

This chapter is organised as follows. In Section 14.2 we review area level models for small area estimation that use spatial information. In Section 14.3 we present model diagnostics that help us build a model for performing small area estimation, and the results of the estimation of the poverty and living conditions indicators for each LLS. In Section 14.3 we also report some diagnostics for evaluating the reliability of the small area estimates.

14.2 Small area estimation models

14.2.1 Fay–Herriot models

Let ϑ be the $m \times 1$ vector of the parameters of inferential interest (typically small area totals y_i, small area means $\hat{y}_i$ with $i = 1, \ldots, m$) and let us assume that the direct estimator $\hat{\boldsymbol{\vartheta}}$ is available and design-unbiased, i.e.:

$$\hat{\boldsymbol{\vartheta}} = \boldsymbol{\vartheta} + \mathbf{e}, \tag{14.1}$$

where $\mathbf{e}$ is a vector of independent sampling errors with mean vector $\mathbf{0}$ and known diagonal variance matrix $\mathbf{R} = diag(\varphi_i)$, φ representing the sampling variances of the direct estimators of the area parameters of interest. Usually φ_i is unknown and estimated according to a variety of methods, including 'generalised variance functions': see Wolter (1985) and Wang and Fuller (2003) for details.

The basic area level model assumes that a $m \times p$ matrix of area-specific auxiliary variables (including an intercept term), $\mathbf{X}$, is linearly related to ϑ so:

$$\hat{\boldsymbol{\vartheta}} = \mathbf{X}\alpha + \mathbf{u}, \tag{14.2}$$

where $\boldsymbol{\alpha}$ is the $p \times 1$ vector of regression parameters, and $\mathbf{u}$ is the $m \times 1$ vector of independent random area-specific effects, with zero mean and $m \times m$ covariance matrix $\sum_u \sigma_u^2 \mathbf{I}_m$, and with $\mathbf{I}_m$ being the $m \times m$ identity matrix.

The combined model (Fay–Herriot, 1979) can be written as:

$$\hat{\boldsymbol{\vartheta}} = \mathbf{X}\alpha + \mathbf{u} + \mathbf{e}, \tag{14.3}$$

and is a special case of the linear mixed model. Under this model, the BLUP $\tilde{\vartheta}_i^{FH}(\sigma_u^2)$ is used extensively to obtain model-based indirect estimators of small area parameters ϑ and associated measures of variability. This approach allows a combination of the survey data with other data sources in a synthetic regression fitted using population area level covariates. More details on the empirical version (EBLUP) of the $\tilde{\vartheta}_i^{FH}(\sigma_u^2)$ predictor can be found in Rao (2003).

14.2.2 Spatial Fay–Herriot models

Salvati (2004), Singh *et al.* (2005) and Petrucci and Salvati (2006) have proposed the introduction of spatial autocorrelation in small area estimation under the Fay–Herriot model. The spatial dependence among small areas is introduced by specifying a linear mixed model with spatially correlated random area effects for ϑ, i.e.:

$$\boldsymbol{\vartheta} = \mathbf{X}\boldsymbol{\alpha} + \mathbf{D}\mathbf{v}, \tag{14.4}$$

where $\mathbf{D}$ is an $m \times m$ matrix of known positive constants and $\mathbf{v}$ is an $m \times 1$ vector of spatially correlated random area effects given by the following autoregressive process, with spatial autoregressive coefficient ρ and $m \times m$ spatial interaction matrix $\mathbf{W}$ (see Cressie, 1993; Anselin, 1992):

$$\mathbf{v} = \rho\mathbf{W}\mathbf{v} + \mathbf{u} \Rightarrow \mathbf{v} = \left(\mathbf{I}_m - \rho\mathbf{W}\right)^{-1}\mathbf{u}. \tag{14.5}$$

The $\mathbf{W}$ matrix describes the spatial interaction structure of the small areas, usually defined through the neighbourhood relationship between areas: generally speaking, $\mathbf{W}$ has a value of one in row i and column j if areas i and j are neighbours. The autoregressive coefficient ρ defines the strength of the spatial relationship among the random effects associated with neighbouring areas. Generally, for ease of interpretation, the spatial interaction matrix is defined in row-standardised form, in which the row elements sum to one; in this case ρ is called a spatial autocorrelation parameter (Banerjee *et al.*, 2004).

Combining (14.1) and (14.4), the estimator with spatially correlated errors can be written as:

$$\hat{\boldsymbol{\vartheta}} = \mathbf{X}\boldsymbol{\alpha} + \mathbf{D}\left(\mathbf{I} - \rho\mathbf{W}\right)^{-1}\mathbf{u} + \mathbf{e}. \tag{14.6}$$

The error term $\mathbf{v}$ has the $m \times m$ SAR covariance matrix:

$$\mathbf{G}(\delta) = \sigma_u^2 \left[\left(\mathbf{I} - \rho \mathbf{W}^T \right) \left(\mathbf{I} - \rho \mathbf{W} \right) \right]^{-1}, \tag{14.7}$$

and the covariance matrix of $\hat{\boldsymbol{\vartheta}}$ is given by:

$$\mathbf{V}(\delta) = \mathbf{R} + \mathbf{DGD}^{\mathrm{T}},$$

where $\delta = (\sigma_u^2, \rho)$.

Under model (14.6), the Spatial Best Linear Unbiased Predictor (SBLUP) estimator of ϑ_i is:

$$\tilde{\vartheta}_i^S(\delta) = \mathbf{x}_i \tilde{\alpha} + \mathbf{b}_i^T \mathbf{GD}^{\mathrm{T}} \left\{ \mathbf{R} + \mathbf{DGD}^{\mathrm{T}} \right\}^{-1} \left(\hat{\vartheta} - \mathbf{X}\tilde{\alpha} \right), \tag{14.8}$$

where $\tilde{\boldsymbol{\alpha}} = (\mathbf{X}^T \mathbf{V}^{-1} \mathbf{X})^{-1} \mathbf{X}^T \mathbf{V}^{-1} \hat{\boldsymbol{\vartheta}}$ and $\mathbf{b}_i^T$ is a $1 \times m$ vector $(0, 0 \ldots 0, 1, \ldots 0)$ with a value of one in the i-th position. The predictor is obtained from Henderson's (1975) results for general linear mixed models involving fixed and random effects. The SBLUP, when $\rho = 0$ and $\mathbf{D} = \mathbf{I}_m$, reduces to the BLUP, i.e. an independent random specific area effects model.

The SBLUP estimator $\tilde{\vartheta}_i^S(\delta)$ in (14.8) depends on δ, that is on the unknown variance component σ_u^2 and spatial autocorrelation parameter ρ. Substituting their asymptotically consistent estimators $\hat{\delta} = (\hat{\sigma}_u^2, \hat{\rho})$, obtained either by Maximum Likelihood (ML) or Restricted Maximum Likelihood (REML) methods, based on the normality assumption of the random effects, the following two stage estimator $\tilde{\vartheta}_i^S(\hat{\delta})$, called the SEBLUP, is obtained:

$$\tilde{\vartheta}_i^S(\hat{\delta}) = \mathbf{x}_i \hat{\alpha} + \mathbf{b}_i^T \hat{\mathbf{G}}\mathbf{D}^{\mathrm{T}} \left\{ \mathbf{R} + \mathbf{D}\hat{\mathbf{G}}\mathbf{D}^{\mathrm{T}} \right\}^{-1} \left(\hat{\vartheta} - \mathbf{X}\hat{\alpha} \right). \tag{14.9}$$

The ML estimators of σ_u^2 and ρ can be obtained iteratively using the 'Nelder–Mead' algorithm (Nelder and Mead, 1965) and the 'scoring' algorithm (Rao, 2003) in sequence. The sequential use of these procedures is necessary because the log-likelihood function has a global maximum as well as some local maximums: for more details see Singh *et al.* (2005) and Pratesi and Salvati (2008).

14.2.3 Semiparametric Fay–Herriot models

An alternative approach to introducing spatial correlation in an area level model is proposed by Giusti *et al.* (2012). We start by considering the Fay and Herriot (1979) model (14.3) again. As already mentioned, this model produces reliable small area estimates by combining the design model and the regression model and then borrowing strength from other domains. It assumes that the direct survey estimators are linear function of the covariates. When this assumption falls down, the Fay–Herriot model can lead to biased estimators of the small area parameters. A semiparametric specification of the Fay–Herriot model, which allows non-linearities in the relationship between the response variable and the auxiliary variables, can be obtained by P-splines.

A semiparametric additive model (hereafter referred to as a semiparametric model) with one covariate x_1 can be written as $f(\mathbf{x}_1)$, where the function $f(\cdot)$ is unknown, but assumed to be sufficiently well approximated by the function:

$$f(\mathbf{x}_1;\beta,\boldsymbol{\gamma}) = \eta_0 + \eta_1 \mathbf{x}_1 + \ldots + \eta_q \mathbf{x}_1^q + \sum_{k=1}^{K} (\mathbf{x}_1 - \kappa_k)_+^q, \tag{14.10}$$

where $\boldsymbol{\eta} = (\eta_0, \eta_1, \ldots, \eta_q)^T$ is the $(q+1)\times 1$ vector of the coefficients of the polynomial function, $\boldsymbol{\gamma} = (\gamma_0, \gamma_1, \ldots, \eta_k)^T$ is the coefficient vector of the truncated polynomial spline basis (P-spline), and q is the degree of the spline $(t)_+^q = t^q$ if $t>0$, 0 otherwise. The latter portion of the model allows for the handling of departures from a q polynomial t in the structure of the relationship. In this portion κ_k for $k=1,\ldots, K$ is a set of fixed knots and, if K is sufficiently large, the class of functions in (14.10) is very large and can approximate most smooth functions. Details on the selection of bases and knots can be found in Ruppert *et al.* (2003).

Since a P-spline model can be viewed as a random effects model (Ruppert *et al.*, 2003; Opsomer *et al.*, 2008), it can be combined with the Fay–Herriot model to obtain a semiparametric small area estimation framework based on linear mixed model regression.

Corresponding to the $\boldsymbol{\eta}$ and $\boldsymbol{\gamma}$ vectors we define:

$$\mathbf{X}_1 = \begin{bmatrix} 1 & x_{11} & \cdots & x_{11}^q \\ \vdots & \vdots & \ddots & \vdots \\ 1 & x_{1m} & \cdots & x_{1m}^q \end{bmatrix}, \quad \mathbf{Z} = \begin{bmatrix} (x_{11}-\kappa_1)_+^q & \cdots & (x_{11}-\kappa_K)_+^q \\ \vdots & \ddots & \vdots \\ (x_{1m}-\kappa_1)_+^q & \cdots & (x_{1m}-\kappa_K)_+^q \end{bmatrix}.$$

Following the same notation already introduced in the previous section, the mixed model representation of the semiparametric Fay–Herriot model (NPFH) can be written as:

$$\hat{\vartheta} = \begin{bmatrix} \mathbf{X} \\ \mathbf{X}_1 \end{bmatrix} [\alpha, \eta] + \mathbf{Z}\boldsymbol{\gamma} + \mathbf{D}\mathbf{u} + \mathbf{e}. \tag{14.11}$$

The $\mathbf{X}_1$ matrix of model (14.11) can be added to the $\mathbf{X}$ effect matrix, and model (14.11) becomes:

$$\hat{\vartheta} = \mathbf{X}\beta + \mathbf{Z}\gamma + \mathbf{D}\mathbf{u} + \mathbf{e} \tag{14.12}$$

where $\boldsymbol{\beta}$ is a $(p+q+1)\times 1$ vector of regression coefficients, and the $\boldsymbol{\gamma}$ component can be treated as a $K\times 1$ vector of independent and identically distributed random variables with mean $\mathbf{0}$ and $K\times K$ variance matrix $\boldsymbol{\Sigma}_\gamma = \sigma_\gamma^2 I_K$. The covariance matrix of model (2.12) is $\boldsymbol{\Sigma}(\psi) = \mathbf{Z}\boldsymbol{\Sigma}_\gamma\mathbf{Z}^T + \mathbf{D}\boldsymbol{\Sigma}_u\mathbf{D}^T + \mathbf{R}$, where $\psi = (\sigma_\gamma^2, \sigma_u^2)$.

Model-based estimation of the small area parameters can be obtained by using best linear unbiased prediction (Henderson, 1975):

$$\tilde{\vartheta}^{NP}(\psi) = \mathbf{X}\tilde{\beta}(\psi) + \Lambda(\psi)\left[\hat{\vartheta} - \mathbf{X}\tilde{\beta}(\psi)\right], \tag{14.13}$$

with $\mathbf{\Lambda}(\psi)=(\mathbf{Z}\mathbf{\Sigma}_\gamma\mathbf{Z}^T+\mathbf{D}\mathbf{\Sigma}_u\mathbf{D}^T)\Sigma(\psi)^{-1}$ and $\tilde{\boldsymbol{\beta}}(\psi) = (\mathbf{X}^T\mathbf{\Sigma}(\psi)^{-1}\mathbf{X})^{-1}\mathbf{X}^T\mathbf{\Sigma}(\psi)^{-1}\hat{\vartheta}$.

The estimator $\hat{\vartheta}^B(\psi)$ depends on the unknown variance components σ_γ^2 and σ_u^2. Replacing them with estimators $\hat{\sigma}_\gamma^2$ and $\hat{\sigma}_u^2$, an empirical best linear unbiased predictor (EBLUP), under the NPFH model, is:

$$\tilde{\vartheta}^{NP}(\hat{\psi}) = \mathbf{X}\hat{\beta}(\hat{\psi}) + \hat{\Lambda}(\hat{\psi})\left[\hat{\vartheta} - \mathbf{X}\hat{\beta}(\hat{\psi})\right], \tag{14.14}$$

where $\hat{\beta}(\hat{\psi}) = (\mathbf{X}^T\hat{\mathbf{\Sigma}}(\hat{\psi})^{-1}\mathbf{X})^{-1}\mathbf{X}^T\hat{\mathbf{\Sigma}}(\hat{\psi})^{-1}\hat{\vartheta}$. Hereafter this estimator is called NPEBLUP. Assuming normality of the random effects, σ_γ^2 and σ_u^2 can be estimated both by the ML and RML procedures (Prasad and Rao, 1990).

When geographically referenced responses play a central role in the analysis and need to be converted to maps, we can utilise bivariate smoothing: $\tilde{f}(\mathbf{x}_1, \mathbf{x}_2)=m(\mathbf{x}_1, \mathbf{x}_2; \boldsymbol{\eta}, \boldsymbol{\gamma})$. This is the case of environment, agricultural, public health and poverty mapping fields of application. P-splines rely on a set of basis functions to handle nonlinear structures in the data. So bivariate basis functions are required for bivariate smoothing. In this chapter we will assume the following model:

$$f(\mathbf{x}_1, \mathbf{x}_2; \eta, \gamma) = \eta_0 + \eta_1\mathbf{x}_1 + \eta_2\mathbf{x}_2 + \mathbf{z}_i\gamma, \tag{14.15}$$

with $\mathbf{z}_i$ being the i-th row of the following $m \times K$ matrix,

$$\mathbf{Z} = \left[C\left(\tilde{\mathbf{x}}_i - \boldsymbol{\kappa}_k\right)\right]_{\substack{1\le i\le m\\ 1\le k\le K}} \left[C\left(\kappa_k - \kappa_{k'}\right)\right]_{1\le k\le K'}^{-1/2}, \tag{14.16}$$

where $C(\mathbf{t})=||\mathbf{t}||^2 \log ||\mathbf{t}||$, $\tilde{\mathbf{x}}_i=(x_{1i}, x_{2i})$ and $\boldsymbol{\kappa}_k$, $k=1,\ldots, K$ are knots. For each area i, $\tilde{\mathbf{x}}_i$ are the coordinates (latitude and longitude) of the centroid of the area. For more details see Opsomer *et al.* (2008), Ruppert *et al.* (2003, Ch. 13), Kamman and Wand (2003) and French *et al.* (2001).

14.2.4 Mean squared error estimation

In practical applications it is important to complement the estimates obtained using the Spatial EBLUP estimator $\tilde{\vartheta}_i^S(\hat{\delta})$ and the semiparametric Fay–Herriot estimator $\tilde{\vartheta}^{NP}(\hat{\psi})$ with an estimate of their variability. For both estimators an approximately unbiased analytical estimator of the MSE is:

$$mse\left[\tilde{\vartheta}_i(\hat{\omega})\right] = g_1(\hat{\omega}) + g_2(\hat{\omega}) + 2g_3(\hat{\omega}), \tag{14.17}$$

where $\tilde{\vartheta}_i(\hat{\omega})$ is equal to $\tilde{\vartheta}_i^S(\hat{\delta})$ when considering the SEBLUP estimator, or to $\tilde{\vartheta}^{NP}(\hat{\psi})$ when the estimator is the semiparametric Fay–Herriot one. The MSE estimator (14.17) is the same as derived by Prasad and Rao (1990); for more details on the specification of the g components under both models see Pratesi and Salvati (2009) and Giusti *et al.* (2012). For a detailed discussion of the MSE and its estimation for the EBLUP based on the traditional Fay–Herriot model (Section 14.2.1) see Rao (2003).

An alternative procedure for estimating the MSE of estimators $\tilde{\vartheta}_i^S(\hat{\delta})$ and $\tilde{\boldsymbol{\vartheta}}^{NP}(\hat{\psi})$ can be based on a bootstrapping procedure proposed by González-Manteiga *et al.* (2007), Molina *et al.* (2009) and Opsomer *et al.* (2008). More in detail, considering the SEBLUP estimator, Molina *et al.* (2009) proposed a non-parametric bootstrap for MSE estimation, in which the bootstrap random effects $(u_1^*,\ldots,u_m^*)^T$ and the random errors $(e_1^*,\ldots,e_m^*)^T$ are obtained by resampling, respectively, from the empirical distribution of the predicted random elements $\hat{\mathbf{u}}=(\hat{u}_1,\ldots,\hat{u}_m)^T$ and the residuals $\hat{\mathbf{r}}=\hat{\boldsymbol{\vartheta}}-\mathbf{X}\alpha-\mathbf{D}\hat{\mathbf{u}}$, both previously standardised. This method avoids the need for distributional assumptions; therefore, it is expected to be more robust to the non-normality of any of the random components of the model.

Under model (14.3) and (14.5), the BLUPs of **u** and **v** are:

$$\tilde{\mathbf{v}}(\delta)=\mathbf{G}(\delta)\mathbf{D}^T\mathbf{V}(\delta)^{-1}\left[\vartheta-\mathbf{X}\alpha(\delta)\right],\quad \tilde{\mathbf{u}}(\delta)=\left(\mathbf{I}-\rho\mathbf{W}\right)\tilde{\mathbf{v}}(\delta)$$

and the covariance matrix of $\tilde{\mathbf{u}}(\boldsymbol{\delta})$ is:

$$\mathbf{V}_u(\delta)=\left(\mathbf{I}-\rho\mathbf{W}\right)\mathbf{G}(\delta)\mathbf{D}^T\mathbf{P}(\delta)\mathbf{G}(\delta)\mathbf{D}^T\left(\mathbf{I}-\rho\mathbf{W}^T\right)$$

where $\mathbf{P}(\boldsymbol{\delta})=\mathbf{V}(\boldsymbol{\delta})^{-1}-\mathbf{V}(\boldsymbol{\delta})^{-1}\mathbf{X}(\mathbf{X}^T\mathbf{V}(\boldsymbol{\delta})^{-1}\mathbf{X})^{-1}\mathbf{X}^T\mathbf{V}(\boldsymbol{\delta})^{-1}$. Moreover, considering the vector of residuals:

$$\tilde{\mathbf{r}}(\delta)=\vartheta-\mathbf{X}\tilde{\alpha}(\delta)-\tilde{\mathbf{v}}(\delta)$$

their covariance matrix is:

$$\mathbf{V}_r(\delta)=\mathbf{R}\mathbf{P}(\delta)\mathbf{R}.$$

Given these quantities, the steps of the nonparametric bootstrap procedure are the following:

1 Fit the model (14.6) to the initial direct estimates $\hat{\vartheta}$, obtaining estimates $\hat{\delta}$ and $\hat{\beta}$. Note that since the covariance matrices $V_u(\delta)$ and $V_r(\delta)$ are not diagonal, the elements of vectors $\tilde{\mathbf{v}}(\delta)$ and $\tilde{\mathbf{r}}(\delta)$ are correlated. Thus, a preliminary standardisation step is crucial, since methods that resample from empirical distributions work well under an ideally *iid* setup. Molina *et al.* (2009) proposed a transformation to make the vectors as close as possible to vectors with uncorrelated and unit variance elements – see their paper for more details on this.

2 With the estimates $\hat{\delta}$ and $\hat{\beta}$ obtained in step 1, calculate predictors of **v** and **u** as follows:

$$\hat{\mathbf{v}}=\mathbf{G}(\hat{\delta})\mathbf{D}^T\mathbf{V}(\hat{\delta})^{-1}\left(\vartheta-\mathbf{X}\hat{\alpha}\right),\quad \hat{\mathbf{u}}=\left(\mathbf{I}-\rho\mathbf{W}\right)\hat{\mathbf{v}}=\left(\hat{u}_1,\ldots,\hat{u}_m\right)^T.$$

Then take $\hat{\mathbf{u}}^S = \hat{\mathbf{V}}_u^{-1/2}\hat{\mathbf{u}}$ where $\hat{\mathbf{V}}_u^{-1/2}$ is the root square of the generalised inverse of $\hat{\mathbf{V}}_u = \mathbf{Q}_u \Delta_u \mathbf{Q}_u^T$, the estimated covariance matrix of $\hat{\mathbf{u}}$, obtained by the spectral decomposition. Here Δ_u is a diagonal matrix, with the eigenvalues of $\hat{\mathbf{V}}_u$ and $\mathbf{Q}_u$ is the matrix with the corresponding eigenvectors in the columns. It is convenient to re-scale the elements $\hat{u}_i^S$ so that they have a sample mean exactly equal to zero and sample variance σ_u^2. This is achieved by the transformation:

$$\hat{u}_i^{SS} = \frac{\hat{\sigma}_u\left(\hat{u}_i^S - m^{-1}\sum_{j=1}^{m}\hat{u}_j^S\right)}{\sqrt{m^{-1}\sum_{d=1}^{m}\left(\hat{u}_d^S - m^{-1}\sum_{j=1}^{m}\hat{u}_j^S\right)^2}}, \quad i = 1,\ldots,m.$$

Construct the vector $\mathbf{u}^* = \left(u_1^*,\ldots,u_m^*\right)^T$ whose elements are obtained by extracting a simple random sample with replacement of size m from the set $\left\{\hat{u}_1^{SS},\ldots,\hat{u}_m^{SS}\right\}$. Then obtain $\mathbf{v}^* = \left(\mathbf{I} - \hat{\rho}\mathbf{W}\right)^{-1}\mathbf{u}^*$ and calculate the bootstrap quantity of interest $\boldsymbol{\vartheta}^{S*} = \mathbf{X}\hat{\boldsymbol{\alpha}} + \mathbf{v}^* = \left(\vartheta_1^{S*},\ldots,\vartheta_m^{S*}\right)^T$.

3 Compute the vector of residuals $\hat{\mathbf{r}} = \boldsymbol{\vartheta} - \mathbf{X}\hat{\boldsymbol{\alpha}} - \hat{\mathbf{v}} = \left(\hat{r}_1,\ldots,\hat{r}_m\right)^T$. Standardise the residuals by $\hat{\mathbf{r}}^S = \hat{\mathbf{V}}_r^{-1/2}\hat{\mathbf{r}} = \left(\hat{r}_1^S,\ldots,\hat{r}_m^S\right)^T$, where $\hat{\mathbf{V}}_r = \mathbf{RP}(\hat{\boldsymbol{\delta}})\mathbf{R}$ is the estimated covariance matrix and $\hat{\mathbf{V}}_r^{-1/2}$ is a root square of the generalised inverse derived from the spectral decomposition. Again, re-standardise these values as:

$$\hat{r}_i^{SS} = \frac{\hat{r}_i^S - m^{-1}\sum_{j=1}^{m}\hat{r}_j^S}{\sqrt{m^{-1}\sum_{d=1}^{m}\left(\hat{r}_d^S - m^{-1}\sum_{j=1}^{m}\hat{r}_j^S\right)^2}}, \quad i = 1,\ldots,m.$$

Construct $\mathbf{r}^* = (r_1^*,\ldots,r_m^*)^T$ by extracting a simple random sample with replacement of size m from the set $\{\hat{r}_1^{SS},\ldots,\hat{r}_m^{SS}\}$. Then take $\mathbf{e}^* = (e_1^*,\ldots,e_m^*)^T$ where $e_i^* = \varphi_i^{1/2} r_i^*, i = 1,\ldots,m$.

4 Construct bootstrap data from the model $\hat{\vartheta}^{S*} = \vartheta^{S*} + \mathbf{e}^* = \mathbf{X}\hat{\alpha} + \mathbf{Du}^* + \mathbf{e}^* = (\hat{\vartheta}_1^{S*},\ldots,\hat{\vartheta}_m^{S*})^T$.

5 Considering $\hat{\alpha}$ and $\hat{\delta}$ as the true $\boldsymbol{\alpha}$ and $\boldsymbol{\delta}$ values, fit model (14.6) to the bootstrap data $\hat{\vartheta}^{S*}$, obtaining the bootstrap estimators $\hat{\alpha}^*$ and $\hat{\delta}^*$.

6 Calculate the bootstrap SBLUP from bootstrap data $\hat{\vartheta}^{S*}$, regarding $\hat{\delta}$ as the true value of $\boldsymbol{\delta}$, obtaining $\tilde{\vartheta}_i^{S*}(\hat{\delta})$. Also calculate the bootstrap SEBLUP using $\hat{\delta}^*$, $\tilde{\vartheta}_i^{S*}(\hat{\delta}^*)$.

7 Repeat steps 2–6 B times. In the b-th bootstrap replication let $\tilde{\vartheta}_i^{S*(b)}$ be the quantity of interest in area i, $\hat{\delta}^{*(b)}$ the bootstrap estimate of $\boldsymbol{\delta}$, $\tilde{\vartheta}_i^{S*(b)}(\hat{\delta})$ the bootstrap SBLUP and $\tilde{\vartheta}_i^{S*(b)}(\hat{\delta}^{*(b)})$ the bootstrap SEBLUP for area i.

A naive bootstrap estimator of the MSE of $\tilde{\vartheta}_i^S(\hat{\delta})$ is given by:

$$mse\left[\tilde{\vartheta}_i^S(\hat{\delta})\right] = B^{-1}\sum_{b=1}^{B}\left[\tilde{\vartheta}_i^{S*(b)}(\hat{\delta}^{*(b)}) - \vartheta_i^{S*(b)}\right]^2. \quad (14.18)$$

A similar bootstrap procedure can be used to obtain an estimate of the MSE of $\tilde{\vartheta}^{NP}(\hat{\psi})$ under the semiparametric Fay–Herriot model:

$$mse\left[\tilde{\vartheta}_i^{NP}(\hat{\psi})\right] = B^{-1}\sum_{b=1}^{B}\left[\tilde{\vartheta}_i^{NP*(b)}(\hat{\psi}^{*(b)}) - \vartheta_i^{NP*(b)}\right]^2. \quad (14.19)$$

For a description of the quantities in (14.19) and of the detailed bootstrap procedure see Giusti *et al.* (2012).

14.3 Results

A central theme of the Horizon 2020 programme is the promotion of innovative and inclusive European societies. In order to achieve inclusive societies, there is a strong need for deeper insight into how social cohesion, solidarity and the reconciliation of differences between social groups or individuals can be achieved. Social and economic inequalities, which include many aspects such as income, wealth, employment, health, environment and wellbeing, have strong implications for social inclusion. Measuring and analysing these phenomena in the right way, particularly at local levels, is extremely relevant to provide policy-makers, and stakeholders in general, with useful information.

In this section we use data from the 2008 EU-SILC in Italy (European Commission, 2009) and from the 2001 Population Census of Italy to estimate the mean equivalised income, the Head Count Ratio (HCR) and the Poverty Gap (PG) for the Local Labour Systems of three Italian regions: Lombardy (Northern Italy), Tuscany (Central Italy) and Campania (Southern Italy). An LLS may include municipalities from different administrative regions. When analysing the LLS of an administrative region, we consider all the LLSs that contain at least one municipality of the region in question. The target small areas are 172 in total: 59 in Lombardy (34 sampled and 25 out-of-sample areas), 57 in Tuscany (32 sampled and 25 out-of-sample areas) and 56 in Campania (18 sampled and 38 out-of-sample areas). By 'sampled areas' we mean areas for which the area-specific sample size is greater than zero; the remaining areas are labelled as 'out-of-sample'. In addition to evaluating the potential poverty dissimilarities within each region, we are also interested in better understanding the so-called north–south divide that characterises Italy in terms of poverty and living conditions.

As explained in the previous sections, an area level model consists of two parts: a 'sampling model' formalising the assumptions concerning direct estimators and their relationships with underlying area parameters (14.1), and a 'linking model', which relates these parameters to area-specific auxiliary information (14.2). Here we considered auxiliary information at the LLSs level of aggregation, using the 2001 Population Census of Italy as data source. A number of variables are considered as candidate covariates: Head of the Household's (HH) occupational status, age, level of education, marital status and gender, ownership of the house and household size. Selection of the covariates

Table 14.1 Covariates used in the small area models for poverty indicators, by indicator and Region. In the table HH means Head of the Household.

Indicators	*Region / Covariates*
	Toscana
Mean Income	HH employed (*vs unemployed*), HH age (*continuous*), HH years in education (*continuous*)
HCR	HH age, Ownership of the house (*vs not ownership*)
PG	Ownership of the house, HH divorced *(vs married)*, HH widowed *(vs married)*
	Lombardia
Mean Income	HH female, HH divorced, Household size (*continuous*)
HCR	HH divorced
PG	Ownership of the house, HH widowed
	Campania
Mean Income	HH divorced
HCR	Ownership of the house
PG	HH divorced, HH employed

Note
HH means Head of the Household.

for each response variable and for each region is based on a regression procedure with stepwise selection. Covariates used in the small area models by indicator and region are reported in Table 14.1.

Concerning the NPFH model, the auxiliary variables enter the model linearly, while the coordinates (latitude and longitude) of the centroid of the LLSs enter through an unknown smooth bivariate function.

To implement the SEBLUP estimator a neighbourhood structure **W** has to be defined for the small areas within the study region. The spatial weight matrix represents the potential interaction between locations. For this reason we have built a spatial weight matrix for each region (i.e. a **W** matrix for the LLSs of Tuscany, one for the LLSs of Lombardy and one for the LLSs of Campania). A general spatial weight matrix can be defined by a symmetric binary contiguity matrix, which can be generated from the topological information given by the Geographical Information System (GIS) based on the criterion of adjacency: the spatial weight is set equal to one if area i shares a boundary with area j and zero otherwise. For easier interpretation, the general spatial weight matrix is defined in row-standardised form, in which the row elements sum to one. In the EU-SILC data we do not have any spatial information. By using an R function (get. Pcent() by library maptools) we obtained the spatial coordinates of the centroids of each LLS.

Before presenting the results of the application of the small area estimators, we show some preliminary data and model diagnostics. In order to detect the spatial pattern (spatial association and spatial autocorrelation) of the chosen poverty indicators, some standard global spatial statistics have been calculated:

Moran's I, Geary's C (Cliff and Ord, 1981), G statistics (Ord and Getis, 1995). The standardised Moran's I is analogous to the correlation coefficient, and its values range from one (strong positive spatial autocorrelation) to minus one (strong negative spatial autocorrelation). Geary's C ranges between zero and two. Positive spatial autocorrelation is found with values ranging from zero to one and negative spatial autocorrelation with values between one and two. Finally, G statistic is an index of the spatial clustering of a set of observations over a defined neighbourhood. The above indexes have been calculated on the equivalised income, HCR and PG for the three regions. They show a positive spatial correlation of income and HCR (except for Lombardy) and negative spatial correlation for PG in Tuscany and Lombardy. The Global *G* values indicate the possibility of clusters of values of the study variable (see Table 14.2).

In Figures 14.1, 14.2 and 14.3 we present maps of EBLUP, SEBLUP and NPEBLUP estimated values of HCR for each LLS in Tuscany, Lombardy and Campania using the Fay–Herriot, spatial Fay–Herriot and nonparametric Fay–Herriot models. We cannot report the maps of the estimates of mean equivalised income and PG by LLS for reasons of space. We have chosen to display the results of the HCR because it is a good indicator of poverty. The results of the estimates of mean equivalised income and PG by LLS are available from the authors upon request.

The maps in Figure 14.1 indicate that the LLSs in the north-western part and northern part of the region of Tuscany, corresponding to the LLSs in the province of Massa-Carrara, and in the northern part of the provinces of Lucca and Prato, are characterised by the highest estimates of HCR, which correspond to the lowest estimates for the mean household equivalised income at LLS level. Hence, these areas can be considered as the most critical in the region. On the other hand, the LLSs in the lowest class of estimated HCR are concentrated in

Table 14.2 Value of Moran's *I*, Geary's *C* and Global *G* statistics

Index	*Value*		
	Income	*HCR*	*PG*
	Toscana		
Moran's *I*	0.171	0.070	–0.158
Geary's *C*	0.635	0.931	1.454
Global *G*	0.076	0.086	0.090
	Lombardia		
Moran's *I*	0.058	–0.041	–0.065
Geary's *C*	0.788	0.709	0.795
Global *G*	0.071	0.080	0.062
	Campania		
Moran's *I*	0.218	0.330	0.287
Geary's *C*	0.606	0.302	0.559
Global *G*	0.071	0.057	0.062

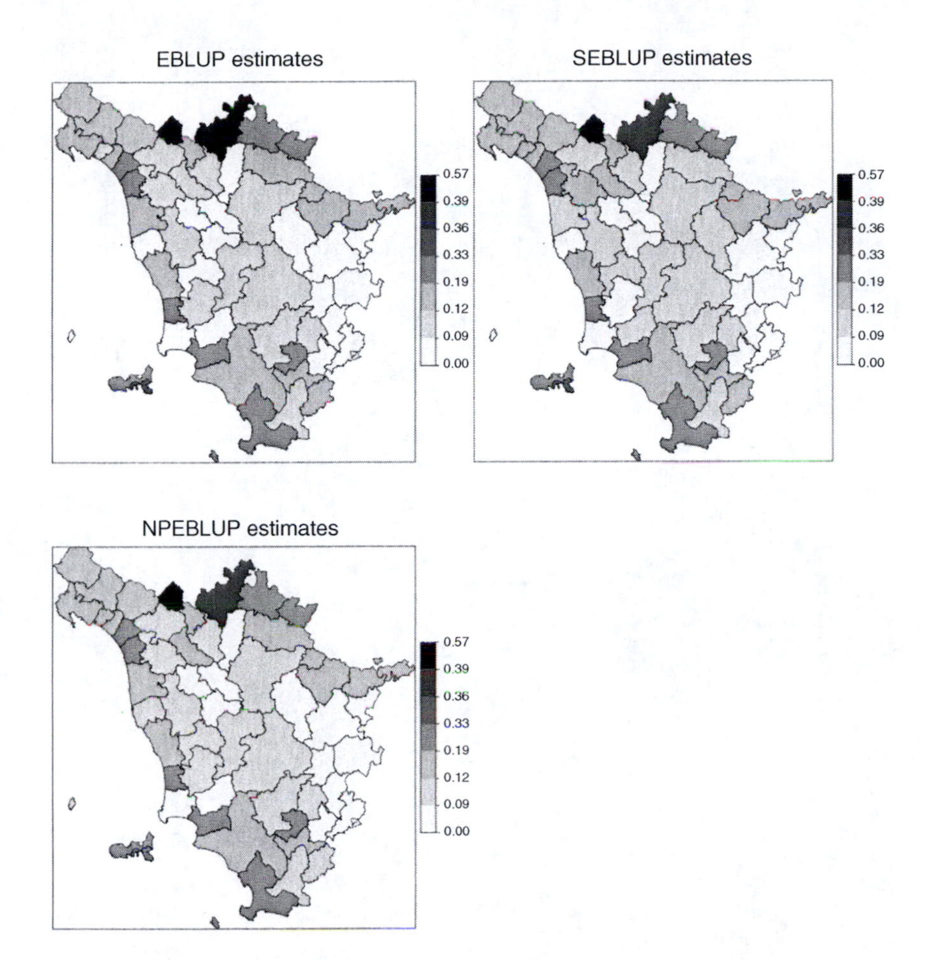

Figure 14.1 Estimated Head Count Ratio for Toscana LLSs: EBLUP, SEBLUP and NPEBLUP estimates.

the provinces of Florence, Siena and Arezzo, in the central-eastern part of the region. These results confirm that the mean household income is usually higher in the LLSs around the biggest cities.

In the maps of Figure 14.2, the LLSs with lower estimated HCR are concentrated in the south-western and south-eastern parts of the Lombardy region. At the opposite end, the LLSs characterised by higher estimates are concentrated in the central and northern parts of the region. These findings are coherent with the estimates at provincial level obtained in the SAMPLE project (SAMPLE project, 2010).

Figure 14.3 highlights that the HCR estimates for the Campania region are more geographically differentiated. Note, for example, how the LLSs characterised by the lowest class of estimated HCR are spread throughout the region.

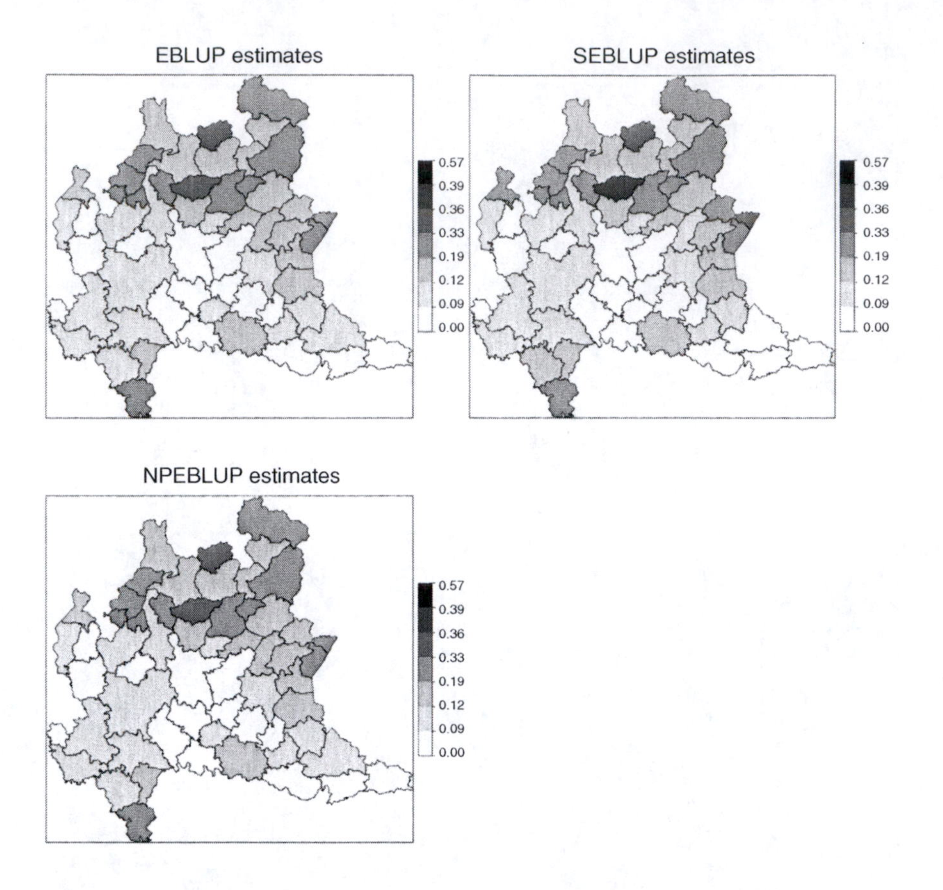

Figure 14.2 Estimated Head Count Ratio for Lombardia LLSs: EBLUP, SEBLUP and NPEBLUP estimates.

The estimates of the HCR at the LLS level allow us to investigate the gap in living conditions between the three regions of Italy. In particular, we can note that the gap between the regions of Lombardy and Tuscany is not very pronounced, in line with the relatively small differences between them in terms of mean equivalised income. Considering the estimates for the region of Campania, on the other hand, we can observe a large difference from the central and northern regions: the lowest estimates of HCR in Campania are comparable to the highest ones in Lombardy and Tuscany. These results confirm the existence of the so-called 'north–south' divide in Italy from the point of view of the wealth of the population. Nonetheless, note that the poverty indicators are calculated assuming a single poverty line at the national level, according to EU guidelines. Accounting for the different average price levels in the calculation of the 'absolute poverty index', as done by ISTAT (ISTAT, 2009), would reduce the evidence of the north–south divide, but naturally not eliminate it.

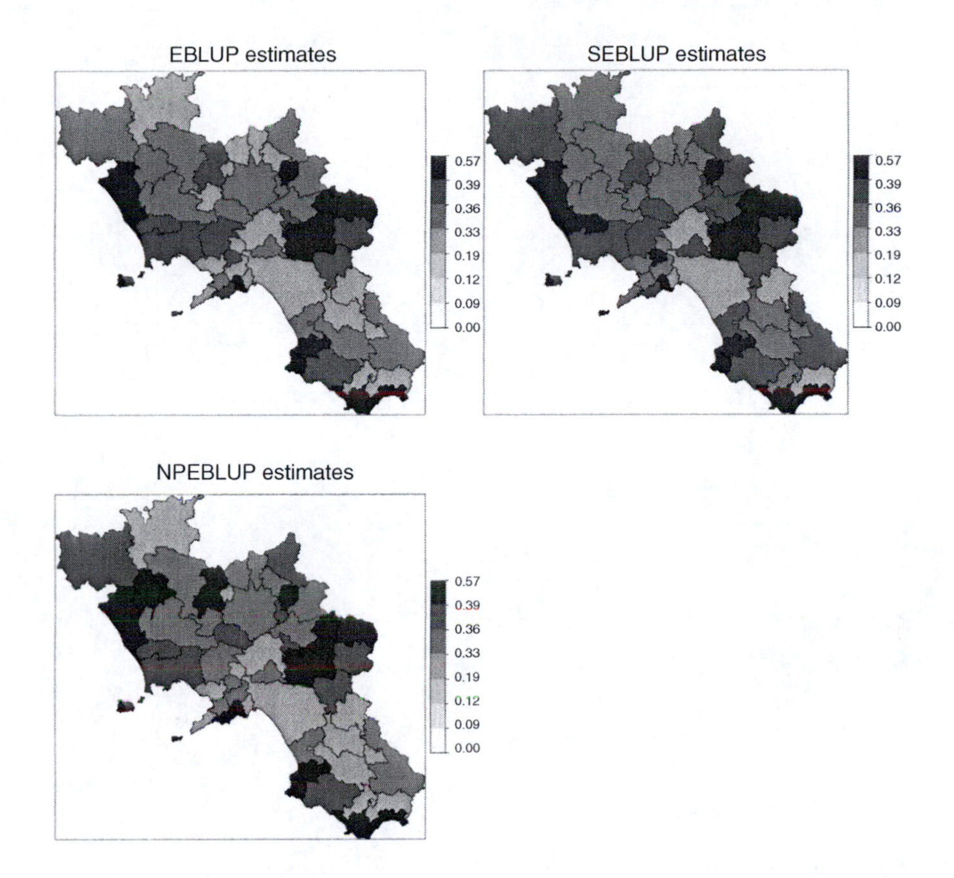

Figure 14.3 Estimated Head Count Ratio for Campania LLSs: EBLUP, SEBLUP and NPEBLUP estimates.

Concerning the results obtained using the different estimators, in general we can note that the spatial distribution of SEBLUP and NPEBLUP estimates appears to be more variable than that obtained with the traditional EBLUP. The range of estimates is larger and there is a wider diversification of the HCR by LLS. Given the same explanatory variables, the result is probably due to the additional spatial information inserted in our estimators. This moderates the smoothing effect resulting from the application of the traditional EBLUP and renders more evident the specific characteristics of the LLSs. This happens without losing precision in the estimates. Figure 14.4 shows the ratios between the coefficients of variation of direct and model-based estimates of mean income, HCR and PG for the region of Tuscany. A ratio greater than one indicates a gain in accuracy of the estimator proposed with respect to the direct estimator. These coefficients of variation are the squared root of the mean squared errors over the corresponding estimates, where the mean squared errors of the model-based

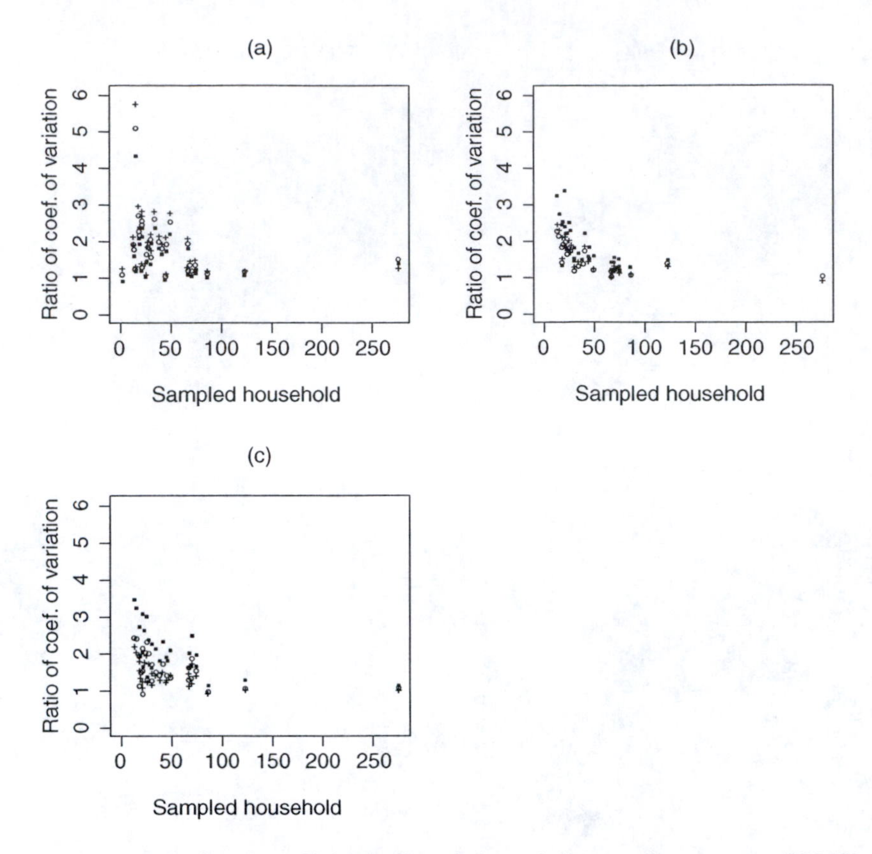

Figure 14.4 Ratio of coefficients of variation of direct estimates over EBLUP (+), SEBLUP (■) NPEBLUP (○) estimates for each LLS of the Toscana region. Ratios of the values of coefficients of variation of (a) mean equivalised income, (b) HCR and (c) PG.

estimators are obtained using the analytical expression for EBLUP and the bootstrap procedure described in Section 14.2.4 for SEBLUP and NPEBLUP. The bootstrap estimator algorithm cannot be applied to the estimation of $mse[\tilde{\vartheta}_i^{S/NP}(\cdot)]$ for out-of-sample areas directly because the point estimator is a synthetic estimator, with a very small variance but a potentially non-negligible bias. To obtain an estimate of $mse[\tilde{\vartheta}_i^{S/NP}(\cdot)]$ for out-of-sample areas we propose a MSE smoothing model:

$$mse\left[\tilde{\vartheta}_i^{S/NP}(\cdot)\right] = b_0 + b_1\tilde{\vartheta}_i^{S/NP}(\cdot) + e_i.$$

For e_i we assume that $E(e_i)=0$, $V(e_i)=\sigma^2$, $E(e_i, e_j)=0$, $\forall i \neq j$. If the assumptions on the residuals hold and the fit is adequate, the parameters b_0 and b_1 can be estimated from the subset of sampled areas (i.e. from those areas for which

n_i>0)). In our case the R^2 of the regression models is always greater than 0.6. The smoothing model is one of the many possible (see Wolter, 2007).

In order to analyse the relationship between gain and sample size, in Figure 14.4 the ratios are plotted against the number of sampled households. We observe that all ratios are larger than one. In particular, SEBLUP and NPEBLUP show larger gains than the EBLUP. Moreover, the gain in accuracy of the estimators proposed increases when the sample size decreases. These results are similar for the regions of Lombardy and Campania, but are not reported here to save space.

The results are consistent with the spatial distribution of the average values of household income, HCR and PG produced by direct estimators. In Figures 14.5, 14.6 and 14.7 we report a diagnostic that can be used to evaluate the small area estimates obtained for the region of Tuscany. Specifically, we plot the model-based estimates on the *X*-axis and the direct ones on the *Y*-axis. It can be of use to look for large divergences of the scatter of points from the regression line $y=x$.

We note that model-based estimates appear to be consistent with the direct estimates of the mean equivalised income and HCR, whereas there are some

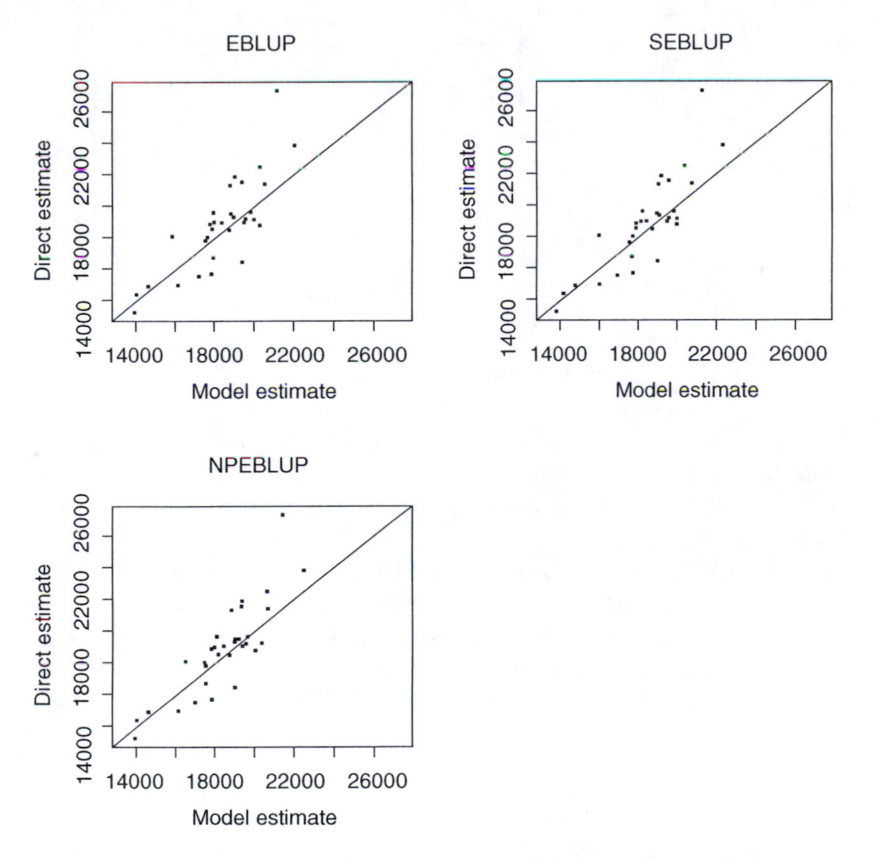

Figure 14.5 Model-based estimates of mean income versus corresponding direct estimates for each area of the Toscana region.

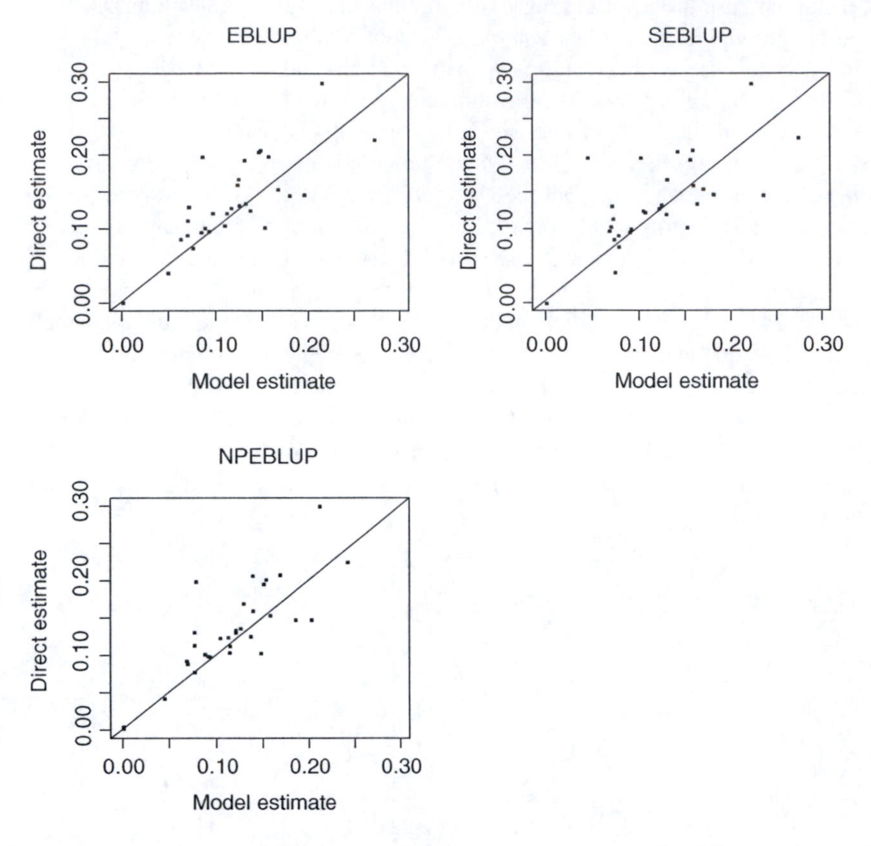

Figure 14.6 Model-based estimates of HCR versus corresponding direct estimates for each area of the Toscana region.

small areas for which the model-based estimates and direct estimates of the PG are notably different. These results indicate that the small area models that allow for more flexible incorporation of the spatial information produce consistent results overall.

The last diagnostic we use to validate the reliability of the model-based small-area estimates is the Goodness of Fit (GoF). This is based on the null hypothesis that the direct and model-based estimates are statistically equivalent. The alternative is that the direct and model-based estimates are statistically different. The GoF diagnostic is computed using the following Wald statistic for every model-based estimator:

$$Wald = \sum_{i=1}^{m} \left\{ \frac{\left(\tilde{\vartheta}^{FH/S/NP}(\cdot) - \hat{\vartheta} \right)^2}{mse\left(\tilde{\vartheta}^{FH/S/NP}(\cdot) \right) + \varphi_i} \right\}.$$

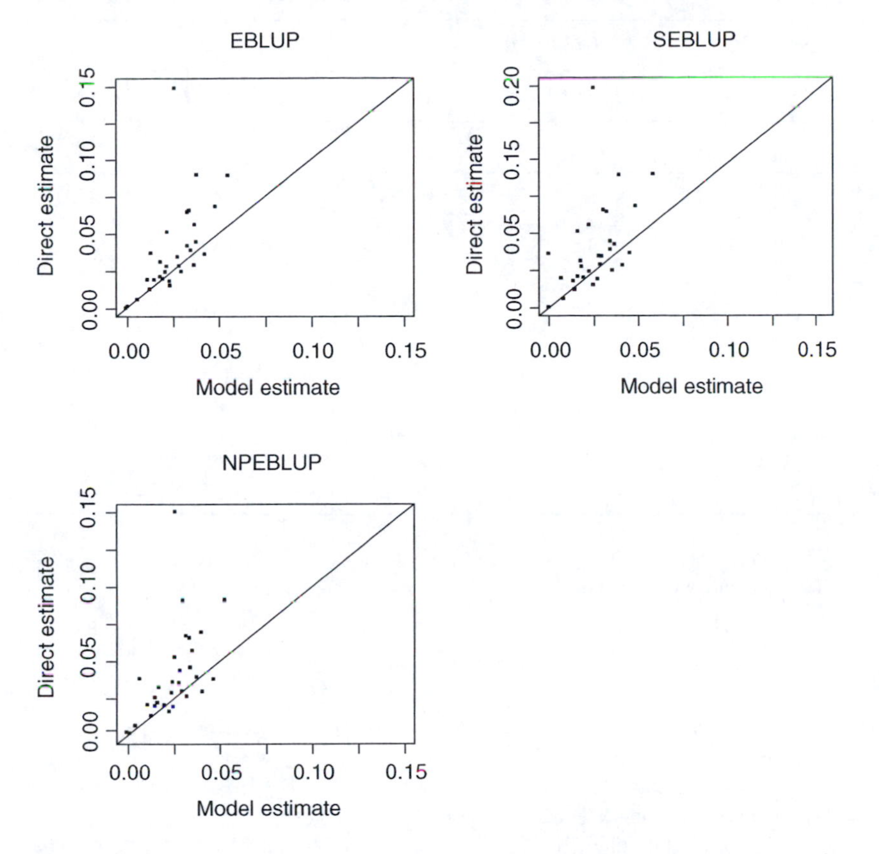

Figure 14.7 Model-based estimates of PG versus corresponding direct estimates for each area of the Toscana region.

Table 14.3 show the results of the GoF diagnostic at the 5 per cent level of significance. The value from the test statistic is compared against the value from a Chi-squared distribution with m degrees of freedom. In our case, this value is 46.19 for Tuscany, 48.6 for Lombardy and 28.8 for Campania. These results indicate that none of the model-based estimates are statistically different from the direct estimates.

The application of spatial SAE models to EU-SILC data in the three regions has shown their potentialities in providing accurate estimates at LLS level. Future developments will possibly focus on estimation at a smaller area level, such as the municipality level. Furthermore, we intend to consider the point and interval estimation of other poverty measures, such as the quantile share ratio and other Laeken indicators, to better investigate poverty and living conditions in the areas of interest.

Table 14.3 Goodness of Fit diagnostic at 5% level of significance

Parameter	*Model-based estimator*		
	EBLUP	*SEBLUP*	*NPFH*
	Toscana		
Mean income	16.76	11.68	12.38
HCR	7.23	10.52	6.09
PG	10.61	17.54	13.41
	Lombardia		
Mean income	12.42	17.83	10.09
HCR	19.77	24.46	19.44
PG	25.82	30.43	25.39
	Campania		
Mean income	4.72	4.46	4.27
HCR	7.13	8.62	8.29
PG	619	7.33	4.93

Note
A smaller value (less than 46.19 for Toscana, 48.6 for Lombardia and 28.8 for Campania) indicates no statistically significant difference between model-based and direct estimates.

References

Anselin, L. (1992). *Spatial Econometrics: Method and Models*, Kluwer Academic Publishers, Boston.

Banerjee, S., Carlin, B.P. and Gelfand, A.E. (2004). *Hierarchical Modeling and Analysis for Spatial Data*, Chapman & Hall, New York.

Battese, G., Harter, R. and Fuller, W. (1988). An Error-Components Model for Prediction of County Crop Areas using Survey and Satellite Data. *Journal of the American Statistical Association*, 83, 28–36.

Cliff, A.D. and Ord, J.K. (1981). *Spatial Processes: Models & Applications*. Pion Limited, London.

Cressie, N. (1991). Small-Area Prediction of Undercount using the General Linear Model. *Proceedings of the Statistical Symposium 90: Measurement and Improvement of Data Quality*. Statistics Canada, Ottawa, 93–105.

Cressie, N. (1993). *Statistics for Spatial Data*, John Wiley & Sons, New York.

European Commission (2005). Regional Indicators to reflect Social Exclusion and Poverty. Report prepared for Employment and Social Affairs DG, with contributions of Betti, G., Lemmi, A., Mulas, A., Natilli, M., Neri, L., Salvati, N., Verma, V. (Research Director). European Commission, Brussels.

European Commission (2009). Description of Silc User Database Variables: Cross-sectional and Longitudinal. Version 2007.1 from 01-03-09. European Commission, Brussels.

Fay, R.E. and Herriot, R.A. (1979). Estimates of Income for Small Places: An Application of James-Stein Procedures to Census Data. *Journal of the American Statistical Association*, 74, 269–277.

French, J., Kammann, E. and Wand, M. (2001). Comment on Paper by Ke and Wang. *Journal of the American Statistical Association*, 96, 1285–1288.

Giusti, C., Marchetti, S., Pratesi, M. and Salvati, N. (2012). Semiparametric Fay–Herriot Model using Penalized Splines. *Journal of the Indian Society of Agricultural Statistics*, 66, 1–14.

González-Manteiga, W., Lombarda, M., Molina, I., Morales, D. and Santamara, L. (2007). Estimation of the Mean Squared Error of Predictors of Small Area Linear Parameters under Logistic Mixed Model. *Computational Statistics and Data Analysis*, 51, 2720–2733.

Henderson, C. (1975). Best Linear Unbiased Estimation and Prediction under a Selection Model. *Biometrics*, 31, 423–447.

ISTAT (2009). La misura della povertà assoluta, *Metodi e norme n. 39*.

Jiang, J. and Lahiri, P. (2006). Mixed Model Prediction and Small Area Estimation. *Test*, 15, 1–96.

Kammann, E.E. and Wand, M.P. (2003). Geoadditive Models. *Journal of the Royal Statistical Society*, C52, 1–18.

Molina, I., Salvati, N. and Pratesi, M. (2009). Bootstrap for Estimating the MSE of the Spatial EBLUP. *Computational Statistics*, 24, 441–458.

Nelder, J.A. and Mead, R. (1965). A Simplex Method for Function Minimization. *Computer Journal*, 7, 308–313.

Opsomer, J.D., Claeskens, G., Ranalli, M.G., Kauermann, G. and Breidt, F.J. (2008). Nonparametric Small Area Estimation using Penalized Spline Regression. *Journal of the Royal Statistical Society, Series B*, 70, 265–286.

Ord, K. and Getis, A. (1995). Local Spatial Autocorrelation Statistics: Distributional Issues and Application. *Geographical Analysis*, 27, 286–296.

Petrucci, A., Pratesi, M. and Salvati, N. (2005). Geographic Information in Small Area Estimation: Small Area Models and Spatially Correlated Random Area Effects. *Statistics in Transition*, 7, 609–623.

Petrucci, A. and Salvati, N. (2006). Small Area Estimation for Spatial Correlation in Watershed Erosion Assessment. *Journal of Agricultural, Biological, and Environmental Statistics*, 11, 169–182.

Pfeffermann, D. (2002). Small Area Estimation: New Developments and Directions. *International Statistical Review*, 70, 125–143.

Prasad, N. and Rao, J. (1990). The Estimation of Mean Squared Error of Small Area Estimators. *Journal of the American Statistical Association*, 85, 163–171.

Pratesi, M. and Salvati, N. (2008). Small Area Estimation: The EBLUP Estimator based on Spatially Correlated Random Area Effects. *Statistical Methods and Applications*, 17, 113–141.

Pratesi, M. and Salvati, N. (2009). Small Area Estimation in the Presence of Correlated Random Area Effects. *Journal of Official Statistics*, 25, 37–53.

Rao, J.N.K. (2003). *Small Area Estimation*, John Wiley & Sons, New York.

Ruppert, D., Wand, M.P. and Carroll, R. (2003). *Semiparametric Regression*. Cambridge University Press, Cambridge/New York.

Saei, A. and Chambers, R. (2005). Small Area Estimation under Linear and Generalized Linear Mixed Models with Time and Area Effects. *Southampton Statistical Sciences Research Institute*, WP M03/15, Southampton.

Salvati, N. (2004). Small Area Estimation by Spatial Models: The Spatial Empirical Best Linear Unbiased Prediction (Spatial EBLUP). *Working Paper no. 2004/04*, 'G. Parenti' Department of Statistics, University of Florence.

SAMPLE project (2010). Deliverable 17: Pilot Applications. Available on the project website: www.sample-project.eu.

Singh, B.B., Shukla, G.K. and Kundu, D. (2005). Spatio-Temporal Models in Small Area Estimation. *Survey Methodology*, 31, 183–195.

Tobler, W.R. (1970). A Computer Movie Simulating Urban Growth in the Detroit Region. *Economic Geography*, 46, 234–240.

Wang, J. and Fuller, W.A. (2003). Mean Squared Error of Small Area Predictors Constructed with Estimated Area Variances. *Journal of the American Statistical Association*, 92, 716–723.

Wolter, K. (1985). *Introduction to Variance Estimation*. Springer-Verlag, New York.

Wolter K.M. (2007). *Introduction to Variance Estimation* (2nd edn). Springer, New York.

15 Outlier robust semi-parametric small area methods for poverty estimation

Nikos Tzavidis, Stefano Marchetti and Steve Donbavand

15.1 Introduction

Estimating economic indicators is crucial to achieving targeted implementation of welfare policies. However, for such policies to be effective policy makers must have access to a detailed picture of deprivation that goes beyond aggregate estimates at the country (national) level, extending to finer geographical levels and to other domains of interest, such as specific groups of individuals. Such a picture can only be constructed by having access to timely and accurate survey and administrative/Census data at appropriate spatial scales. One possible solution for obtaining accurate indicators at finer spatial scales is the use of small area estimation methodologies. The term 'small areas' is typically used to describe domains (e.g. geographic areas) whose sample sizes are not large enough to allow sufficiently precise direct estimation, i.e. estimation that is based only on the sample data from the domain (Rao, 2003). Small area-specific sample sizes often also hamper the use of conventional design-based estimators. In such cases model-based estimation procedures can be considered to improve the precision of the direct estimates.

Small area estimation is conventionally concerned with the estimation of small area averages and totals. More recently, some research effort has been shifted towards methods for estimating poverty (deprivation) indicators at the small area level, also known as poverty mapping. Poverty mapping can offer a detailed description of the spatial distribution of poverty and inequality within a country. It combines individual and household survey data with Census/administrative data, with the objective of estimating welfare indicators for geographic areas or domains of interest. In recent years a range of alternative model-based small area methodologies for poverty mapping have been proposed.

The seminal paper by Elbers, Lanjouw and Lanjouw (2003, ELL) proposed a methodology for estimating poverty indicators at the small area level. This methodology, widely described in Chapter 12 of this book, consists of a nested error regression (random effects) model with cluster random effects, which is estimated by using survey data. The response variable, which is not available in the Census, is the logarithm of a welfare variable, e.g. income or consumption, while the explanatory variables, used for modelling the welfare variable, are available in both the survey and the Census datasets. Once the model has been estimated using the survey data, the estimated model parameters are combined with Census micro-data to form unit-level synthetic Census predictions of the welfare variable. The synthetic values of income/consumption in relation to a defined poverty line are then used to estimate indicators of deprivation, such as the incidence of poverty; Head Count Ratio (HCR), Poverty Gap (PG) and poverty severity (Foster *et al.*, 1984).

More recently, Molina and Rao (2010) proposed an Empirical Best Prediction (EBP) approach for estimating poverty indicators at the small area level, which is similar to the ELL approach but generates Census predictions of income/consumption by using the conditional predictive distribution of the out-of-sample data, given the sample data. Molina and Rao (2010) demonstrated the superior performance of the EBP approach, compared to the ELL approach, under the nested error regression model.

Economic data usually involve long-tail distributions and outliers. To reduce the effect of outliers and hence make the Gaussian assumptions of the nested error regression model more plausible, both the ELL and EBP approaches use a logarithmic transformation of outcome variables. An alternative approach for checking the effect of outliers involves using outlier robust estimation. An outlier robust methodology for small area poverty estimation is based on the M-quantile approach (Chambers and Tzavidis, 2006; Tzavidis *et al.*, 2010; Marchetti *et al.*, 2012; Chambers *et al.*, 2013). The M-quantile approach to poverty mapping also utilises a model for a welfare variable, but estimation of the 'random' effects makes no explicit parametric assumptions. In this case model-based Census predictions and the corresponding estimates of poverty indicators are produced using a smearing-type estimator (Duan, 1983).

The aim of this chapter is to review recently proposed small area methodologies for poverty estimation, with specific emphasis on outlier robust estimation. The chapter is organised as follows. In Section 15.2 we present the ELL and EBP approaches to small area poverty estimation. In Section 15.3 we present the outlier robust, semi-parametric M-quantile approach to poverty estimation and discuss the estimation of the Mean Squared Error (MSE) of poverty estimates. In Section 15.4 we present results of a Monte-Carlo simulation study that contrasts the ELL and M-quantile approaches under a range of parametric assumptions. Although similar comparisons between the EBP and M-quantile approaches are not reported in the present chapter, we do comment on some preliminary results. Using data from the survey of income and living conditions (EU-SILC) and Census micro-data from Italy, in Section 15.5 we apply the M-quantile approach to derive HCR and PG estimates for ten provinces in Tuscany. These estimates are validated against direct estimates of HCR and PG. In Section 15.6 we provide information about the availability of software, in the form of R functions, that implements the outlier robust small area estimation procedures, and in Section 15.7 we conclude the chapter with some final remarks and areas for future research.

15.2 Small area estimation of poverty indicators using the nested error regression model

In what follows we assume that a vector of p auxiliary variables $\mathbf{x}_{ij}$ is known for each population unit i in small area $j=1,\ldots, m$ and that values of the welfare variable of interest y_{ij} are available from a random sample, s, that includes units from all the small areas of interest. We denote the population size, sample size, sampled part of the population and non-sampled part of the population in area j respectively by N_j, n_j, s_j, r_j. We assume that the sum over the areas of N_j and n_j is equal to N and n respectively. We further assume that, conditional on covariate information, for example design variables, the sampling design is ignorable.

All poverty mapping methods we describe in this chapter assume the availability of survey data on a welfare variable (income/consumption) and explanatory variables that can be used to model the outcome variable. In addition, the methods assume the availability of Census/administrative data on the same set of explanatory variables. Some methods further assume that the Census and survey data are linked. However, this assumption is fairly unrealistic as, in most cases, the link between the survey and the Census data is unknown. Having said this, the estimation methods can be modified so that the linkage assumption is not necessary. Finally, the methods based on the nested error regression model (ELL and EBP) conventionally use a logarithmic transformation of the welfare variable. In contrast, the M-quantile approach uses the raw values of the welfare variable. Nevertheless, before proceeding to small area estimation, it is always advisable to use model diagnostics. Depending on the results of the model diagnostics, all methods can be implemented using either the raw or the transformed values of the welfare variable. In this chapter we focus on the estimation of HCR and PG as defined by Foster *et al.* (1984). Using t to denote the poverty line, different poverty indicators are defined by the area-specific mean of the variable derived:

$$z_{ij}(\alpha,t)=\left(\frac{t-y_{ij}}{t}\right)^{\alpha} I\left(y_{ij}\le t\right),\ \mathrm{i}=1,\cdots,\mathrm{N}_{\mathrm{j}}.$$

Setting $\alpha=0$ defines the HCR, $z_{ij}(0, t)$, whereas setting $\alpha=1$ defines the PG, $z_{ij}(1, t)$.

15.2.1 The ELL method

As introduced in Chapter 12, the most widely used method for small area poverty mapping is the so-called ELL method. In its simplest form, and assuming a non-informative sampling design, the ELL method assumes a nested error regression model on the logarithmically transformed values of y_{ij},

$$y_{ij}=\mathbf{x}_{ij}\beta^{T}+u_j+\varepsilon_{ij},\ u_j\sim N\left(0,\sigma_u^2\right),\varepsilon_{ij}\sim N\left(0,\sigma_\varepsilon^2\right). \tag{15.1}$$

For notational simplicity we sometimes refer to y_{ij} as the income variable or the logarithm of the income variable. The method starts by estimating (15.1) using the sample data. Once estimates of the fixed effects, $\hat{\beta}$, of the variance components, $\hat{\sigma}_u^2$, $\hat{\sigma}_\varepsilon^2$, and of the area random effects, $\hat{u}_j$, have been obtained, the ELL method uses the following bootstrap population model to generate L synthetic Censuses,

$$y_{ij}^* = x_{ij}\hat{\beta}^{\mathrm{T}} + u_j^* + \varepsilon_{ij}^*, u_j^* \sim N\left(0, \hat{\sigma}_u^2\right), \varepsilon_{ij}^* \sim N\left(0, \hat{\sigma}_\varepsilon^2\right). \tag{15.2}$$

The exact steps of the Monte-Carlo simulation are as follows. Start by estimating (15.1) using the sample data; draw L population vectors of y_{ij}^* using (15.2); using the synthetic values of the welfare variable, y_{ij}^*, compute the ELL estimate from the 1th synthetic Census, $\hat{z}_j^{*WB(1)}(\alpha,t)$; last, average the results over L Monte Carlo simulations.

Using the bootstrap population model (15.2), one can further compute the MSE of the estimated poverty indicators,

$$\mathrm{MSE}\left[\hat{z}_j^{WB}(\alpha,t)\right] = \mathrm{L}^{-1}\sum_{l=1}^{\mathrm{L}}\left[\hat{z}_j^{*WB(l)}(\alpha,t) - E\left(\hat{z}_j^{*WB}(\alpha,t)\right)\right]^2.$$

One distinguishing aspect of the ELL method is that the random effect is specified at the cluster (e.g. primary sampling units) level and not at the level of the target small area. This is in contrast to the alternative methodologies we describe later in this chapter. For the sake of simplicity, in this chapter we assume that clusters and target small areas coincide. As Molina and Rao (2010) pointed out, when small areas and clusters coincide, and since $E(u_j^*) = 0$, $E(\varepsilon_{ij}^*) = 0$, in the simplest case of estimating a small area mean, and denoting by U_j the population of units in domain j, the ELL method leads to

$$E(y_{ij}^*) = N_j^{-1}\sum_{k \in U_j} \mathbf{x}_{kj}\hat{\beta}^{\mathrm{T}},$$

which is a synthetic regression estimator. It may be reasonable to assume that in many cases a synthetic regression estimator will be less efficient than competing model-based estimators.

15.2.2 The EBP method

The EBP method was proposed by Molina and Rao (2010). Like the ELL approach, the EBP approach also uses a nested error regression model on the logarithmically transformed welfare variable. Let us start our description of the method by decomposing the population-level small area-specific poverty indicator as follows,

$$z_j(\alpha,t) = N_j^{-1}\left[\sum_{i \in s_j} z_i(\alpha,t) + \sum_{k \in r_j} z_k(\alpha,t)\right]. \tag{15.3}$$

The first component in (15.3) is observed in the sample, whereas the second component is unknown and should be estimated by using a small area model, which is itself estimated with the sample data. Similarly to the ELL method, the EBP method starts by estimating the nested error regression model to obtain estimates of the fixed effects, $\hat{\beta}$, of the variance components, $\hat{\sigma}_u^2$, $\hat{\sigma}_\varepsilon^2$, and of the small area random effects $\hat{u}_j$.

The EBP method then simulates out-of-sample data from the conditional distribution of the out-of-sample data, given the sample data. This is done by using the following bootstrap population model for generating L synthetic Censuses:

$$y_{ij}^* = x_{ij}\hat{\beta}^{\mathrm{T}} + \hat{u}_j + u_j^* + \varepsilon_{ij}^*, u_j^* \sim N\left(0, \hat{\sigma}_u^2(1-\gamma_j)\right), \varepsilon_{ij}^* \sim N\left(0, \hat{\sigma}_\varepsilon^2\right), \tag{15.4}$$

$$\gamma_j = \frac{\hat{\sigma}_u^2}{\left(\hat{\sigma}_u^2 + \hat{\sigma}_\varepsilon^2 / n_j\right)}.$$

The exact steps of the Monte-Carlo simulation are as follows. Start by estimating (15.1) using the sample data; draw L out of sample vectors of y_{ij}^* using (15.4); combine the sample y_{ij} with the out-of-sample y_{ij}^* values; compute the EBP estimate for the 1th synthetic Census, $\hat{z}_j^{*EPB(1)}(\alpha, t)$; average the results over L Monte-Carlo simulations. Focusing again on the simplest case of estimating a small area mean, and since $E(u_j^*) = 0$, $E(\varepsilon_{ij}^*) = 0$, the EBP approach leads to

$$E(y_{ij}^*) = N_j^{-1}\left[\sum_{i \in s_j} y_i + \sum_{k \in r_j} x_{kj}\hat{\beta}^T + \hat{u}_j\right],$$

which is expected to be more efficient than the synthetic regression estimates obtained with the ELL approach.

MSE estimation for the EBP estimates relies on a parametric bootstrap scheme (see also González-Manteiga *et al.*, 2008). In particular, using the bootstrap population model (15.2) B bootstrap populations are generated and the target population parameters are computed for each bootstrap population. From each bootstrap population a sample is selected and the EBP approach is implemented using the sample and bootstrap population data. MSE estimates of the EBP estimates are computed over the B bootstrap replications.

15.3 Outlier robust small area estimation of poverty indicators

A recently proposed approach to small area estimation is based on the use of a quantile/M-quantile regression model (Chambers and Tzavidis, 2006). The classical regression model summarises the behaviour of the mean of a random variable y at each point in a set of covariates $\mathbf{x}$. Instead, quantile regression summarises the behaviour of different parts of the conditional distribution of y at each point in the set of $\mathbf{x}$'s. Let us for the moment and for notational simplicity drop the area-specific subscript j. Suppose that $(\mathbf{x}_i, y_i)$, $i = 1, \ldots, n$, denotes the values observed for a random sample consisting of n units, where $\mathbf{x}_i$ are row p-vectors of a known design matrix $\mathbf{x}$, and y_i is a scalar response variable corresponding to the realisation of a continuous random variable with an unknown continuous cumulative distribution function F. A linear regression model for the qth conditional quantile of y_i given $\mathbf{x}_i$ is:

$$Q_{y_i}(q|\mathbf{x}_i) = \mathbf{x}_i^T \beta(q).$$

Estimates of the qth regression parameters, $\boldsymbol{\beta}(q)$, are obtained by minimising:

$$\sum_{i=1}^{n} \left\{ \left| y_i - x_i^T \beta(q) \right| \left[(1-q) I\left(y_i - x_i^T \beta(q) \leq 0\right) + qI\left(y_i - x_i^T \beta(q) > 0\right) \right] \right..$$

Quantile regression presents a generalisation of median regression whilst expectile regression (Newey and Powell, 1987) presents a 'quantile-like' generalisation of mean regression. M-quantile regression (Breckling and Chambers, 1988) integrates these concepts within a framework defined by a 'quantile-like' generalisation of regression based on influence functions (M-regression). The M-quantile of order q for the conditional density of y given the set of covariates $\mathbf{x}$, $f(y|\mathbf{x})$, is defined as the solution $MQ_y(q|\mathbf{x};\, \psi)$ of the estimating equation $\int \psi_q(y - MQ_y(q|\mathbf{x};\, \psi)) f(y|\mathbf{x}) dy = 0$, where ψ_q denotes an asymmetric influence function, which is the derivative of an asymmetric loss function ρ_q. A linear M-quantile regression model for y_i given $\mathbf{x}_i$ is one in which we assume that:

$$MQ_{y_i}(q \mid x_i; \psi) = x_i^T \beta_\psi(q), \tag{15.5}$$

and estimates of $\boldsymbol{\beta}_\psi(q)$ are obtained by minimising:

$$\sum_{i=1}^{n} \rho_q\left(y_i - \mathbf{x}_i^T \boldsymbol{\beta}_\psi(q)\right). \tag{15.6}$$

Different regression models can be defined as special cases of (15.6). In particular, by varying the specifications of the asymmetric loss function ρ_q we obtain the expectile, M-quantile and quantile regression models as special cases. When ρ_q is the square loss function, we obtain the linear expectile regression model if $q \neq 0.5$ (Newey and Powell, 1987) and the standard linear regression model if $q = 0.5$. When ρ_q is the loss function described by Koenker and Bassett (1978) we obtain the linear quantile regression. Throughout this chapter we will take the linear M-quantile regression model to be defined by using as ρ_q the Huber loss function (Breckling and Chambers, 1988). Setting the first derivative of (15.6) equal to zero leads to the following estimating equation:

$$\sum_{i=1}^{n} \psi_q\left(r_{iq}\right) x_i = 0,$$

where $r_{1q} = y_i - x_i^T \beta_\psi(q)$,

$$\sum_{i=1}^{n} \{2\psi(s^{-1} r_{iq})[(1-q) I(r_{iq} \leq 0) + q I(r_{iq} > 0)]$$

where s is an estimate of scale, such as the Mean Absolute Deviation. Since the focus of our chapter is on M-type estimation, we use the Huber Proposal 2 influence function. Provided that the tuning constant of the influence function is strictly greater than zero, estimates of model parameters are obtained using iteratively weighted least squares (IWLS).

Chambers and Tzavidis (2006) extended the use of M-quantile regression models to small area estimation. Following this development, these authors characterised the conditional variability across the population of interest by the M-quantile coefficients of the population units. For unit i with values y_i and $\mathbf{x}_i$, this coefficient is the value θ_i, such that $MQ_y(\theta_1|\mathbf{x}_i; \psi)) = y_i$. The M-quantile coefficients are determined at the population level. Consequently, if a hierarchical structure does explain part of the variability in the population data, then we expect units within clusters (domains) defined by this hierarchy to have similar M-quantile coefficients. An area-specific semi-parametric (empirical) random effect, θ_j, can be computed by the expected value of the M-quantile coefficients in area j.

Having presented the M-quantile small area model, we now focus on the estimation of the HCR and PG. Similarly to the EBP approach, we start by decomposing the population-level small area-specific poverty indicator, as follows:

$$z_j(\alpha,t) = N_j^{-1}\left[\sum_{i \in s_j} z_i(\alpha,t) + \sum_{k \in r_j} z_k(\alpha,t)\right]. \tag{15.7}$$

The first component in (15.7) is observed in the sample, whereas the second component is unknown and predicted values can be obtained by using a small area model. The EBP approach (Molina and Rao, 2010) makes explicit parametric Gaussian assumptions for the error terms of the nested error regression model. The same is true for the ELL approach. If it is known that the true error distribution is normal, the EBP will offer the optimal approach for estimating poverty indicators for small areas. However, what if the true error distribution is unknown?

A non-parametric approach to estimating (15.7) is offered by using a smearing-type estimator, which can be motivated by the work of Duan (1983). More specifically:

$$z_j(\alpha,t) = N_j^{-1}\left[\sum_{i \in s_j} z_i(\alpha,t) + \sum_{k \in r_j} E(z_k(\alpha,t))\right]. \tag{15.8}$$

We are now interested in finding an estimator for $E[z_k(\alpha, t)]$. For simplicity, let us focus on the simplest case, i.e. that of estimating the HCR, $z_k(0, t)$. In this case, $z_k(0, t)=I(y_k \leq t)$. The y_k values are unknown and hence we can use the M-quantile small area model to predict these values. It follows that:

$$E\left[z_k(0,t)\right] = \int I\left(x_i^T \beta_\psi(\theta_j) + \varepsilon \leq t\right) dF(\varepsilon). \tag{15.9}$$

Since we make no assumptions about the error distribution, $F(\varepsilon)$, we can estimate $F(\varepsilon)$ from the empirical distribution of the residuals:

$$\hat{F}(\varepsilon) = n^{-1} \sum_{i=1}^{n} I(\hat{\varepsilon}_i \leq \varepsilon).$$

It follows that:

$$\begin{aligned}\hat{E}\left[z_k(0,t)\right] &= \int I\left(x_k^T \hat{\beta}_\psi(\hat{\theta}_j) + \hat{\varepsilon}_i \leq t\right) d\hat{F}(\varepsilon) \\ &= n_j^{-1} \sum_{k \in r_j} \sum_{i \in s_j} I\left(x_k^T \hat{\beta}_\psi(\hat{\theta}_j) + \hat{\varepsilon}_i \leq t\right),\end{aligned} \tag{15.10}$$

where $\hat{\varepsilon}_i$ are the estimated residuals from the M-quantile fit. An estimator of $z_k(0, t)$ is then obtained by substituting (15.10) into (15.7) leading to:

$$\hat{z}_j(0,t) = N_j^{-1} \left[\sum_{i \in s_j} z_i(0,t) + \hat{E}\left[z_k(0,t)\right] \right]. \tag{15.11}$$

The same approach can be followed to estimate $z_k(1, t)$ or any other of the FGT poverty measures. Since robust estimation for the M-quantile model is automatic, the M-quantile model is conventionally estimated using the raw values of the welfare survey variable. However, the decision as to whether to use transformed or untransformed values depends on what the model diagnostics suggest.

15.3.1 MSE estimation

Mean Squared Error estimation for (15.11) is discussed in detail in Marchetti *et al.* (2012) and is based on a non-parametric bootstrap scheme. Here we recall the main steps of this scheme. Starting from sample s, selected from a finite population U without replacement, we fit the M-quantile small area model and obtain estimates of $\hat{\theta}_j$ and $\hat{\beta}_\psi(\hat{\theta}_j)$, which are used to compute the model residuals. We then generate B bootstrap populations, U^{*b}. From each bootstrap population we select L bootstrap samples, using simple random sampling within the small areas and without replacement, such that $n_j^* = n_j$. Using the bootstrap samples we obtain estimates of the target poverty indicators by using the methodology described in Section 15.3. Bootstrap populations are generated by sampling from the empirical distribution of the residuals, or a smoothed version of this distribution, conditionally or unconditionally on the small areas. Bootstrap estimators of the bias and variance of the estimated target small area parameter, $\hat{\tau}_j$, are defined respectively by:

$$\hat{B}\left(\hat{\tau}_j\right) = B^{-1}L^{-1}\sum_{b=1}^{B}\sum_{l=1}^{L}\left(\hat{\tau}_j^{*bl} - \tau_j^{*b}\right)$$

and

$$\hat{V}\left(\hat{\tau}_j\right) = B^{-1}L^{-1}\sum_{b=1}^{B}\sum_{l=1}^{L}\left(\hat{\tau}_j^{*bl} - \bar{\hat{\tau}}_j^{*bl}\right)^2,$$

where τ_j^{*b} is the small area parameter of the bth bootstrap population, $\hat{\tau}_j^{*bl}$ is the small area parameter estimated by using the 1th sample from the bth bootstrap population, and

$$\bar{\hat{\tau}}_j^{*bl} = L^{-1}\sum_{l=1}^{L}\hat{\tau}_j^{*bl}.$$

The bootstrap MSE estimator of the estimated small area target parameter is then defined as:

$$\hat{M}\left(\hat{\tau}_j\right) = \hat{V}\left(\hat{\tau}_j\right) + \hat{B}\left(\hat{\tau}_j\right)^2. \tag{15.12}$$

15.4 An empirical study

We designed a model-based simulation study to compare the performance of the ELL and M-quantile (MQ) methods in estimating small area FGT poverty measures. Both area-specific values of HCR and PG are estimated. Similar comparisons between the EBP and the MQ approaches will be reported elsewhere. In addition to model-based estimates of poverty indicators, we further evaluate the properties of the direct estimator of the HCR and the PG.

Three super-population models are used in order to generate populations with different characteristics. In each population individuals, $i=1,\ldots, N_j$, are clustered within 30 areas, $j=1,\ldots, 30$. The response variable, y_{ij}, which reflects a welfare indicator, is generated for each individual in the population under three different scenarios for the nested error regression model. Under the first scenario a single covariate is drawn from a normal distribution $x_{ij} \sim N(\mu_j, 1)$ with the mean varying across areas, $3 \leq \mu_j \leq 10$. The intercept and the slope take the values $\boldsymbol{\beta}=(3000, -150)$. Area effects and individual errors are drawn from normal distributions, $u_j \sim N(0, 200^2)$, $\varepsilon_{ij} \sim N(0, 800^2)$. Population values of the welfare variable are constructed using $y_{ij}=x_{ij}\boldsymbol{\beta}^T+u_j+\varepsilon_{ij}$. In total 9580 population units are generated. A sample is taken from each of the populations generated so that the sample size of each area is 10 per cent of its total size. For this scenario the ELL and MQ methods are implemented by estimating the working model using the raw values of the outcome variable.

Under the second scenario we use two covariates with values drawn from the following distributions, $x1_{ij} \sim N(0, 4)$, $x2_{ij} \sim N(0, 1)$, which are combined with the following intercept and slope terms, $\boldsymbol{\beta}=(20, -1, 0.5)$. Area effects and individual errors are again drawn from normal distributions, $u_j \sim N(0, 0.64^2)$, $\varepsilon_{ij} \sim N(0, 4^2)$. Population values of the welfare variable in this case are constructed using $y_{ij}=\exp(x_{ij}\boldsymbol{\beta}^T+u_j+\varepsilon_{ij})$. The population size is 7500 and the sample size is 480. The sample size for a given area is not necessarily proportional to the population size, which varies between eight and 34 with N_j equal to 250 for all areas. In this scenario the ELL and MQ methods are implemented by estimating the working small area model using the logarithmic values of the outcome variable.

For the third scenario the intercept and slope parameter values and the parametric form of the distributions, from which the covariate and area effects are drawn, remain the same as in the first scenario. In 16 of the 30 areas individual errors are also drawn from identical distributions, as in the first scenario. However, in the remaining areas a small number of units (between ten and 20, depending on the area) are drawn from a distribution with a higher variance, $\varepsilon_{ij} \sim N(0, 4000^2)$, thus introducing contamination, which invalidates the parametric assumptions of the nested error regression model. As it is not immediately obvious why a logarithmic transformation will alleviate the effects of contamination in this case, the ELL and MQ methods are implemented by estimating the working model, using the raw values of the outcome variable.

Each super-population model is used to simulate $H=500$ populations. The true empirical values of the FGT poverty measures are calculated for each area from the corresponding Monte-Carlo population as:

$$z_j(\alpha,t)=N_j^{-1}\sum_{i=1}^{N_j} z_{ij}(\alpha,t),$$

where the poverty status of an individual is calculated as:

$$z_{ij}(\alpha,t)=\left(\frac{t-y_{ij}}{t}\right)^{\alpha} I\left(y_{ij}\leq t\right).$$

The Bias and Root MSE (RMSE) of the estimates for each area are calculated over simulations using:

$$\text{Bias}\left(\hat{z}_j(\alpha,t)\right)=H^{-1}\sum_{h=1}^{H}\left(\hat{z}_j(\alpha,t)-z_j(\alpha,t)\right),$$

$$\text{RMSE}\left(\hat{z}_j(\alpha,t)\right)=\sqrt{H^{-1}\sum_{h=1}^{H}\left(\hat{z}_j(\alpha,t)-z_j(\alpha,t)\right)^2}.$$

The results for each scenario are presented in Tables 15.1–15.3.

Table 15.1 Scenario 1. Across area distribution of the bias and RMSE of estimates of HCR and PG

	Min.	*25th*	*Median*	*Mean*	*75th*	*Max.*
Bias – HCR						
MQ	–0.0091	–0.0066	–0.0056	–0.0055	–0.0045	0.0016
ELL	–0.0071	–0.0041	0.0033	0.0017	0.0055	0.0127
Direct	–0.0054	–0.0023	–0.0001	–0.0005	0.0017	0.0033
RMSE – HCR						
MQ	0.0232	0.0310	0.0354	0.0382	0.0443	0.0657
ELL	0.0336	0.0449	0.0532	0.0583	0.0725	0.0921
Direct	0.0404	0.0518	0.0541	0.0583	0.0647	0.0912
Bias – PG						
MQ	–0.0059	–0.0033	–0.0026	–0.0027	–0.0018	–0.0010
ELL	–0.0033	–0.0016	0.0010	0.0008	0.0022	0.0077
Direct	–0.0053	–0.0009	–0.00004	–0.0002	0.0007	0.0014
RMSE – PG						
MQ	0.0083	0.0122	0.0151	0.0182	0.0224	0.0446
ELL	0.0118	0.0170	0.0219	0.0268	0.0345	0.0610
Direct	0.0157	0.0217	0.0240	0.0275	0.0317	0.0559

Note
Results are averaged over Monte-Carlo simulations.

Table 15.2 Scenario 2. Across area distribution of the Bias and RMSE of estimates of HCR and PG

	Min.	*25th*	*Median*	*Mean*	*75th*	*Max.*
Bias – HCR						
MQ	–0.0076	–0.0039	–0.0016	–0.0016	0.00094	0.0037
ELL	–0.0184	–0.0064	0.000007	0.0030	0.0114	0.0305
Direct	–0.0115	–0.0061	–0.000081	–0.0013	0.0035	0.0089
RMSE – HCR						
MQ	0.055	0.060	0.064	0.064	0.067	0.079
ELL	0.102	0.106	0.109	0.109	0.111	0.119
Direct	0.076	0.107	0.121	0.121	0.132	0.171
Bias – PG						
MQ	–0.0094	–0.0063	–0.0037	–0.0041	–0.0018	0.0007
ELL	–0.0165	–0.0060	0.000096	0.0024	0.0110	0.0264
Direct	–0.0118	–0.0041	–0.0002	–0.0005	0.0034	0.0081
RMSE – PG						
MQ	0.048	0.053	0.055	0.056	0.058	0.069
ELL	0.089	0.092	0.095	0.095	0.097	0.103
Direct	0.061	0.087	0.100	0.099	0.110	0.137

Note
Results are averaged over Monte-Carlo simulations.

Table 15.3 Scenario 3. Across area distribution of the Bias and RMSE of estimates of HCR and PG

	Min.	*1st Qu.*	*Median*	*Mean*	*3rd Qu.*	*Max.*
Bias – HCR						
MQ	–0.0139	–0.0076	–0.0048	–0.0052	–0.0027	7E–07
ELL	0.0004	0.0139	0.0199	0.0182	0.0233	0.0294
Direct	–0.0061	–0.0028	–0.0002	–0.0005	0.0018	0.0053
RMSE-HCR						
MQ	0.0223	0.0300	0.0357	0.0377	0.0419	0.0596
ELL	0.0432	0.0510	0.0572	0.0619	0.0714	0.0918
Direct	0.0399	0.0540	0.0584	0.0607	0.0655	0.0874
Bias – PG						
MQ	–0.0237	–0.0119	0.0039	–0.0029	0.0059	0.0067
ELL	–0.0125	–0.0013	0.0110	0.0070	0.0143	0.0202
Direct	–0.0031	–0.0016	–0.0004	–0.0002	0.0006	0.0036
RMSE – PG						
MQ	0.0117	0.0151	0.0205	0.0223	0.0257	0.0454
ELL	0.0178	0.0223	0.0271	0.0316	0.0371	0.0640
Direct	0.0164	0.0231	0.0399	0.0388	0.0537	0.0698

Note
Results are averaged over Monte-Carlo simulations.

Examining the simulation results, we note that the M-quantile estimates of HCR and PG are more efficient than the corresponding ELL estimates. As we described earlier, the ELL approach provides synthetic regression estimators, which can be significantly less efficient than indirect small area estimates. On the other hand, both the ELL and M-quantile approaches are generally more efficient than the direct estimates, implying that a small area model improves estimation in this case.

15.5 An application: poverty mapping for provinces in Tuscany

In Italy, the European Survey on Income and Living Conditions (EU-SILC) is carried out yearly by the Italian National Statistical Institute (ISTAT), with the aim of producing estimates of poverty and living conditions at both national and regional (NUTS-2) levels. Regions are planned domains, for which EU-SILC estimates are published, while the provinces (LAU-1 level) are unplanned domains. Provinces are further partitioned into municipalities (LAU-2 level). The regional samples are based on a stratified two-stage sample design. In each province the municipalities are the Primary Sampling Units (PSUs) and the households the Secondary Sampling Units (SSUs). The PSUs are divided into strata according to their population sizes, with the SSUs being selected by means of systematic sampling from each PSU. Data are collected for each household sampled by means of a questionnaire. Official estimates are provided only at regional level because direct estimates may have large sampling errors for most provinces. Hence, one way of obtaining reliable estimates at the province level is by using small area estimation techniques. In this section we describe the application of small area methodologies for the estimation of small area poverty indicators (HCR and PG) regarding provinces in Tuscany. For the purposes of this application we used survey data from the 2008 EU-SILC. This survey collected information from 1495 households in Tuscany. Table 15.4 presents the population and number of households sampled for each province.

The small area methods presented in the previous sections require auxiliary information for all units (households) in the population. This information is available from Italy's 2001 Population Census. Access to Census data from Tuscany was given by ISTAT to researchers from the University of Pisa, under the condition that the data are kept and analysed in a secure setting at the University of Pisa. Census variables, also available in the survey dataset, included: the gender, age, occupational status, civil status and years of education of the head of the household, the ownership status of the house, the square metres of the house and the number of household members. These explanatory variables are included in the small area model that is used to model equivalised household income. Although the 2008 EU-SILC data were collected six years after the Census, the 2001–2007 period was one of relatively slow growth and low inflation in Italy, so it is reasonable to assume that the structural relationship between income and the covariates remained stable during this period.

In Table 15.5 we present the parameter estimates of the M-quantile (median) model. Since the main target of small area estimation is prediction, we decided to include in the model as many explanatory variables as possible. In a preliminary data analysis step we also estimated a nested error regression model (on the logarithm of equivalised income) with random effects specified at the province level. The estimated intra-province coefficient, although small, appeared to be significant, hence justifying the use of indirect small area estimators. In addition, the model residuals exhibited some departures from the assumption of normality.

Table 15.4 Number of total households and number of sampled households in Tuscan provinces

Province	*Total households*	*Sampled households*
Massa and Carrara	90052	105
Lucca	164526	150
Pistoia	121192	136
Firenze	427172	415
Livorno	151722	105
Pisa	170865	149
Arezzo	141821	143
Siena	116101	104
Grosseto	102284	65
Prato	97171	123
Tuscany	1582906	1495

Table 15.5 Estimated M-quantile regression parameters using the 2008 EU-SILC data from Italy

Variable	*Value*	*P-value*
Constant	2932.704	0.064
House dimension (m^2)	43.226	0.000
Ownership status	2058.885	0.001
Age of the head of the household	–15.800	0.323
Employment status of the head of the household	4654.613	0.000
Gender of the head of the household	2066.183	0.000
Years of education of the head of the household	511.733	0.000
Household size	8.952	0.966
Civil status of the head of the household single	–950.662	0.154
Civil status of the head of the household divorced	–1370.146	0.116
Civil status of the head of the household widow	1626.224	0.041

Using the 2008 EU-SILC survey data, the Census 2001 data and the M-quantile model parameter estimates (Table 15.5), we obtained estimates of HCR and PG using the methodology outlined in Section 15.3. The poverty line used to compute estimates of HCR and PG is set to 9381.91 euros, corresponding to 60 per cent of the median equivalised Italian household income. This estimate is derived using the 2008 EU-SILC income values, which are available for the entire sample in Italy with weights equal to the cross-sectional EU-SILC household weights. In the region of Tuscany the direct estimates of HCR and PG are respectively 0.127 (0.010) and 0.040 (0.004). In parentheses we report the estimated root mean squared errors, which have been adjusted for the complex survey design of the EU-SILC. Direct and model-based (MQ) estimates of HCR and PG for each province in Tuscany are presented in Table 15.6. Model-based estimates are also presented in Figure 15.1. For each province we report the estimated HCR and PG, as well as the estimated root mean squared error in parentheses.

Table 15.6 M-quantile (MQ) and direct estimates of the HCR and the PG in provinces of Tuscany

Province	*MQ*		*Direct*	
	HCR (RMSE)	*PG (RMSE)*	*HCR (RMSE)*	*PG (RMSE)*
Massa-Carrara (MS)	0.221 (0.027)	0.095 (0.016)	0.185 (0.040)	0.043 (0.013)
Lucca (LU)	0.174 (0.019)	0.070 (0.010)	0.171 (0.034)	0.049 (0.014)
Pistoia (PT)	0.143 (0.017)	0.054 (0.009)	0.148 (0.035)	0.048 (0.017)
Firenze (FI)	0.132 (0.011)	0.050 (0.006)	0.124 (0.019)	0.047 (0.010)
Livorno (LI)	0.185 (0.021)	0.076 (0.012)	0.104 (0.032)	0.026 (0.009)
Pisa (PI)	0.133 (0.015)	0.050 (0.008)	0.110 (0.027)	0.044 (0.015)
Arezzo (AR)	0.146 (0.016)	0.056 (0.009)	0.086 (0.023)	0.018 (0.006)
Siena (SI)	0.139 (0.017)	0.053 (0.009)	0.122 (0.037)	0.029 (0.014)
Grosseto (GR)	0.194 (0.026)	0.081 (0.015)	0.191 (0.052)	0.067 (0.025)
Prato (PO)	0.124 (0.017)	0.045 (0.009)	0.074 (0.031)	0.012 (0.005)

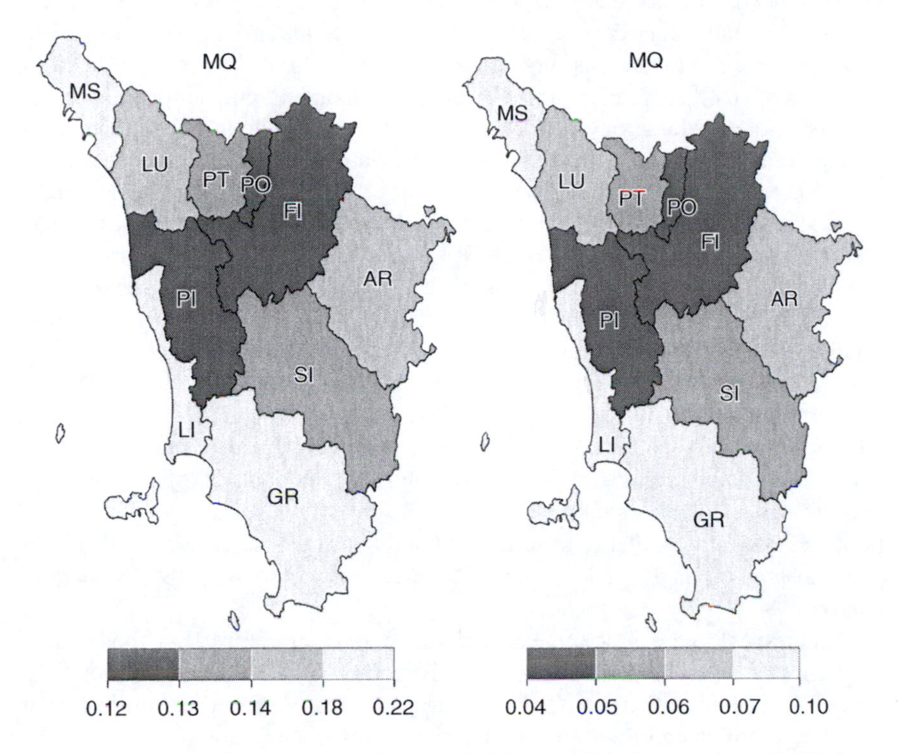

Figure 15.1 Model-based (MQ) estimates of HCR and PG for provinces in Tuscany using the 2008 EU-SILC data and 2001 Census micro-data from Italy.

According to the M-quantile estimates, the province with the highest number of poor households and the highest poverty gap is Massa-Carrara. According to the direct estimates, the province with the highest number of poor households and the highest poverty gap is Grosseto, with Massa-Carrara being very close. One way of evaluating the validity of the model-based estimates is by using external diagnostics, e.g. expert opinions concerning local poverty. There is strong evidence that the provinces of Grosseto and Massa-Carrara are the poorest in Tuscany. The model-based estimates for both provinces stand out, indicating that the MQ estimates are plausible. Moreover, the RMSE of the model-based estimates is lower than the corresponding RMSE of the direct estimates, indicating the efficiency gains of using model-based estimation in this case.

15.6 R software for poverty mapping with the M-quantile approach

The importance of having access to software for implementing the estimation approaches we have described in this chapter is now widely recognised. Marchetti and Tzavidis have developed open source R functions for small area estimation with the M-quantile approach. More specifically, estimation of small area poverty indicators can be achieved with the function *mq.sae.poverty*. The packages required to use this function are MASS and np. The function provides estimates of small area HCRs and PGs and corresponding bootstrap MSE estimates using the methodology described in Section 15.3. The arguments required are the response variable y, the matrix of covariates x, covariate information for out-of-sample units x outs and values relating to the convergence of the algorithm. L specifies the number of Monte-Carlo runs (synthetic Censuses generated) to estimate HCR and PG. If MSE=TRUE, bootstrap MSE estimates are produced. B denotes the number of bootstrap populations and R denotes the number of bootstrap samples to be selected from each bootstrap population. Finally, the method defines the type of residuals used to generate the bootstrap population: 'su' (smoothed unconditional), 'eu' (empirical unconditional), 'sc' (smoothed conditional), 'ec' (empirical unconditional). The default is set to 'eu', which is computationally faster. Similar functions for estimating small area averages and the small area quantiles of a distribution function with the M-quantile approach are available using the functions *mq.sae* and *mq.sae.quant*. These functions are currently available from the authors upon request. However, there are plans to prepare an R package for small area estimation with the M-quantile approach.

With respect to the alternative estimation approaches, the ELL method can be implemented using software available from the World Bank. For the EBP approach, an R function for implementing point and MSE estimation has been written by Molina and Rao (2010). More information about this function and the M-quantile functions can be obtained from Molina *et al.* (2010).

15.7 Concluding remarks and areas for future research

In this chapter we have reviewed some recently proposed model-based methods for poverty estimation, with particular emphasis on outlier robust semi-parametric approaches. The outlier robust approach we advocate in this chapter is based on the use of a smearing-type estimator that can be used when the parametric assumptions of the nested error regression model are questionable. Hence, the smearing-type estimators are expected to offer efficiency gains when the model assumptions are not valid, but are equally expected to be less efficient when the assumptions of the nested error regression model hold.

In our empirical study we compared the ELL and M-quantile approaches for poverty estimation. The results are in line with the evidence already presented by Molina and Rao (2010) in the case of the EBP approach. That is, the ELL method offers a synthetic estimator and indirect estimators may offer substantial gains in efficiency. Empirical comparisons between the EBP and the M-quantile methods are not presented in this chapter, but we expect the EBP to be more efficient than the M-quantile approach when the model assumptions hold.

One key difference between the ELL method and the alternative methods is that the random effect is specified at the level of cluster (e.g. psu) and not at the level of the target small area. The reason for specifying the random effect at the cluster level is because the ELL approach aims to account for the clustered design of household surveys. However, the drawback of this approach is the resulting synthetic ELL estimator for the target small areas. An alternative approach to accounting for the effect of the sampling design involves including as many design variables as possible in the small area model and then assuming an ignorable sampling design. A further alternative approach, and one that the authors are currently investigating, is to allow for more complex hierarchical structures that better reflect the sampling design. This is perhaps a more realistic approach, as in most applications we expect clusters and small areas not to coincide. In this chapter we further assumed that all target small areas are part of our sample data. However, in most small area estimation applications we expect a number of the small areas to be out-of-sample. In this case, small area estimation is performed using synthetic estimation. All methods we presented in this chapter can be adapted for this purpose.

The M-quantile approach is not the only outlier robust estimation method that can be used. An alternative approach to outlier robust poverty estimation involves using an outlier robust version of the EBP method, in which the variance components are estimated by using the robust EBLUP (REBLUP) approach proposed by Sinha and Rao (2009). Nevertheless, the REBLUP approach is computationally intensive, making its use with large datasets extremely difficult, especially since MSE estimation requires the use of bootstrap. Therefore, more research is needed to develop outlier robust approaches to poverty estimation.

Finally, in this chapter we focused on the estimation of fairly simple deprivation indicators. In many instances practitioners are interested in more complex measures of poverty that extend well beyond income deprivation and incorporate other dimensions of deprivation, such as health, education, social security and cohesion. In our experience extending the small area methods to the simultaneous estimation of multidimensional deprivation indicators is something that policy makers and practitioners look forward to and further research is needed in order to achieve this goal.

References

Breckling, J. and Chambers, R. (1988). M-quantiles. *Biometrika*, 75, 761–771.

Chambers, R. and Tzavidis, N. (2006). M-quantile models for small area estimation. *Biometrika*, 93, 255–268.

Chambers, R., Chandra, H. and Tzavidis, N. (2011). On bias-robust mean squared error estimation for pseudo-linear small area estimators. *Survey Methodology*, 37, 153–170.

Chambers, R., Chandra, H., Salvati, N. and Tzavidis, N. (2013). Outlier robust small area estimation. *Journal of the Royal Statistical Society Series B* (in press).

Duan, N. (1983). Smearing estimate: a nonparametric retransformation method. *Journal of the American Statistical Association*, 78, 605–610.

Elbers, C., Lanjouw, J.O. and Lanjouw, P. (2003). Micro-level estimation of poverty and inequality. *Econometrica*, 71, 355–364.

Foster, F., Greer, J. and Thorbecke, E. (1984). A class of decomposable poverty measures. *Econometrica*, 52, 761–766.

González-Manteiga, W., Lombardia, M.J., Molina, I., Morales, D. and Santamaria, L. (2008). Bootstrap mean squared error of a small-area EBLUP. *Journal of Statistical Computation and Simulation*, 78, 443–462.

Koenker, R. and Bassett, G. (1978). Regression quantiles. *Econometrica*, 46, 33–50.

Marchetti, S., Tzavidis, N. and Pratesi, M. (2012). Non parametric bootstrap mean squared error estimation for M-quantile estimators of small area averages, quantiles and poverty indicators. *Computational Statistics and Data Analysis*, 56, 2889–2902.

Molina, I. and Rao, J.N.K. (2010). Small area estimation of poverty indicators. *Canadian Journal of Statistics*, 38, 369–385.

Molina, I., Morales, D., Pratesi, M. and Tzavidis, N. (eds) (2010). Final small area estimation development and simulation results. Vol. Deliverable 12 and 16 – S.A.M.P.L.E. Project. European Union, 7th Framework Programme.

Newey, W. and Powell, J. (1987). Asymmetric least squares estimation and testing. *Econometrica*, 55, 819–847.

Rao, J.N.K. (2003) *Small Area Estimation*. Wiley, London.

Sinha, S. and Rao, J.N.K. (2009). Robust small area estimation. *Canadian Journal of Statistics*, 37, 381–399.

Tzavidis, N., Marchetti, S. and Chambers, R. (2010). Robust prediction of small area means and distributions. *Australian and New Zealand Journal of Statistics*, 52, 167–186.

16 Poverty and social exclusion in 3D

Multidimensional, longitudinal and small area estimation

Gianni Betti and Achille Lemmi

16.1 Introduction

This final chapter is not merely a conclusion of the book. It seeks to connect the three previous parts of the book, or the three dimensions of poverty and social exclusion: Multidimensionality, Longitudinal poverty and Small area estimation. It constitutes the "fil rouge" of the whole book, and examines and proposes methods that combine these three dimensions.

The chapter is composed of six sections: after this introduction, we present and propose new and old approaches to measure multidimensional poverty at longitudinal level (Section 16.2), multidimensional poverty at local level (Section 16.3) and longitudinal poverty at local level (Section 16.4). Finally, Section 16.5 presents a new method for estimating poverty at local level using longitudinal or cross-sectional data: the cumulation approach; here we propose applying this approach to fuzzy multidimensional measures of poverty, to form a 3D picture.

16.2 Multidimensional and longitudinal poverty

Two of the first attempts to study multidimensional poverty in a longitudinal context are the works of Betti and Verma (1999) and Cheli and Betti (1999), which make use of fuzzy set theory previously proposed by Cerioli and Zani (1990) and later formalised in the so-called Totally Fuzzy and Relative (TFR) approach by Cheli and Lemmi (1995).

Betti and Verma (1999) denote with μ_i^t and μ_i^{t+1} individual multidimensional degrees of poverty of unit i at two successive points of time, t and $t+1$. These are two fuzzy sets and hence they define measures of persistence or otherwise of poverty at the individual level, as follows: persistent poverty, i.e. present at both time points, as the intersection of the two sets:

$$\mu_i^{(P)} = \min\left(\mu_i^t, \mu_i^{t+1}\right); \tag{16.1}$$

any-time poverty, that is at one or both of the time points, as the union of the two fuzzy sets:

$$\mu_i^{(S)} = \max\left(\mu_i^t, \mu_i^{t+1}\right). \tag{16.2}$$

Transient poverty, i.e. at one but not both times, is the difference between the above, while non-poverty is the complement of (16.1).

The method can be extended to any number T of periods (years). The indices of persistent poverty and any-time poverty are defined as follows:

$$\mu_i^{(P)} = \min\left(\mu_i^1, \mu_i^2, ..., \mu_i^t, ..., \mu_i^T\right) \tag{16.3}$$

$$\mu_i^{(S)} = \max\left(\mu_i^1, ..., \mu_i^t, ...\mu_i^T\right). \tag{16.4}$$

More indices can be calculated in the case of T periods. Let $\mu_i^{[j]}$ be the ranked value for individual i so that $\mu_i^{[1]} \le \mu_i^{[2]} \le \ldots \le \mu_i^{[T]}$. Then $\mu_i^{[1]} = \min(\mu_i^1, \mu_i^2, \ldots, \mu_i^t, \mu_i^t, \ldots, \mu_i^T) = \mu_i^{(P)}$ is the propensity to always be poor throughout the period T, i.e. the index of persistent poverty; $\mu_i^{[T]}$ is the index of any-time poverty during T. In general, $\mu_i^{[j]}$ can be seen as the propensity to be poor for at least $(T+1-j)$ of the T periods.

Since we can identify the propensity to persist in poverty to varying degrees, it could be very instructive to compare $\bar{\mu}^{[1]}, \bar{\mu}^{[2]}, \ldots, \bar{\mu}^{[T]}$ against the average $\bar{\mu}$ over T periods.

Cheli and Betti (1999) denote $\bar{\mu}_i^t$ and $\bar{\mu}_i^{t+1}$ of Betti and Verma (1999) with

$$\mathbf{g}_\mathbf{i}^{(t)} = [g_{i0}^{(t)}, g_{i1}^{(t)}], \quad i = 1, \ldots, n; \quad t = 1, \ldots, T;$$

the vector whose components represent the degrees of membership in the two multidimensional fuzzy states.

One of the fundamental tools with which to analyse the dynamics between two discrete or fuzzy states is represented by the so-called transition matrix (T); they define such a matrix in the following way:

$$t_{kl}^{(1,2)} = \frac{\mathrm{E}[g_{ikl}^{(1,2)}]}{\mathrm{E}[g_{ik}^{(1)}]} \tag{16.5}$$

where $g_{ikl}^{(1,2)}$ represents the joint membership function for the i-th unit in states k and l at times 1 and 2, respectively, whereas $E\left[g_{kl}^{(1,2)}\right]$ and $E\left[g_k^{(1)}\right]$ represent the mean values of the corresponding memberships in the population and are calculated by the corresponding mean values in the sample. The joint membership in equation (16.5) is defined as follows:[1]

$$g_{ikl}^{(1,2)} = \min[g_{ik}^{(1)}, g_{il}^{(2)}] \tag{16.6}$$

under marginal constraints:

$$\sum_l g_{ikl}^{(1,2)} = g_{ik}^{(1)}, \ \sum_k g_{ikl}^{(1,2)} = g_{il}^{(2)} \text{ and } \sum_k \sum_l g_{ikl}^{(1,2)} = 1.$$

When considering more than two times (T), and considering a model with first order memory, after a few simple algebraic manipulations, Cheli and Betti (1999) obtain:

$$\mathrm{E}[g^{(1,2,\ldots,T)}_{k_1k_2\ldots k_T}] = \mathrm{E}[g^{(1,2,3)}_{k_1k_2k_3}]\, t^{(3,4|2)}_{k_3k_4|k_2} \ldots t^{(T-1,\,T|T-2)}_{k_{T-1}k_T|k_{T-2}}. \quad (16.7)$$

Betti *et al*. (2002) study multidimensional poverty dynamics and the influence of socio-demographic factors. A fuzzy and multidimensional approach is chosen here in order to define two different poverty measures: one unidimensional and the other multidimensional. A panel regression model is proposed to deal with the unobservable heterogeneity among longitudinal units. The specified model combines autoregression with variance components. The empirical analysis is conducted using the data set of the British Household Panel Survey (BHPS) from 1991 to 1997.

The theory of multidimensional and fuzzy poverty in a longitudinal context has been further developed by Betti *et al.* (2006) and empirically applied by Betti and Verma (2009).

Betti *et al.* (2006) defines the fuzzy multidimensional or supplementary poverty measure FS, as introduced in Chapter 5, formula (5.4): in the longitudinal context a person's multidimensional membership over T period is defined as $(\mu_1, \mu_2, \ldots, \mu_T)$, $\mu_t \in [0, 1]$. The complement of the above at each time is defined as $\bar{\mu} = 1 - \mu_t$. The above cross-sectional measures generate 2^T longitudinal sequences of length T, in which any element t can take one of two values, μ_t and its complement $\bar{\mu}_t = (1 - \mu_t)$.

Let $S(1, 2, \ldots, T)$ be a particular pattern of T "poor" and "non-poor" sets, for which the membership function (m.f.) is required. Let the *elements* (cross-sectional sets) of this pattern be grouped into two parts, $S_1 = (\ldots, t_1, \ldots)$ and $S_2 = (\ldots, t_2, \ldots)$, where t_1 indicates any T_1 elements of the same type (say, "poor") in the first group, and t_2 any T_2 elements of the opposite type ("non-poor") in the same group, with $T_1 + T_2 = T$. Let $m_1 = \min(\ldots, \mu_{t1}, \ldots)$; $M_2 = \max(\ldots, \mu_{t2}, \ldots)$. The joint m.f. (JMF) required for the particular pattern of interest is given by the following:

$$JMF = \max\left(0, m_1 - M_2\right). \quad (16.8)$$

Different types of longitudinal measures correspond to, or can be simply derived from, different patterns S. A number of applications are described later. As an example, for the propensity to be poor at time 1, non-poor at time 2 and then re-entering poverty at time 3, Betti *et al.* (2006) define:

$$S_1 = (1,3),\; S_2 = (2),\; JMF = \max\left(0, \min\left(\mu_1, \mu_3\right) - \mu_2\right). \quad (16.9)$$

On the basis of the above, a general procedure is formulated in the following terms. Consider any sequence of cross-sectional propensities for poverty or deprivation. This can always be expressed in the form $(\ldots, \mu_{t1}, \ldots)$, $(\ldots, \mu_{t2}, \ldots)$, where t_1 indicates T_1 elements of the same type in one group, and t_2 indicates T_2 elements of the opposite type in the other group.

Sort the elements into two groups by type, for instance all T_1 elements of one type followed by all T_2 elements of the other type. Construct the intersection for each group involving elements of the same type, using the standard operator. Finally, construct the intersection of the two results of the operation above. Since the temporal order of cross-sectional propensities is immaterial in the construction of their intersection using this rule, the application of this rule can be seen as being without memory. More precisely, Betti *et al.* (2006) designate it as a procedure "without chronology": the outcome depends on the whole "history" (i.e. the specified type of cross-sectional sets in the time sequence $t=1$ to T, and the associated membership functions), but it does not depend on the actual chronology, or the temporal sequence, of those cross-sections.[2]

Betti and Verma (2009) further develop the work of Betti *et al.* (2006) and apply the methodology to the eight waves of the European Community Household Panel (ECHP) survey for 15 countries. They define $r_i=\min(a_j: j=1 \text{ to } i)$ as the minimum value of the membership functions for the first i periods, and $R_i=\max(a_j: j=1 \text{ to } i)$ as the corresponding maximum value. Then for row i, the aggregation of all the sets it represents is given, separately for each panel, by the intersection of the first $(i+1)$ specified cells, as shown at the bottom of Table 16.1.[3] Summing over the first $(T-1)$ rows gives:

$$\begin{aligned}&\sum_{i=1}^{T-1}\left(r_i - r_{i+1}\right)+\sum_{i=1}^{T-1}\left(R_{i+1} - R_i\right) = \left(r_1 - r_T\right)+\left(R_T - R_1\right)\\ &= R_T - r_T = \max\left(a_i\right) - \min\left(a_i\right)\end{aligned} \quad (16.10)$$

The result is pleasantly simple. For any number of periods with propensities to poverty (or a more general form of deprivation) as (a_i), the propensity to continuous is $C_i=\min(a_i)$, and the propensity to any-time is $A_i=\max(a_i)$.

Recently, Nicholas and Ray (2012) have expanded the literature on static multidimensional deprivation by proposing dynamic deprivation measures that incorporate both the persistence and duration of deprivation across multiple dimensions. Their article also illustrates the usefulness of this extension by applying it to Australian panel data for the period 2001–2008.

Finally, the most recent attempt to study multidimensional poverty and material deprivation from an intertemporal perspective is Chapter 7 of this volume. In it, Bossert *et al.* employ the EU-SILC panel data, which includes information on different aspects of well-being over time, to compare EU countries based on measures that take this additional intertemporal information into consideration. Following the path of material deprivation experienced by each individual over time, the results gained in this chapter provide a different picture from the annual results. Since the measurement of material deprivation is used by the EU member states and the European Commission to monitor national and EU progress in the fight against poverty and social exclusion, these results suggest that time cannot be neglected. Countries should not only be compared based on their year by year results, but additional information needs to be gained by following individuals over time and producing an aggregate measure once time is taken into account.

Table 16.1 Membership function for interaction sets for T time periods

	Panel 1												*Panel 2*											
Time	1	2	3	4	5	6	..		T-3	T-2	T-1	T	1	2	3	4	5	6	..		T-3	T-2	T-1	T
Set group																								
1	+	–	?	?	?	?	..	?	?	?	?	?	–	+	?	?	?	?	..	?	?	?	?	?
2	+	+	–	?	?	?	..	?	?	?	?	?	–	–	+	?	?	?	..	?	?	?	?	?
3	+	+	+	–	?	?	..	?	?	?	?	?	–	–	–	+	?	?	..	?	?	?	?	?
4	+	+	+	+	–	?	..	?	?	?	?	?	–	–	–	–	+	?	..	?	?	?	?	?
..	..	..	..	..	..	..	..	..	..	..	..	..	..	..	..	..	..	..	..	..	..	..	..	
T-3	+	+	+	+	+	+	..	+	+	–	?	?	–	–	–	–	–	–	..	–	–	+	?	?
T-2	+	+	+	+	+	+	..	+	+	+	–	?	–	–	–	–	–	–	..	–	–	–	+	?
T-1	+	+	+	+	+	+	..	+	+	+	+	–	–	–	–	–	–	–	..	–	–	–	–	+
T	+	+	+	+	+	+	..	+	+	+	+	+	–	–	–	–	–	–	..	–	–	–	–	–

Source: Betti and Verma (2009).

16.3 Multidimensional poverty at local level

We can define two categories of works that study multidimensional poverty at local level: in the first category, approaches are based on counting some simple or composite indicators related to the whole population in a specific country, i.e. using census data. Although this approach provides estimates that are not affected by sampling error, it can only be repeated every ten years, i.e. when the subsequent census is conducted; in the second category, approaches are based on so-called small area estimation (SAE) techniques, and do not necessarily require the use of census data. In this section we are more interested in and focused on the latter category; however, we may recall (among many others) one particularly interesting work based on census data, i.e. the World Bank (2005) socio-economic atlas of Tajikistan. This includes information on a range of different indicators of the population's well-being, including education, health, economic activity and the environment. A unique feature of the atlas is the inclusion of estimates of material poverty at the *Jamoat* local level.

The first seminal paper in the second category mentioned above is that of Berman and Phillips (2000). This paper explores the domains and indicators of social inclusion and exclusion and their interaction at national and *community* level. Here, social inclusion/exclusion is dynamically conceptualised within the overarching construct of social quality. Micro and macro aspects of social quality are discussed, along with the relationships between organisations, institutions and communities, groups and individuals. These interactions between social inclusion and exclusion at national and community level are then exemplified, ranging from inclusion to exclusion at both community and national level, via the intermediate stages of inclusion in one realm and exclusion in another. The social policy implications of the relationship between national and community exclusion are drawn, both for macro/institution and organisational levels (in relation to legislation and society-wide service provision) and for micro, group and citizen levels (in relation to social work).

Verma *et al.* (2005), in their report to the European Commission on *Regional indicators to reflect social exclusion and poverty*, subsequently improved by Betti *et al.* (2012), describe and illustrate a statistical methodology for generating comparative indicators of well-being and extend their use to the level of sub-national regions. The choice of the methodology very much depends on the type of data available for its implementation. The main data sources for EU countries are the ECHP and EU-SILC surveys, supplemented by rich data from diverse sources in the Eurostat database "NewCronos". The strategy Verma *et al.* (2005) and Betti *et al.* (2012) propose for the construction of regional indicators of well-being has three fundamental aspects: (a) making the best use of available data from national sample surveys, such as by cumulating and consolidating the information to obtain more robust measures that permit greater spatial disaggregation, (b) exploiting "meso" data to the maximum, such as the highly disaggregated tabulations available in the Eurostat data source "NewCronos", for the purpose of constructing regional indicators and (c) using the two sources in combination to produce more precise estimates for regions using appropriate SAE techniques, as described in Chapters 11–15 of this book.

The most recent work analysing multidimensional poverty at local level is by Ferretti and Molina (2012); this paper studies small area estimation of computationally complex poverty indicators, such as Fuzzy Monetary and Fuzzy Supplementary measures, which are defined in Chapter 5 as (5.1) and (5.4) respectively. For these indicators, a faster version of the empirical best/bayes (EB) method of Molina and Rao (2010) is proposed. The new method allows feasible estimation of such computationally complex indicators in large populations, and can still considerably reduce the computation time when the original EB method is feasible.

16.4 Longitudinal poverty at local level

In this section we limit our focus to two approaches proposed in literature for the estimation of longitudinal or repeated measures of poverty and inequality at local level. We start with the pioneering work of Elbers, Lanjouw and Lanjouw (2003, ELL), to which Chapter 12 of this volume is devoted.

The first approach, initially proposed by Dabalen and Ferrè (2008) for updating poverty mapping between two periods, starts from the hypothesis that there is only one census, and two household surveys: a *new* and an *old* one. When the old household survey is conducted about the same time as the census, the ELL method can be applied and welfare estimates obtained for that year. The problem is how to update the poverty estimates for small areas when there is a new survey but no new census, especially when the new survey may contain either very few observations, or none at all in most of the small areas. In this case, Dabalen and Ferrè (2008) propose constructing a counterfactual consumption distribution of the old household survey, using information from both the old and new household surveys and matching the corresponding estimates with the old census data, following the methodology proposed by Lemieux (2002). Betti *et al.* (2013) have further updated the poverty mapping to cover three periods.

Here, using the notation of Dabalen and Ferrè (2008), we show how to construct the counterfactual distribution for the old data (t_0), which will permit us to link the census data with the new survey data (t). Let us consider a consumption model using the new survey data:

$$\ln(y_{t,i}) = \beta_t' X_{t,i} + \varepsilon_{t,i} \tag{16.10}$$

where y_t denotes consumption in year t, i indexes the household, β_t is a parameter (that captures the returns to covariates in t), X_t is a vector of covariates that are in common between the old and the new surveys, and ε_t is the unobserved component of consumption. Using the new survey without additional adjustment and applying the ELL estimator would be problematic, because the returns to covariates, the parameter β, may have changed between t_0 and t. Moreover, the profile of the population – that is covariates such as education levels, age composition and so on – may also have changed. Finally, the returns to unobserved covariates may also have changed. To recreate a consumption distribution that resembles consumption of t_0, these changes would need to be accounted for. Therefore, construction of the counterfactual consumption distribution proposed by Dabalen and Ferrè (2008) is based on three basic steps (not to be confused with the Stage zero, Stage one and Stage two of ELL; see Chapter 12). The first step consists in creating the consumption distribution that would have prevailed in t_0 if the parameters were as in t. This can be seen as:

$$\ln(y^{p}_{t_0,i}) = \hat{\beta}'_{t,i} X_{t_0,i}. \tag{16.11}$$

Equation (16.11) accounts for changes in the parameters of covariates by using the estimated parameters from the t survey to estimate consumption distribution in the t_0 survey. However, in addition to these parameters, levels of covariates may have changed because the population is t much more educated compared to t_0. When the covariates of interest have a small number of categories (i.e. the education variable has only two values, primary and higher education) then a simple reweighting of each cell would be sufficient. But when changes in multiple covariates are of interest, as they are in this case, it is not feasible to perform cell-by-cell reweighting. Instead, a score that reduces the dimension of the data can be created, by stacking the new and old surveys and then running a probit model of the form:

$$P_{iT} = \text{Prob}(survey = t \mid Z_{iT}, M_{iT}) = \alpha'_z Z_{iT} + \alpha'_m M_{iT}. \tag{16.12}$$

In principle, a large set of observable household-level characteristics can be included, Z_{iT}, as well as the migration status of the household, M_{iT}, or any suitable variables that capture the scale of migration, which is a crucial concern when trying to update poverty maps. Equation (16.12) makes it possible to obtain a propensity score – the predicted probability of being in period $T=\{t_0, t\}$ – that is conditional upon the observable characteristics:

$$\psi_{iT} = \frac{1-P_{iT}}{P_{iT}} \Big/ \frac{P_T}{1-P_T}, \tag{16.13}$$

where P_T is the unconditional probability that an observation belongs to period T or the share of year t observations in the total observations (that is, both years). In this framework, accounting for changes in the distribution of the observable characteristics is equivalent to reweighing the consumption distribution estimated in equation (16.12), thus obtaining:

$$\ln(y^{r}_{t_0,i}) = \ln(y^{p}_{t_0,i}) \times \psi_{t_0,i}. \tag{16.14}$$

The only step remaining is to add a measure of the unobserved component of consumption. If the dispersion in unobserved consumption were due to random events which were unrelated to systematic differences across households, then there would be nothing more to say about the error term. However, one reason to add a measure of the unobserved consumption is that the residual is unlikely to be just a random component of consumption. Instead, it may reflect systematic (albeit unexplained) differences between households. Therefore, for these two reasons, the consumption in equation (16.13) can be adjusted with counterfactual residuals. This can be done by first estimating a consumption model for the t_0 data and ranking all the households on the basis of the residual distribution for that year. Then assign to each household in t_0 the value of the ranking from the empirical distribution of residuals in the new survey in year t, which corresponds to the year t_0 rank. In this way the counterfactual consumption is calculated so that the consumption is what would have been observed in t_0 if the parameters, the distribution of covariates and the unmeasured determinants of consumption were as in t. From equations (16.10) and (16.14), the counterfactual wealth distribution can be rewritten as:

$$\ln(y^c_{t_0,i}) = \psi_{t_0,i}(\ln(y^p_{t_0,i} + \varepsilon^r_{t,i})) = \psi_{t_0,i}(\hat{\beta}'_t X_{t_0,i} + \varepsilon^r_t), \tag{16.15}$$

where $\varepsilon^r_{t,i}$ denotes the value of the ranking in t assigned to a household with the same residual rank in year t_0. Dabalen and Ferrè (2008) apply this approach to update poverty mapping in Albania using census data (2001) and two LSMS surveys, conducted in 2002 and 2005; Betti *et al.* (2013) further update the poverty mapping in Albania over a decade, using LSMS 2002, 2005 and 2008.

A second approach to updating poverty mapping without using census is data is proposed by Emwanu *et al.* (2006); this approach has one additional constraint with respect to that of Betti *et al.* (2013), since it needs panel data; in particular it requires the estimation of a relationship between per capita consumption in the year of interest and household characteristics in the census year.

16.5 The cumulation of fuzzy measures: combining the 3D

The method proposed by Verma *et al.* (2010) seeks to improve the sampling precision of poverty indicators for sub-national regions, through the cumulation of data in a rotational panel design. The reference data used is that of the EU-SILC, as it is the most important source of comparative data on poverty in Europe to date. This dataset involves a rotational panel that replaces a quarter of the sample each year, hence allowing a unit to be monitored for a maximum of four years. Cross-section and longitudinal databases of various durations are derived.

The measures can be calculated by aggregating information regarding individual elementary units into average measures – such as means, rates and proportions – or distributional measures – such as variation or dispersion measures. It is interesting to note that a set of average measures at regional level can serve as a dispersion measure and can be a valid tool in identifying geographical disparity.

When using survey data, one can adopt different approaches to construct regional indicators:

1 direct estimation – as done at national level;
2 alternative indicators – with indicators that use data more intensively;
3 cumulation of data over time – by changing the temporal reference period;
4 synergy with other sources – adopting small area estimation techniques.

16.5.1 Cumulation in rotational panel design

Based on a common blueprint, the EU-SILC survey has been developed in each country in order to adapt to individual countries' various needs. Most countries use the standard rotational panel design, in which each cross-section at time *t* contains units that have been in the survey from one to four years. Figure 16.1 shows how the first sample has been in the study for four years, and is present in the cross-section data at time *t*. The same can be said for the second, third and fourth samples active in the survey at time *t*.

Direct estimation would identify an estimator at time *t* by calculating it with respect to the cross-section dataset, independently of the values in adjacent time frames. The cumulation method, on the contrary, relies on the integration of information in multiple time frames. The sample overlap must be taken into account, and assumptions need to be made to guarantee a correct estimation of the indicator, which will refer to an intermediate time frame.

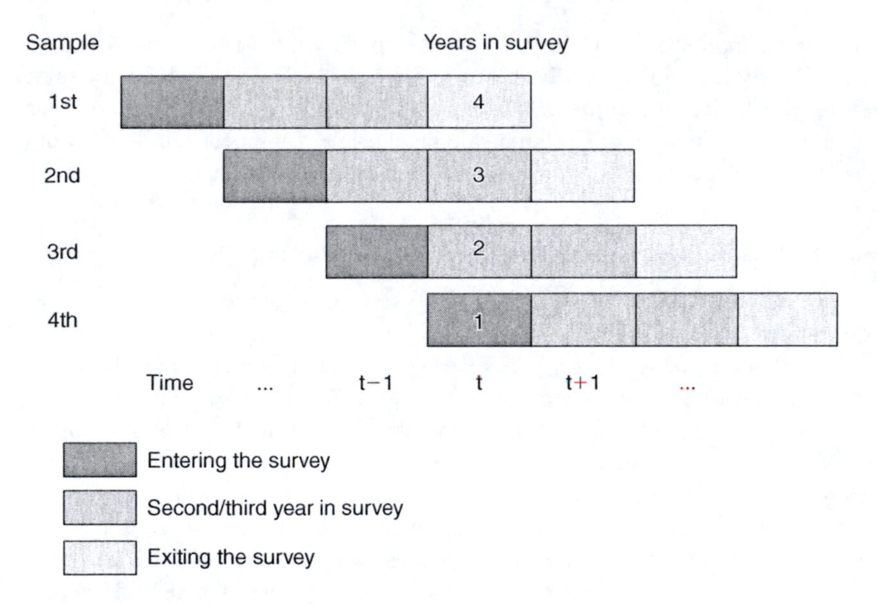

Figure 16.1 Rotational panel diagram: four sub-samples and four (yearly) time periods.

16.5.2 Pooling data versus pooling estimates

To pool data from two or more data sources the analysis must be conducted on the same type of units, such as households or individuals, and the variables must be measured in a comparable way. The pooling of information can be done by:

1 combining source estimates;
2 pooling data at the micro level.

The technical details and relative efficiencies of the procedures depend on the characteristics of the situation. The two approaches may give numerically equivalent results, one may provide more accurate estimates than the other, or only one may be appropriate/feasible.

Linear statistics, such as totals pooling country estimates $\{\varphi_i\}$ with some appropriate weights $\{\omega_i\}$, give the same result as pooling data at the micro level with weights $\{\omega_{ij}\}$ rescaled as

$$w'_{ij} = w_{ij}\left(\omega_i \Big/ \Sigma w_{ij} \right).$$

For ratios of the form

$$\varphi_i = \Sigma w_{ij} v_{ij} \Big/ \Sigma w_{ij} u_{ij},$$

the two approaches give very similar, although not identical, results, corresponding to the "separate" and "combined" ratio estimates respectively.

Cumulation is concerned with an equally common issue, in which the pooling of different sources is computed on the same population. In the case of a rotational panel design, the pooling of different sources describes similar and largely overlapping populations. In a national survey this allows for pooling over survey waves, which increases the precision due to the additional information. This is of critical importance in regional estimates where the information available is scarce.

Estimates from samples of the same population are most efficiently pooled with weights in proportion to their variances. This implies that they are efficiently pooled when the population has been sampled with similar sampling designs and in direct proportion to its sample size.

Alternatively the samples may be pooled at the micro level, with unit weights being inversely proportional to the probability of being selected in any single sample. This latter procedure may be more efficient, but may be not feasible as it requires the unit probability of selection in each of the samples, irrespective of whether the unit appears in the particular sample. An additional difficulty must be accounted for when dealing with complex sampling designs, as the structure of the resulting pooled sample can become too complex for variance estimation.

Different waves of a rotational panel survey increase the difficulty, due to the fact that the waves do not correspond to exactly the same population. This is analogous to combining samples from multiple frames, for which it has been noted that micro level pooling is generally not the most efficient method. For the reasons above, pooling of wave-specific estimates rather than micro data sets is generally the most appropriate approach to aggregation over time from surveys such as EU-SILC.

16.5.3 Variance estimation of cumulated measures

Standard poverty analysis determines the statistical unit's poverty status (poor or not-poor) based on the income distribution for each separate wave. As a benchmark indicator Verma *et al.* (2010) take the Head Count Ratio (HCR), which identifies the proportion of poor people within the sample. These proportions are then averaged over a number of consecutive waves. The objective is to quantify the gain in sampling precision from such pooling, given that data from different waves of a rotational panel are highly correlated. In other words, suppose a person's poverty status is determined from his income position in the income distribution separately for each year; once the poverty measure has been computed for each wave in the dataset, it is averaged over the number of consecutive waves. The cumulation method measures how much gain in sampling precision derives from this pooling, given that consecutive years generally present highly correlated data.

To estimate variances while properly taking complexity into account, one can use, for example, the Jack-knife Repeated Replications (JRR) variance estimation methodology. The sample of interest is formed by the union of all cross-sectional samples being compared or aggregated, hence the JRR method defines a set of replications by excluding a unit simultaneously from every wave in which the unit appears. For each replication the measure required is constructed for each of the cross-sectional samples involved. These measures are then used to obtain the average required for the replication, according to the standard JRR procedure (Betti *et al.*, 2007; Verma and Betti, 2011). Originally introduced as a technique for bias reduction, the JRR method has now been widely tested and used for the variance estimation of complex statistics, such as cumulative and longitudinal measures.

JRR relies on replications generated through repeated re-sampling of the parent sample, and the following algorithm applies to statistics of any complexity:

$$\text{var}(Q) = \sum_k \left[(1 - f_k) \frac{a_k - 1}{a_k} \sum_j (Q_{kj} - Q_k)^2 \right]. \tag{16.16}$$

In the application of JRR, a replication is formed by:

1 eliminating one Primary Selection Unit (PSU) from the parent sample;
2 compensating the weight given to the remaining units in the stratum.

This generates a number of replications equal to the number of PSUs in the sample. Verma *et al.* (2010) have applied this method to estimate variances in the EU-SILC rotational panel design for:

1 sub-populations (including geographical domains);
2 longitudinal measures, such as persistent poverty rates;
3 measures of net change and averages over cross-sections.

The union of all cross-sectional samples being compared or aggregated forms the parent sample. Using the common basis as structure of this total sample, a set of JRR replications is defined. Each replication is formed so that the replication-identifying unit is excluded in its construction, hence being excluded in every wave in which the unit appears. For each replication, the measure required is constructed for each of the cross-sectional samples involved, and these are used to obtain the averaged measure required for the replication. The variance of the statistic of interest is estimated from the replication estimates in the usual way.

In conclusion, the poverty measures estimates calculated through the cumulation method proposed by Verma *et al.* (2010) are able to incorporate information from multiple time frames. This additional information can substantially increase the sampling precision – depending on the overlap between different waves – hence the method is successful in its principal objective.

This method, applied to the fuzzy measure FS, combines the three dimensions (3D): Multidimensionality, Longitudinal poverty and Small area estimation.

Notes

1 The reasons that justify the choice of this specification are given in Manton *et al.* (1992) and Cheli (1995).
2 This procedure has certain similarities with that proposed by Betti *et al.* (2004). However, the present procedure is more general and more consistent.
3 The following rules are used in computing the intersection of a sequence of cross-sectional sets of the form appearing in rows of Table 16.1. In accordance with the Composite rules defined in Betti *et al.* (2006), Betti and Verma (2009) first take the standard intersection of the sequence of similar states (such as all "+") over i periods, and then take the intersection of the result with the dissimilar state which immediately follows it using the bounded operation. Each of the remaining $(T-i-1)$ cells marked "?" represents all states, i.e. membership function identically equal to 1, so that it makes no difference to the intersection being considered.

References

Berman, Y., Phillips, D. (2000), Indicators of social quality and social exclusion at national and community level, *Social Indicators Research*, 50(3), pp. 329–350.

Betti, G., Cheli, B., Cambini, R. (2004), A statistical model for the dynamics between two fuzzy states: theory and an application to poverty analysis, *Metron*, 62, pp. 391–411.

Betti, G., Cheli, B., Lemmi, A., Verma, V. (2006), Multidimensional and longitudinal poverty: an integrated fuzzy approach, in Lemmi, A., Betti, G. (eds), *Fuzzy Set Approach to Multidimensional Poverty Measurement*. New York: Springer, pp. 111–137.

Betti, G., D'Agostino, A., Neri, L. (2002), Panel regression models for measuring multidimensional poverty dynamics, *Statistical Methods and Applications*, 11(3), pp. 359–369.

Betti, G., Dabalen, A., Ferré, C., Neri, L. (2013), Updating poverty maps between censuses: a case study of Albania, in Laderchi, C.R., Savastano, S. (eds), *Poverty and Exclusion in the Western Balkans*, Economic Studies in Inequality, Social Exclusion and Well-Being 8. Springer Science+Business Media, New York, pp. 55–70.

Betti, G., Gagliardi, F., Lemmi, A., Verma, V. (2012), Sub-national indicators of poverty and deprivation in Europe: methodology and applications, *Cambridge Journal of Regions, Economy and Society*, 5(1), pp. 149–162.

Betti, G., Gagliardi, F., Nandi, T. (2007), Jackknife variance estimation of differences and averages of poverty measures, Working Paper no. 68/2007, DMQ, University of Siena.

Betti, G., Verma, V. (1999), Measuring the degree of poverty in a dynamic and comparative context: a multi-dimensional approach using fuzzy set theory, Proceedings of the ICCS-VI, Lahore, Pakistan, 27–31 August, 11, pp. 289–301.

Betti, G., Verma, V. (2009), Fuzzy measures of the incidence of relative poverty and deprivation: a longitudinal and comparative perspective, *Advances and Applications in Statistics*, 12(2), pp. 235–273.

Cerioli, A., Zani, S. (1990), A fuzzy approach to the measurement of poverty, in Dagum, C., Zenga, M. (eds), *Income and Wealth Distribution, Inequality and Poverty*. Berlin: Springer Verlag, pp. 272–284.

Cheli, B. (1995), Totally fuzzy and relative measures of poverty in dynamics context, *Metron*, 53(1), pp. 183–205.

Cheli, B., Betti, G. (1999), Totally fuzzy and relative measures of poverty dynamics in an Italian pseudo panel, 1985–1994, *Metron*, 57(1–2), pp. 83–104.

Cheli, B., Lemmi, A. (1995), A totally fuzzy and relative approach to the multidimensional analysis of poverty, *Economic Notes*, 24, pp. 115–134.

Dabalen, A., Ferrè, C. (2008), Updating poverty maps: a case study of Albania, mimeo, World Bank.

Elbers, C., Lanjouw, J.O., Lanjouw, P. (2003), Micro-level estimation of poverty and inequality, *Econometrica*, 71(1), pp. 355–364.

Emwanu, T., Hoogeveen, J., Okwi, P.O. (2006), Updating poverty maps with panel data, *World Development*, 34(12), pp. 2076–2088.

Ferretti, C., Molina, I. (2012), Fast EB method for estimating complex poverty indicators in large populations, *Journal of the Indian Society of Agricultural Statistics*, 66, pp. 105–120.

Lemieux, T. (2002), Decomposing changes in wage distributions: a unified approach, *Canadian Journal of Economics*, 35(4), pp. 646–688.

Manton, K.G., Woodbury, M.A., Stallard, E. and Corder, L.S. (1992), The use of the grade-of-membership technique to estimate regression relationships, *Sociological Methodology*, 22, pp. 321–381.

Molina, I. and Rao, J.N.K. (2010), Small area estimation of poverty indicators. *Canadian Journal of Statistics*, 38, pp. 369–385.

Nicholas, A. and Ray, R. (2012), Duration and persistence in multidimensional deprivation: methodology and an Australian application, *Economic Record*, 88, pp. 106–126.

Verma, V., Betti, G. (2011), Taylor linearization sampling errors and design effects for poverty measures and other complex statistics, *Journal of Applied Statistics*, 38(8), pp. 1549–1576.

Verma, V., Betti, G., Lemmi, A., Mulas, A., Natilli, M., Neri, L. and Salvati, N. (2005), Regional indicators to reflect social exclusion and poverty, Final report, Project VT/2003/45, European Commission, Employment and Social Affairs DG.

Verma, V., Gagliardi, F., Ferretti, C. (2010), Cumulation of poverty measures to meet new policy needs, Proceedings of the Italian Statistical Society, Padua, June.

World Bank (2005), The socio-economic atlas of Tajikistan, World Bank Office of Dushanbe, Dushanbe, Tajikistan.

Index

Page numbers in *italics* denote tables, those in **bold** denote figures.